Advance Praise for
Saving Species on Private Lands:
Unlocking Incentives to Conserve Wildlife and Their Habitats
By Lowell E. Baier

Praise from Government

This straightforward handbook will rally the ecological power of private landowners. By clarifying multiple confusing laws, amendments, and regulations, it spreads a map for landowners and advocates to travel together. As a guide to new progress, Lowell Baier's *Saving Species on Private Lands* is a welcome hope.

Dirk A. Kempthorne
Secretary of the Interior (2006–2009)
Governor of Idaho (1999–2006)
U.S. Senator (1993–1999)

This book takes on the complex issues of how landowners can conserve wildlife, access public and private support for doing so, and avoid regulation under the Endangered Species Act. This practical, well organized book is a valuable resource for landowners, partner organizations, government officials, students, and policymakers alike.

Thomas J. Vilsack
Secretary of Agriculture (2009–2017)
Governor of Iowa (1999–2007)

This book demystified the Endangered Species Act, explains how it affects private land, and provides landowners with the knowledge they need to navigate it and thrive.

Frank H. Murkowski
Governor of Alaska (2002–2006)
U.S. Senator (1980–2002)

Private lands provide an incredibly rich and vibrant tapestry of habitats, across our nation, that serve as home to tens of thousands of at-risk species, ranging from invertebrates to plants, from birds to mammals. Private landowners can make a serious conservation difference through voluntary, planned management actions in concert with, and supported by, myriad state, federal, and non-governmental partners. Lowell Baier is a preeminent, indefatigable legal scholar about conservation and endangered species, and I can think of no better person to provide such a timely and important resource as *Saving Species on Private Lands.*

Ronald J. Regan
Executive Director
Association of Fish and Wildlife Agencies

County governments have partnered with federal and state agencies, tribal governments, non-profits and private landowners to pursue innovative solutions for species conservation, leading to improved outcomes on the ground. Voluntary conservation efforts create buy-in from all impacted parties, more effectively establishing the Endangered Species Act within communities. In *Saving Species on Private Lands: Unlocking Incentives to Conserve Wildlife and Their Habitats*, Lowell Baier uses real world examples of voluntary conservation programs that benefit species and landscapes without jeopardizing landowners' needs. This book serves as an important guide for better implementation of the Endangered Species Act.

Jonathan Shuffield
Associate Legislative Director
National Association of Counties

Praise from Industry

Ranchers are perhaps our most important resource for the conservation of wildlife. *Saving Species on Private Lands* arms these conservationists with critical information to aid them in securing the technical assistance, regulatory assurances, and financial incentives necessary to navigate the Endangered Species Act. These tools can help ranchers continue to provide the large unbroken landscapes needed for healthy wildlife populations.

Ethan L. Lane
Vice President, Government Affairs
National Cattlemen's Beef Association
Executive Director, Public Lands Council (2015–2019)

Home builders and developers face daunting ESA rules and regulations. Lowell Baier's book demystifies the Endangered Species Act and provides a well-structured guide for private landowners who want to enjoy their property rights while making a difference through wildlife conservation.

Greg Ugalde
2019 Chairman of the Board
National Association of Home Builders

In *Saving Species on Private Lands*, Lowell Baier provides a roadmap for voluntary conservation by private landowners to help endangered, threatened, and at-risk species. This book explains how existing financial and regulatory incentives for landowners can make a difference—for the good of both people and the environment.

Marc A. Brinkmeyer
Chairman of the Board
Idaho Forest Group

Today, a new conservation paradigm is sweeping across America. Federal, state, local government agencies, foundations, non-profits, and willing landowners are joining together to find win-win solutions that keep working lands productive and American workers employed while conserving more of our habitat and wildlife than ever before. This book explains how and why they work, and how they can be put to work by landowners across the nation. I recommend this as a must read for private landowners desiring to successfully navigate the ESA!

George Emmerson
President
Sierra Pacific Industries

Wildlife conservation in America is at a crossroads. Will we continue to lose wildlife, or will we roll up our sleeves and work together to save it? In this powerful book, Lowell Baier explains how private landowners can secure a portfolio of incentives to conserve wildlife on their property, while navigating the ESA and related laws and regulations.

Marc C. Mondavi
Co-Proprietor
C. Mondavi & Family/Charles Krug Winery

Saving Species on Private Lands is a timely and valuable resource for landowners willing to engage in the greatest conservation effort of all time. Lowell Baier deciphers voluntary conservation and helps landowners use proven tools to make smart decisions leading to real conservation. His vision of the natural harmony that can exist between people and wildlife on working lands will inspire and empower the rising generation of conservation leaders.

Dave Tenny
President & CEO
National Alliance of Forest Owners

Respected authority Lowell Baier has prepared a valuable resource for landowners and managers. He has consolidated into one book a discussion of the challenges faced by all of us in the land management business. This well written book covers a wide range of topics, identifying resources and options for managing our private property for sustainable agriculture and wildlife habitat, especially at risk, threatened and endangered species.

Thomas Price
President, Price Cattle Company
Chairman, InterWest Ranch and Farm Management, Inc.
Northwest Rangeland Trust

Saving Species on Private Lands: Unlocking Incentives to Conserve Wildlife and Their Habitats documents 40 years of innovations in wildlife conservation with examples and resources incentivizing landowners and land managers, especially in dealing with at-risk, threatened and endangered species.

Scott Manley, Ph.D.
Director, Rice Stewardship Partnership

Saving Species on Private Lands: Unlocking Incentives to Conserve Wildlife and Their Habitats is a comprehensive and accessible handbook that lays out a win-win blueprint to promote wildlife conservation, with an emphasis on at-risk, threatened and endangered species, and at the same time improve the value of private land. This is a must read for the entire agricultural community.

Steve J. Hageman
Chairman of the Board, Hageman Group LLC
President, Hageman Farms, Inc
1983 Founding Partner, Remington Hybrid Seed Co. Inc.

Praise from Conservation Organizations

This book is an important resource to those who want to protect birds and the places they need while also managing their land without sacrificing profits. Ranchers, farmers, and even people who just have a large piece of property will find something in here, like how to tap into government incentives, find conservation partners, or navigate compliance with federal laws like the Endangered Species Act. A great new tool for all landowners.

David O'Neill
Chief Conservation Officer
National Audubon Society

At a time when more than one-third of all U.S. species are at heightened risk of extinction, incentivizing habitat restoration on private lands has never been more important. That's easy to say, but incredibly hard to do. Thankfully, Lowell Baier has charted a clear path forward that will unleash the collaborative conservation desperately needed to recover the full diversity of fish, wildlife, and plants for future generations.

Collin O'Mara
President and Chief Executive Officer
National Wildlife Federation
Secretary of Natural Resources and Environmental Control for Delaware (2009–2014)

Private lands hold the key to the future of wildlife conservation, with two thirds of all threatened and endangered species living on privately owned lands. This book shows that private landowners are true champions of wildlife conservation and provides much needed information to help them understand the ESA and better engage in effective species conservation work on the ground. This book fills a tremendous need for private landowners.

Nick Wiley
Chief Operating Officer
Ducks Unlimited, Inc.

America's private lands represent the future for wildlife conservation. Healthy agricultural landscapes conserve wildlife by focusing on habitats they depend upon while benefitting all of society through environmental services. *Saving Species on Private Lands: Unlocking Incentives to Conserve Wildlife and Their Habitats* will serve as a keystone resource for farmers, ranchers, and all landowners interested in wildlife conservation, at-risk and endangered species, and healthy sustainable landscapes for future generations.

Howard K. Vincent
President and CEO
Pheasants Forever and Quail Forever

As we look to the future of wildlife conservation, we must build on the success that we have seen from collaboration, and engage private landowners as full partners. *Saving Species on Private Lands: Unlocking Incentives to Conserve Wildlife and Their Habitats* is a comprehensive book that will help landowners understand the resources available to them to promote wildlife conservation on private land. This is an excellent resource to shape a new paradigm in wildlife conservation

Rebecca Humphries
Chief Executive Officer
National Wild Turkey Federation

This comprehensive, exhaustively researched book contains everything a farmer, rancher, forester, or other landowner needs to understand how government, private funding and technical assistance can help them improve the wildlife conservation value of their land.

Blake Henning
Chief Conservation Officer
Rocky Mountain Elk Foundation

Saving Species on Private Lands: Unlocking Incentives to Conserve Wildlife and Their Habitats is a must-read book for anyone who wants to do something to help our declining wildlife. With information on financial resources for conservation, technical assistance for landowners, and the legal and regulatory landscape of conservation, the book clearly lays out opportunities for working landowners to step up and conserve wildlife—without hurting their businesses in the process.

Timothy C. Brady
President
Boone and Crockett Club

Saving Species on Private Lands: Unlocking Incentives to Conserve Wildlife and Their Habitats is an essential read for those that value wildlife and its habitat. Private landowners provide the majority of wildlife habitat for our nation's wildlife and this book provides a practical approach for sailing through the rocky shoals of federal law and regulations. It's a powerful guide for landowners to achieve both their economic and wildlife conservation goals."

Steve Williams, Ph.D.
President of the Wildlife Management Institute
Director of the U.S. Fish and Wildlife Service (2002–2005)

Saving Species on Private Lands is a very important book for conservation in America for the twenty-first century. For more than forty-five years, since its passage by Congress in 1973, the Endangered Species Act has been primarily the province of federal and state wildlife agencies for implementation of conservation strategies. As a result not much progress has been made on species recovery. This is not surprising since 80 percent of endangered species habitat nationwide is on private lands, and public agencies have largely ignored private land owners for decades. Progress on species recovery will only be made by encouraging the stewardship by private land owners on a scalable basis. This core constituency of 13 million Americans is the key to species recovery in America. Lowell Baier's book is a major contribution to breaking through this barrier to engage and encourage private land owners to participate in species recovery. I encourage any landowner hosting endangered species or having species in their vicinity of their properties to read this important book. It will advance the conservation cause in America more than any and all environmental advocacies.

Amos S. Eno
President & Founder
Land Conservation Assistance Network
Executive Director, National Fish and Wildlife Foundation (1991–1999)

Aldo Leopold once wrote, "Conservation will ultimately boil down to rewarding the private landowner who conserves the public interest." Lowell Baier's *Saving Species on Private Lands: Unlocking Incentives to Conserve Wildlife and Their Habitats* will reinvigorate this truth by decoding what has become a complex labyrinth of laws and policies that disincentivize landowners and distance them from the agencies that actually have the tools and resources to help them achieve successful conservation on their lands. Without a pathway, it's hard to get anywhere. This text provides a pathway and couldn't be more timely.

Gray N. Thornton
President & CEO
Wild Sheep Foundation

Every farmer, rancher, or small businessperson knows that to do a job, you need the right tools. In this well-written book, Lowell Baier lays out the tools that private landowners can use to carry out voluntary conservation on their land to benefit at-risk species. This guide explains how landowners can prepare a conservation plan, sign up for payments under the federal Farm Bill, avoiding the pitfalls of the Endangered Species Act, and everything in between.

James L. Cummins
Executive Director
Wildlife Mississippi

Private land and landowners are critical to healthy wildlife populations. Partnerships between landowners and other wildlife conservationists are key to conserving declining wildlife species, and necessary to ensure working landscapes remain productive for human communities as well. This book explains programs and approaches to partnership-driven voluntary conservation on private lands and is a valuable resource for landowners and others who want to learn more about wildlife conservation opportunities on private land.

Steve Jester
Executive Director
Partners for Conservation

Private landowners play a critical role for species and habitat conservation. *Saving Species on Private Lands* documents decades of work with the private sector and illustrates how wildlife conservation can integrate with operations—from forestry and energy development to farming and ranching—on private lands.

Edward B. Arnett, Ph.D.
Chief Scientist
Theodore Roosevelt Conservation Partnership

Whether working to preserve species on the brink of extirpation or aspiring to keep common species common, overcoming landowner uncertainty about their options can be even more important than providing direct technical or financial assistance for conservation. This well-researched, comprehensive volume is an excellent resource for landowners, non-profit organizations, foundations, and agencies to navigate the bureaucratic terrain and deliver on-the-ground conservation.

Brent A. Rudolph, Ph.D.
Chief Conservation & Legislative Officer
Ruffed Grouse Society
American Woodcock Society

Praise from the Academic Community

Conservation happens when we stop fighting those who hold different opinions about endangered species. It happens when we invest that time and energy to design and implement management actions to help wildlife. This book tells us how to do that.

William F. Porter, Ph.D.
Boone and Crockett Professor of Wildlife Conservation
Michigan State University

As the world confronts the sixth great extinction crisis in its history, thousands upon thousands of landowners are working voluntarily to reverse the decline of countless at-risk species across the United States. This book examines their efforts, documents their successes, and provides a roadmap for more to follow in their footsteps. The importance of voluntary conservation on private lands cannot possibly be overstated, and Lowell Baier's book could not be more timely.

Chad J. Bishop, Ph.D.
Director of Wildlife Biology, University of Montana
President, National Association of University Fisheries and Wildlife Programs

Public debate in America is too often about what should happen, or what the government should do, and not about what can happen and what we all can do, and wildlife conservation is no exception. In *Saving Species on Private Lands*, Lowell Baier provides a guide that every landowner can follow to conserve at-risk species and other wildlife, without placing an undue burden on themselves and their families. The purposeful alignment of wildlife conservation and private land management will prove essential to effectively address our most pressing conservation challenges.

Joshua J. Millspaugh, Ph.D.
Boone and Crockett Professor of Wildlife Conservation
University of Montana

Aldo Leopold wrote, in the 1930 American Game Policy, that we must recognize the landowner as the custodian of wildlife, and that only the landowner can practice management efficiently. Yet, in spite of broad acceptance of this principle, there has not been a comprehensive guide as to how landowners can become wildlife stewards and use their lands for personal enjoyment and profit at the same time. The need for such guidance has become critical, with expanding legions of at-risk wildlife species, most of which are on private lands. Lowell Baier, one of the most respected conservationists in the Nation, has stepped forward once again to take a pressing issue head-on. He has produced a remarkable book that should be read by all in the conservation institution, from academics to wildlife managers, industrialists to the small landholder. This splendidly written, exhaustively referenced work, is the most thorough and clear treatment of the 1973 Endangered Species Act as it relates to private lands I have read, and his comprehensive guidance on programs and tools available to landowners should contribute most significantly in efforts to incentivize private lands conservation.

John F. Organ, Ph.D., CWB
Chief Emeritus, U.S. Geological Survey
Cooperative Fish and Wildlife Research Units
Former Chief, Northeast Region USFWS Wildlife
and Sport Fish Restoration Program
Past President, Fellow, and Honorary Member, The Wildlife Society
Adjunct Associate Professor, University of Massachusetts, Amherst
and Michigan State University

A great barometer of a healthy society is the health and well-being of wild things and wild places, and Lowell Baier's *Saving Species on Private Lands* clearly explains how landowners can secure financial incentives and navigate regulatory obstacles to successfully conserve wild species for generations to come. This is a must read for all private landowners and conservationists.

Steven Leath, Ph.D.
President, Auburn University (2017-2019)
President, Iowa State University (2012-2017)

Saving Species on Private Lands: Unlocking Incentives to Conserve Wildlife and Their Habitats would make a great choice as a comprehensive and accessible textbook for courses that focus on the tools available for resource managers whose objectives include influencing management decisions on private lands.

Christine L. Thomas, Ph.D.
Dean, Professor of Resource Management
College of Natural Resources
University of Wisconsin, Stevens Point

Saving Species on Private Lands is much more than a book about species at risk of extinction, it is a guide for empowering landowners. The book briefly describes the history, legal authority and operation of institutions that support voluntary conservation by landowners. *Saving Species on Private Lands* is essential reading for practitioners of conservation in the United States at educational institutions, extensions and research organizations.

Perry S. Barboza, Ph.D., Professor
Wildlife and Fisheries Sciences Boone & Crockett Chair
in Wildlife Conservation & Policy
Texas A&M University

SAVING SPECIES ON PRIVATE LANDS

Unlocking Incentives to Conserve Wildlife and Their Habitats

LOWELL E. BAIER
With Christopher E. Segal

ROWMAN & LITTLEFIELD
Lanham • Boulder • New York • London

Published by Rowman & Littlefield
An imprint of The Rowman & Littlefield Publishing Group, Inc.
4501 Forbes Boulevard, Suite 200, Lanham, Maryland 20706
www.rowman.com

6 Tinworth Street, London SE11 5AL, United Kingdom

British Library Cataloguing in Publication Information Available

Library of Congress Cataloging-in-Publication Data Is Available

ISBN 978-1-5381-3937-0 (cloth)
ISBN 978-1-5381-3938-7 (paperback)
ISBN 978-1-5381-3939-4 (electronic)

™

Contents

List of Illustrations

Guide to Acronyms and Terms

ABC: American Bird Conservancy
ACEP: Agricultural Conservation Easement Program
ACEP-ALE: Agricultural Conservation Easement Program–Agricultural Land Easements
ACEP-WRE: Agricultural Conservation Easement Program–Wetland Reserve Easements
AFWA: Association of Fish and Wildlife Agencies
AGCF: Arkansas Game and Fish Commission
ARP: Acreage Reserve Program
BCCP: Balcones Canyonlands Conservation Plan
BLM: Bureau of Land Management
CCA: Candidate Conservation Agreement
CCAA: Candidate Conservation Agreement with Assurances
CCC: Commodity Credit Corporation
CE: Categorial Exclusion
CELCP: Coastal and Estuarine Land Conservation Program
CFOSC: Community Forest and Open Space Conservation program
CLEAR: Clean Lakes, Estuaries and Rivers Initiative
CPW: Colorado Parks and Wildlife
CREP: Conservation Reserve Enhancement Program
CRGP: Coastal Resilience Grants Program
CRP: Conservation Reserve Program
CRU: U.S. Geological Survey Cooperative Research Units
CSeP: Conservation Security Program
CSP: Conservation Stewardship Program
CWA: Clean Water Act of 1972

CWHP: Colorado Wildlife Habitat Program
DJ: Dingell-Johnson Sport Fish Restoration Act of 1950
DNR: Department of Natural Resources
DoD: Department of Defense
DU: Ducks Unlimited
EA: Environmental Assessment
EIS: Environmental Impact Statement
EPA: Environmental Protection Agency
EQIP: Environmental Quality Incentives Program
ESA: Endangered Species Act of 1973
Federal Nexus: Any connection that a private landowner has with the federal government through an agricultural or other government program, such as federal crop subsidies, crop insurance, flood insurance, federal loans, disaster relief payments, permits under federal anti-pollution laws, or enrollment in federal conservation programs under the Farm Bill or other laws
FLPMA: Federal Land Policy and Management Act of 1976
FONSI: Finding of No Significant Impact
FSA: Farm Service Agency
FSP: Forest Stewardship Program
FWCA: Fish and Wildlife Coordination Act of 1934
FWS: U.S. Fish and Wildlife Service
GEBF: Gulf Environmental Benefit Fund
HCP: Habitat Conservation Plan
HFRP: Healthy Forests Reserve Program
INRMP: Integrated Natural Resources Management Plan
IRC: Internal Revenue Code
IRS: Internal Revenue Service
ITP: Incidental Take Permit
JV: Migratory Bird Joint Venture
LandCAN: Land Conservation Assistance Network
LPCI: Lesser Prairie Chicken Initiative
LSR: Landscape Scale Restoration
MDF: Mule Deer Foundation
NAWCA: North American Wetlands Conservation Act
NAWMP: North American Waterfowl Management Plan
NCCP: California Natural Community Conservation Planning Act of 1991
NCIF: Natural Capital Investment Fund
NEPA: National Environmental Policy Act of 1969
NFWF: National Fish and Wildlife Foundation
NGO: Non-governmental organization
NMFS: National Marine Fisheries Service

NOAA: National Oceanic and Atmospheric Administration
NRCS: Natural Resources Conservation Service
NWF: National Wildlife Federation
NWTF: National Wild Turkey Federation
ODFW: Oregon Department of Fish and Wildlife
PECE: Policy for Evaluation of Conservation Efforts When Making Listing Decisions
PF: Pheasants Forever and Quail Forever
PHU: Panther Habitat Unit
PR: Pittman-Robertson Wildlife Restoration Act of 1937
Producer: A farmer, rancher, or other agricultural landowner
PWCI: Pecos Watershed Conservation Initiative
QDMA: Quality Deer Management Association
R&D: Research and Development
RCPP: Regional Conservation Partnership Program
REPI: Readiness and Environmental Protection Integration program
RGS/AWS: Ruffed Grouse Society & American Woodcock Society
RIBITS: Regulatory In-Lieu Fee and Bank Information Tracking System
RMEF: Rocky Mountain Elk Foundation
RMS: Resource Management Service, LLC
SAFE: State Acres for Wildlife Enhancement
SGI: Sage Grouse Initiative
SHA: Safe Harbor Agreement
SWAP: State Wildlife Action Plan
SWG: State Wildlife Grant
TCF: The Conservation Fund
TNC: The Nature Conservancy
TPW: Texas Parks & Wildlife
TSP: Technical Service Provider
TU: Trout Unlimited
USDA: U.S. Department of Agriculture
USGS: U.S. Geological Survey
VPA-HIP: Voluntary Public Access and Habitat Incentive Program
WAFWA: Western Association of Fish and Wildlife Agencies
WHIP: Wildlife Habitat Incentives Program
WLFW: Working Lands for Wildlife
WPA: Waterfowl Production Area
WRP: Wetlands Reserve Program
WSFR: Wildlife and Sport Fish Restoration Funds

Foreword

AMERICANS LOVE WILDLIFE and want their lives to be part of ours. But we share a persistent concern.

We are not accomplishing as much for today's imperiled species as we have for the American bald eagle and the American peregrine falcon, and other species that have been retrieved from near extinction.

Many of us have worked diligently on this problem with heartfelt concern and hopeful visions. We have yet to surmount it.

Our efforts too often have bogged down into conflict, not only across the western United States, but across the entire country. Futile disagreements have clouded a shared vision of abundant wildlife held by individuals, property owners, and often whole communities.

With this book, my friend and lifelong conservationist Lowell Baier seeks to solve this problem by turning the efforts on the best target: the problem. Most of this book profiles federal, state, and private programs that seek to eliminate regulatory and practical obstacles that discourage landowners from improving wildlife habitat and conserving wildlife. These programs provide financial assistance, technical support, and regulatory assurances to landowners engaged in voluntary conservation.

Lowell is the right person to write this book. Lowell is a lawyer, successful entrepreneur, award-winning author, and has been an advisor to President George H.W. Bush and subsequent administrations on conservation policy. Lowell is the first President Emeritus of the Boone and Crocket Club, the oldest wildlife conservation organization in the United States. In 2016, Lowell received the National Wildlife Federation's highest honor for a lifetime of conservation service.

We should never get to the point of needing to list species. When we must list species as endangered, we must not leave them there as catalog entries. We must act to recover species with equal or better zeal than we devote to protections.

The strategy set by the Endangered Species Act in 1973 remains one of prohibiting harm. Today we need to produce benefits. Restrictive rules raised conflicts between landowners and advocates. Each side responded with defenses that put up opposing blockades against what they could lose. This closed many paths forward.

In the forty-seven years since the Endangered Species Act was passed, amendments and rules shimmed and tweaked its administration. Some provisions conflicted with each other. Others confused the effort. The Endangered Species Act was twelve pages when enacted in 1973. Today it is thirty-six pages long. The accompanying regulations initially were thirty pages. Today the regulations fill five thousand pages.

There is a deeper problem, too. The U.S. Constitution leaves to the states all powers not granted to the federal government. Care for wildlife was one of these. The Endangered Species Act strained this principle by giving the federal government the upper hand over the everyday job of states. The gray wolf is such a clear example of this tension that when the late Peter Jennings began what became the Project for Journalists and the Constitution, he chose the wolf issue as a case study. He came to interview me in 2002 about state and federal struggles in Idaho.

I am one of many who have tried to reconcile this emotional and polarizing issue. Years earlier, in 1997, as a member of the U.S. Senate, and with the help of many, I had advanced a bill that turned adversaries into advocates. It was better at saving species. It was better for people and communities. After passing committee on a bipartisan vote, the bill was not called up to be considered by the full U.S. Senate.

As a guide to new progress, Lowell Baier's *Saving Species on Private Lands* is a welcome hope.

Here is a manual for something that can be done today, without needing complicated changes to policy. Here is a change of strategy. American species conservation can move from the elusive vision of being a backstop against extinction . . . into a driver of recovery.

The book explores a solution that—although widely recognized—has not been realized. Rely on landowners. Make species into assets, instead of liabilities, on private land. On average, 74 percent of each of the lower forty-eight states is privately owned land, and 75 percent of all endangered species rely on private land habitat for survival.

This straightforward handbook will rally the ecological power of private landowners. By clarifying multiple confusing laws, amendments, and regulations, it spreads a map for landowners and advocates to travel together with the support of public funding.

There are incentives available to landowners to make them partners in species conservation by giving them more reason to act, rather than restrictions to fear. The many public programs adding speed to this progress date back to the 1934 Duck Stamp Act and up to the most recent 2018 Farm Bill.

Please read this book. No one will agree with everything Lowell writes, but Lowell is the first conservationist to offer a comprehensive inventory of the many options to conserve species. Take courage to persevere toward abundant success. The idea of America is to do better and achieve more. It is the hope of our wildlife legacy.

DIRK KEMPTHORNE
Secretary of the Interior (2006–2009)
Governor of Idaho (1999–2006)
U.S. Senator (1993–1999)

"Conservation is getting nowhere because it is incompatible with our Abrahamic concept of land. We abuse land because we regard it as a commodity belonging to us. When we see land as a community to which we belong, we may begin to use it with love and respect."

–ALDO LEOPOLD, *A SAND COUNTY ALMANAC*

Introduction and Chapter Summaries

ACCORDING TO THE WORLD'S LEADING SCIENTISTS, mankind today finds itself on the brink of an ecological catastrophe of unimaginable proportions. The fifth extinction crisis sixty-five million years ago killed all the dinosaurs. Today, we are in the midst of the sixth mass extinction crisis in the Earth's history. The 2019 United Nations Global Assessment Report on Biodiversity and Ecosystem Services reports that 60 percent of the world's mammals, birds, fish, and reptiles have gone extinct since 1970. Moreover, one million of the remaining eight million animal and plant species are at risk of extinction. The International Union for the Conservation of Nature reports there are twenty-eight thousand species threatened with extinction. According to the World Wildlife Fund, man has severely altered 75 percent of the earth's surface and 66 percent of its oceans.

In October 2019, leading researchers from seven institutions published in the journal *Science* a multi-year study of 529 bird species that demonstrated the United States and Canada have lost three billion birds since 1970, despite high reproduction rates. This represents 29 percent of the total bird population of North America. Birds are an indicator species (like a barometer) that infer the health of an entire ecosystem. Driving this dramatic decline are the loss of their habitat as land is cleared for agricultural production; use of neonicotinoid insecticides that poison birds and insects, one of their primary food sources; and cats, both domestic and feral, which kill one billion birds in the United States each year. The manifest causes for all of the above are deforestation, pollution and fossil fuel emissions, overgrazing, bush meat hunting and poaching, invasive species, land fragmentation and habitat destruction, ocean acidification, and overfishing. The continuing threat of climate change promises to make summers hotter, winters drier, and storms more severe. The facts and statistics are overwhelming both intellectually and emotionally. We can't grasp the enormity of this, and the challenges before us can make us feel hopeless. What can any one of us, or a few of us, do to combat such unimaginable forces? The answer is plenty, as history has demonstrated.

America's farmers, ranchers, foresters, and other private landowners have been showing us the way for generations, leading the continuing crusade to improve America's lands and its diverse wildlife populations. Voluntary private land conservation efforts have the potential, as has been repeatedly demonstrated, to restore lost wildlife populations, improve water quality and restore wetlands, reduce and eliminate air pollution, and even sequester carbon emissions while still profitably producing grain, beef, timber, etc. Where landowners are leaning into this work, a host of federal assistance programs, state government agencies, non-governmental organizations, and both public and private financial backers are stepping up to support their conservation efforts. In fact, a dramatic new paradigm is emerging in America focused on incentives for private working landowners to engage in bold wildlife conservation activities with an emphasis on at-risk, threatened, or endangered species.

This book documents these emerging new programs, their historical successes, and the growing number of current opportunities for voluntary conservation on private lands. It is a field guide for landowners and the government, non-governmental organizations, and foundation staff who work with them. Moreover, it is a handbook for the public at large, which needs to better understand both how much is being accomplished and how much more is still possible. Rural America continues in a period of perpetual transition as urban sprawl spreads, the energy industry's footprint grows (not just oil and gas but also renewables), and family farms are handed down to the next generation. For every private landowner who wants to help conserve wildlife and our other natural resources, there is an assistance program, conservation partner, or creative conservation solution out there. This book will help you find them.

Part I introduces the concept of proactive voluntary conservation of fish and wildlife on private land. Chapter 1 provides a brief history of private land in America, an explanation of the federal Endangered Species Act of 1973 (ESA), and a call to action for private landowners. Chapter 2 introduces the greater sage-grouse and the red-cockaded woodpecker, two species used as examples throughout the book, and explains their conservation history and current circumstances. Chapter 3 provides landowners with information on conservation planning resources. Together with appendix A, it shows landowners interested in conservation whom to call for assistance.

Part II provides an introduction to the tools of voluntary conservation. Chapter 4 explains the primary mechanisms used to enroll land in conservation programs, including grants, cost-sharing payments, incentive payments, rental contracts, and conservation easements. It also discusses the tax consequences, most positive but some negative, of each. Chapter 5 expands upon these concepts and shows how they fit into regulatory programs such as compensatory mitigation under the ESA and Clean Water Act. It describes how landowners can work with industry, investors, and government agencies to develop conservation banks and other mitigation tools.

Part III addresses the federal Farm Bill, the single largest source of conservation funding in the nation. The Farm Bill requires landowners to enroll in a conservation program for a fixed term, and applications are generally competitive. This part provides landowners with information on the various Farm Bill programs, including requirements, benefits, and how to enroll in each. Chapter 6 introduces the Farm Bill, its history, and the organizations that implement Farm Bill programs and assist landowners interested in enrolling. Chapter 7 addresses Farm Bill incentive and rental contract programs, including the Environmental Quality Incentives Program, Conservation Stewardship Program, and Conservation Reserve Program. Chapter 8 addresses Farm Bill easement programs, including the Agricultural Conservation Easement Program and the Healthy Forests Reserve Program. Chapter 9 details Farm Bill partnership programs, which include the Regional Conservation Partnership Program and the Voluntary Public Access and Habitat Incentive Program, as well as Working Lands for Wildlife, a Farm Bill–wide strategy that uses partnerships to improve efficiency. Additional information on the historic impact of the various Farm Bill conservation programs can be found in appendix B.

Part IV covers additional incentives and resources for private landowners beyond the federal Farm Bill. Chapter 10 addresses other federal conservation assistance programs, including programs operated by the Fish and Wildlife Service, Forest Service, Department of Defense, National Oceanic and Atmospheric Administration, Environmental Protection Agency, Department of Agriculture, Bureau of Land Management, National Fish and Wildlife Foundation, and U.S. Geological Survey. Chapter 11 documents non-federal conservation assistance programs, including those operated by state fish and wildlife agencies, state forestry agencies, other state and local government entities, non-governmental organizations, and land trusts.

Finally, part V addresses the ESA, explaining how it impacts landowners and how landowners can navigate its rules. Chapter 12 explains the operation of the ESA and provides landowners with resources to determine if and how the ESA impacts them. Chapter 13 documents tools developed over the years by Congress, the Fish and Wildlife Service, and the National Marine Fisheries Service that help landowners deal with ESA regulations. These include conservation actions landowners can take and permit programs landowners can enroll in, such as Habitat Conservation Plans, Safe Harbor Agreements, Candidate Conservation Agreements with Assurances, and regional or programmatic versions of these tools. The chapter closes with tools that federal agencies can use to reduce the ESA's regulatory burden, including 4(d) rules, critical habitat exclusions, experimental populations, and the Policy for Evaluation of Conservation Efforts When Making Listing Decisions.

INTRODUCTION TO PROACTIVE WILDLIFE CONSERVATION ON PRIVATE LAND

I

"The land ethic simply enlarges the boundaries of the community to include soils, waters, plants, and animals, or collectively: the land. In short, a land ethic changes the role of Homo sapiens from conqueror of the land-community to plain member and citizen of it. It implies respect for his fellow-members, and also respect for the community as such."

–ALDO LEOPOLD, *A SAND COUNTY ALMANAC*

Private Land and the Endangered Species Act

1

“CONSERVATION WILL ULTIMATELY boil down to rewarding the private landowner who conserves the public interest” and “the thing to be encouraged is the use of private land in such a way as to combine the public and the private interest to the greatest degree possible.” These words were written in 1934 by Aldo Leopold, the father of scientific wildlife management and a man decades ahead of his time in his understanding of the complex and mutually dependent relationship of wildlife and humankind. In the same essay, Leopold called himself a “political and economic dreamer,” acknowledging that our society lacked both the appetite and the tools for rewarding private landowners for conserving wildlife.[1] Leopold is best known for the “land ethic” he laid out in *A Sand County Almanac*, the idea that the relationships between people and land are intertwined and inseparable.

Eighty-six years later, America has begun to understand and act on the lessons that Leopold was trying to teach. Today, successful wildlife conservation depends on the adoption of Leopold’s land ethic not only by landowners such as farmers and ranchers, but by government agencies, non-profit organizations that fund conservation, and the public at large that supports conservation programs. To reach this point, we have weathered four periods of exploitation of our natural resources, including the early settlement of the United States and human use of its abundant natural resources, the industrial revolution, the crushing despair of the Great Depression, and the burgeoning postwar period when an expanding population demanded jobs, housing, and recreation, each drawing upon our natural resource reserves. Each period was characterized by different challenges and different responses from those dedicated to conservation. Following the early settlement of the

Figure 1.1. Aldo Leopold, 1887–1948. Widely recognized as the father of scientific wildlife management, Leopold was also a fierce conservationist and the creator of what became the National Wilderness Preservation System. His most famous work, *A Sand County Almanac*, published posthumously, has sold millions of copies and ranks among the likes of Henry David Thoreau's *Walden* and Rachel Carson's *Silent Spring* in its influence on conservation. Courtesy of Aldo Leopold Foundation.

United States, thinkers like Henry David Thoreau and John Muir called for a human connection to nature. The industrial revolution, in turn, gave rise to the U.S. Forest Service, the National Wildlife Refuges, the Migratory Bird Treaty Act, and state fish and wildlife agencies working to propagate wildlife through breeding, translocations, and releases. The Great Depression and the catastrophic Dust Bowl gave rise to modern soil management and many agricultural assistance programs that exist to this day.[2] And the postwar era, of course, gave rise to the Endangered Species Act of 1973 (ESA) and sixty-six other pieces of legislation dedicated to environmental and consumer protection. It also led to the 1985 Farm Bill and the development of modern Farm Bill conservation programs,[3] and a new focus on conservation on "working lands"—lands that are economically productive and support an individual's livelihood, such as farms, ranches, and timberlands.[4]

Although many of our historic battles over wildlife management have focused on our public lands, the current frontier of conservation is on private, and especially working, lands. In the continental United States, 74 percent of all land is privately owned.[5] Among species listed as threatened or endangered under the ESA, two-thirds can be found on privately owned land,[6] as can hundreds more species at risk of being listed. These include both greater sage-grouse in the west and red-cockaded woodpecker in the forests of the southeast. Even in the public land states of the American West, private land is essential to conservation. Although public land comprises, for example, 61.6 percent of Idaho, 63.1 percent of Utah, and a staggering 79.6 percent of Nevada,[7] private land provides critical wildlife habitat in every state, including

Figure 1.2. Map of public lands in the United States.
Courtesy of the U.S. Geological Survey National Geospatial Program.

much of the most imperiled and ecologically valuable areas. For example, over 75 percent of remaining wetlands and over 80 percent of remaining grasslands in the United States are located on private land.[8]

The concentration of public land in the West is a product of history. Beginning in 1781, the federal government took possession of all land west of the borders of the original thirteen states, which was then expanded through the Louisiana Purchase (1804), the end of the Mexican-American War (1848), the purchase of Alaska (1867), and other events. Altogether, over 1.8 billion acres of land—75 percent of the entire United States—have passed through the hands of the federal government, and some 520 million acres remain in public ownership.[9] The vast majority of that land is in the West and Alaska.

Notwithstanding the use of the Oregon Trail as early as the 1830s and the California gold rush of 1849, settlement of the West did not begin in earnest until passage of the Homestead Act in 1862, which authorized disposal of federal land through the granting of 160-acre claims to homesteaders. Over the decades, 287,500,000 acres were so granted.[10] Land that was claimed was that most suitable for settlement: land with access to water, good soil for crops, and other characteristics that also make land ecologically valuable.[11]

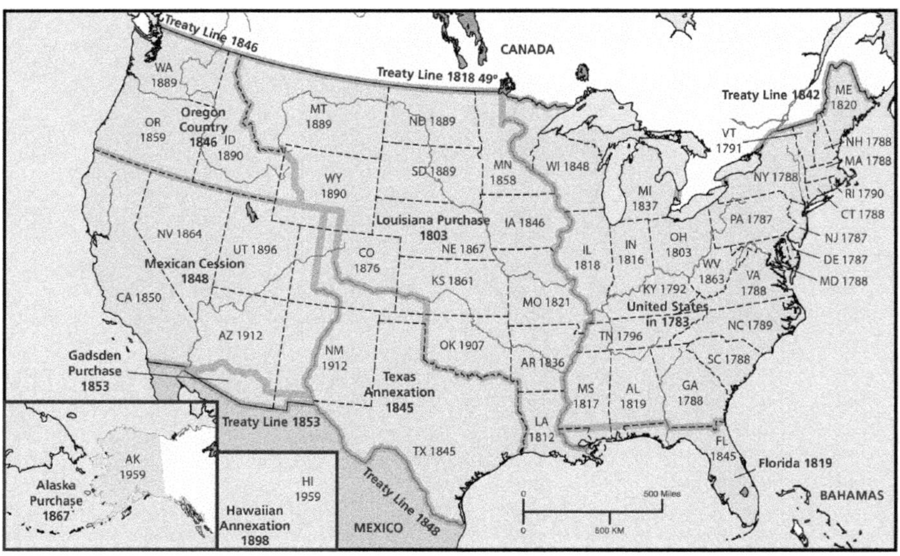

Figure 1.3. Map of westward expansion.
Courtesy of the Choices Program, Department of History, Brown University.

Other land was disposed of through grants to railroads in a checkerboard pattern, with 94,400,000 acres of the public domain granted to railroads in 640-acre sections of land alternating with sections retained by the government.[12] The theory was that construction of the railroad would increase the value of both the land granted to the railroad and the land retained by the government. In fact, much of the land retained by the government was never sold and is still public land today, interwoven with privately owned land.

Leasing government land for grazing and timber and mineral extraction moreover fell under the federal government's land disposal policy, beginning with the General Mining Act of 1872, which remains in effect today. Although the Homestead Act and similar laws were repealed in 1976, leasing continues. For example, as of the end of fiscal year 2017, the Bureau of Land Management had under lease a total of 25,742,991 acres for oil and gas operations and 458,669 acres for coal mining.[13]

Passage of the Federal Land Policy and Management Act of 1976 caused a shift away from land disposal. As a result, much of our wildlife in the West today exists primarily on public land. But many species also rely on private land for certain life-cycle functions such as seasonal food sources, nesting, nurseries, or overwintering. For other species, private lands that meander through valleys and along stream beds can provide important migratory corridors and enhance connectivity between populations on public land. Even for non-migratory populations, limiting protections to public land creates isolated "habitat islands" that lack connectivity and are less effective than landscapes for conserving species.[14]

The historic approach to conserving endangered species focused on public land and utilized federal legislation and command-and-control policies. Statutes such as the ESA and the Clean Water Act prohibit actions that would harm listed wildlife or pollute public waterways, and land management statutes such as Federal Land Policy and Management Act of 1976 and the National Forest Management Act require public land managers to apply these laws. This has resulted in both conservation successes and failures on public land, but the history of efforts to bring this approach onto private land is unequivocally one of failure, punctuated by conflict, litigation, recriminations, and distrust. At its worst, federal wildlife management created perverse incentives that encouraged landowners to mismanage their land in order to prevent the appearance of any endangered species upon their property.

In the last forty years, policymakers have increasingly understood and acted upon what Leopold was trying to tell us in 1934, that when private landowners conserve wildlife they are serving the public interest and should be rewarded accordingly. Congress, federal agencies such as the U.S. Fish and Wildlife Service (FWS), state governments, and non-governmental organizations ranging from hunting groups to foundations and land trusts have created a multitude of incentive and assistance programs. These programs form a constellation both expansive and dense, and in the best designed programs considerable effort goes into matching the right people and the right land with the right program at the right time.

This book is a guide to the programs presently available to private landowners interested in conserving wildlife on their land. It identifies and summarizes the different conservation mechanisms, programs, and partners available to them. It lays out the benefits and limitations of programs, and identifies resources for landowners to connect with programs for which they may be eligible.

Conservation Efforts Should Target All At-Risk Species

Under the ESA, species are listed as either "endangered," meaning "in danger of extinction throughout all or a significant portion of its range," or "threatened," meaning "likely to become an endangered species within the foreseeable future throughout all or a significant portion of its range." When a species is listed, the ESA grants jurisdiction over the species to either FWS or the National Marine Fisheries Service. Although these listed species are usually the most imperiled, they are far from the only species that are declining and in need of conservation. On the contrary, wildlife of virtually every kind is declining sharply worldwide, in what has been called the "sixth great extinction." Listing a species under the ESA, or the possibility of such a listing, can drive conservation of an entire ecosystem. Habitat conservation efforts generally focus on a single keystone or apex species that acts as an indicator species for many others that utilize the same habitat. Hence they collectively create a rich abundance of biodiversity within a given habitat. For example, the greater sage-grouse is both a keystone and indicator species in the vast sagebrush sea of the Great Plains for 353 species that share the same habitat and ecosystems.

Many species that are not listed as threatened or endangered are nonetheless at risk of decline and eventual extinction. Some of these species are formally defined by FWS as either "candidate" species or "at-risk" species. "Candidate" species are species "being considered . . . for listing as an endangered or threatened species, but not yet the subject of a proposed rule." Some of these are species that were found to warrant listing under the ESA, but were not listed due to a lack of agency resources (these species are also called "warranted but precluded"). Others are species "for which [FWS] . . . has on file sufficient information on biological vulnerability and threats to support a proposal to list as endangered or threatened under the Endangered Species Act."[15] "At-risk" species include "candidate species and other unlisted species that are declining and are at risk of becoming . . . candidate[s] for listing under the Endangered Species Act. Th[ese] may include, but [are] not limited to, State listed species, species identified by States as species of greatest conservation need, or species with State heritage ranks of G1 or G2."[16]

In this book, species in all of these different categories are collectively referred to as "at-risk species," meaning species that are at risk of decline and eventual extinction unless humans engage in active conservation efforts to preserve them. Such efforts may include captive breeding and reintroduction, restrictions on hunting and commercial uses, or protection from pollutants such as pesticides. Most often, however, what species need is for their habitat to be protected, restored, and expanded. This work can most easily be accomplished by the people who own that habitat.

Habitat conservation, as well as other conservation actions, is most efficient when directed toward at-risk species that are not yet listed under the ESA. Healthier populations are more resilient to threats such as wildfire, disease, and climate change. In most cases, healthier populations are also found in larger, less fragmented habitat. It is usually cheaper to preserve existing habitat than to restore degraded habitat, and to preserve populations that do not require translocations or captive breeding and reintroduction. Finally, regulatory oversight of species not listed under the ESA is limited to state fish and wildlife agencies, simplifying the legal framework within which conservation programs operate.

As Leopold understood, in our system of private property, the cost of conserving land must fall in large part upon the owners of that land, be they family farmers or ranchers, towns, corporations, or government agencies. The conservation programs discussed in this book are intended to meet Leopold's challenge to reward private landowners for conserving their land for the benefit of the public's interest in maintaining and restoring healthy wildlife populations. Species listed under the ESA must not be ignored, but neither should other at-risk species be overlooked.

The Impact of the Endangered Species Act on Private Landowners

The idea of focusing conservation efforts on habitat and including non-ESA-listed species is neither new nor inconsistent with the ESA. The link between habitat and population health was well-understood in the 1970s, and the ESA embraces it. The purpose of the ESA, in the words of Congress in the Preface of the Act, is "to provide a means whereby the ecosystems upon which endangered species and threatened species depend may be conserved, [and] to provide a program for the conservation of such endangered species and threatened species."[17] All of the tools available under the ESA, from listing species and designating critical habitat to developing recovery plans, using lawsuits to enforce the act, and even delisting species, are simply means to achieve the purpose of conserving ecosystems and species.

When conserving at-risk species, habitat is often the central issue. In the vast majority of cases, species are in decline and perhaps in danger of future ESA listing because of loss of habitat due to development. Development takes many forms, including forestry (especially in the southeast), oil and gas exploration and extraction (especially in the west), roadbuilding, and urbanization nationwide. To protect species, we must protect habitat. To protect habitat, we must restrict development. And most of this development is on private land. Thus, as Leopold understood, landowners ultimately bear much of the cost of conservation. In fact, many people criticize the ESA for functioning as a regulatory land use law. The tension between commercial development and wildlife values has been the root of most bitter conflicts over wildlife, from the Tellico Dam that threatened the snail darter in the late 1970s all the way to energy development that threatens the greater sage-grouse today.

Despite these very public controversies, the ESA's practical ability to protect habitat and drive conservation efforts on private land is surprisingly limited. Private landowners making changes to the habitat on their land are only subject to ESA regulations when a) they jeopardize the continued existence of a listed species, or adversely modify the officially designated critical habitat of a listed species; *and* b) federal funding or authorization is involved in the habitat modification.

As a threshold matter, the ESA only takes effect when a species is officially listed as threatened or endangered. The ESA's primary enforcement mechanism, Section 9, imposes powerful and rigid protections for individual members of a species through restrictions on "take" of species, punishable by civil and criminal penalties under Section 11. (To "take" a species is defined as to "harass, harm, pursue, hunt, shoot, wound, kill, trap, capture, or collect, or to attempt to engage in any such conduct.") In contrast to these protections for individual species, the ESA protects habitat in a more limited and roundabout away.[18]

The ESA's only explicit protection for habitat appears in Section 7(a)(2), which provides that "[e]ach Federal agency shall . . . insure that any action authorized, funded, or carried out by such agency . . . is not likely to jeopardize the continued existence of any endangered species or threatened species or result in the destruction or adverse modification of [critical] habitat of such species." Critical habitat may not be destroyed or "adversely modified," but the

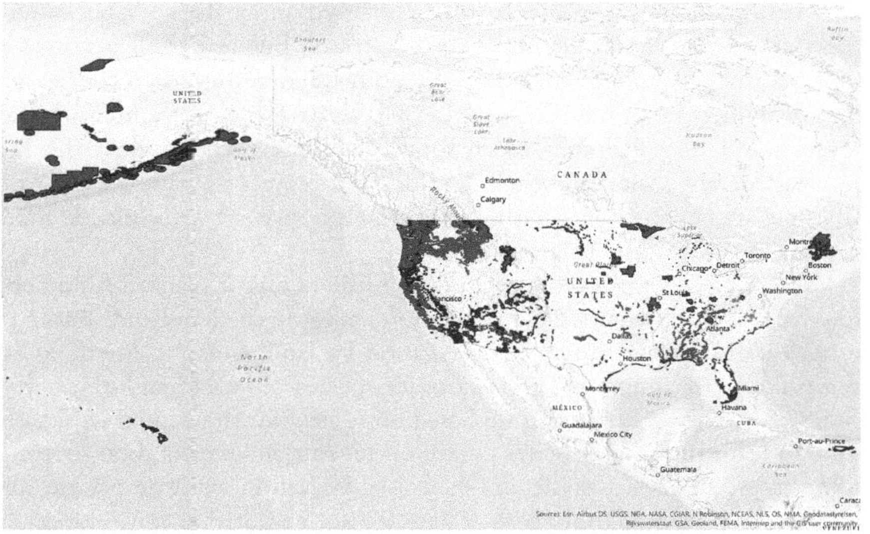

Figure 1.4. Map of designated and proposed critical habitat under the Endangered Species Act of 1973, as of September 2019.
Courtesy of the U.S. Fish and Wildlife Service.

reach of Section 7 is limited to critical habitat on public land and on those private lands where there is an action authorized, funded, insured, subsidized, or carried out by a federal agency, otherwise known as a "federal nexus."

Section 7 is a powerful and effective means of protecting species and their habitat on public lands. On private lands, its impact is more limited, although a federal nexus can be created by enrolling land in federal crop subsidy, crop insurance, or flood insurance programs, receiving a federal loan, receiving federal disaster relief, applying for a permit under federal anti-pollution laws, or enrolling in a voluntary federal conservation program under the Farm Bill or other laws.

The taking protections of Sections 9 and 11 are also limited in their effectiveness. The penalties imposed by these sections are retroactive and do not create incentives for affirmative actions to improve habitat, increase species' numbers, and ultimately recover species. Furthermore, the actual imposition of penalties for taking listed species is rare. In short, the ESA inconveniences a relatively small number of private landowners, greatly worries many more, and fails to encourage the conservation actions that at-risk species actually need.

Voluntary Conservation for All At-Risk Game and Non-Game Species

A better solution is available in the form of voluntary conservation undertaken by willing landowners. Much recent attention has focused on prelisting conservation for at-risk species that may be listed under the ESA, because the threat of a listing drives attention to these cases. But nearly all private land conservation is voluntary, including for both listed and unlisted species, and when intended to promote ecosystem health at a broader, landscape scale. Conservation may feel coerced in some situations, such as when the ESA requires federal consultation on a development project in critical habitat of a listed species, but these projects nearly always go forward with suitable modifications to protect listed species.[19]

Voluntary conservation has expanded greatly in recent years. As Aldo Leopold wrote, "The implements for restoration lie not in the legislature, but in the farmer's toolshed."[20] Voluntary conservation by landowners is emerging as a powerful tool for improving conditions for species and avoiding listings under the ESA. Voluntary conservation is based on respect for the natural world, on a desire to leave the land in better condition for our children, and on successful partnerships. Effective wildlife and habitat conservation work on private lands increasingly relies on partnerships of state, federal, and private conservation entities working alongside willing landowners.[21] Voluntary conservation requires buy-in from all participants, and that buy-in yields, in turn, commitment, follow-through, and innovative solutions to challenging problems.

Today there are a growing number of programs and initiatives to incentivize voluntary conservation. Some of them are under the umbrella of the ESA and designed by Congress or the FWS and National Marine Fisheries Service to make the ESA less burdensome by offering regulatory incentives and assurances to landowners. Far more operate entirely outside the context of the ESA. Many of these programs are focused on providing funding for conservation on private land. Programs under the federal Farm Bill provide the majority of this funding, but there are also other federal grant programs and non-governmental organization programs, as well as programs that employ market-based solutions to drive private capital toward conservation. Other programs provide scientific expertise and technical assistance to landowners. Many combine funding with technical assistance and engage multiple partners. Collectively, these tools within the conservation toolbox stand ready to be picked up and used by private landowners across the nation.

For many private landowners, the greatest challenge may not be deciding they want to manage some or all of their land for conservation, but identifying the programs and partners available to them. The purpose of this book is to help landowners understand the universe of voluntary conservation programs. For each, it identifies purposes and requirements, advantages and disadvantages, and implementing organizations. It includes stories of successes as well as challenges that had to be overcome. It shares its ultimate purpose with the programs it describes: To realize Aldo Leopold's vision and reward private landowners for conserving the ecosystems and wildlife that sustain our civilization, as well as life on earth itself.

Notes

1. Aldo Leopold, "Conservation Economics," in *The River of the Mother of God and Other Essays by Aldo Leopold*, ed. Susan L. Flader and J. Baird Callicott (Madison, WI: University of Wisconsin Press, 1991), 200, 202.

2. See chapter 6.

3. John J. Morgan et al., "A State Assessment of Private Lands Wildlife Conservation in the United States," *Wildlife Society Bulletin* 43, no. 3 (2019): 328–37. For information on Farm Bill programs, see chapters 6 through 9.

4. Steven P. Riley, Robert Abernethy, and Donald F. McKenzie, "Case Studies of Scalable Wildlife Conservation in Working Landscapes," *Wildlife Society Bulletin* 43, no. 3 (2019): 363–71.

5. Morgan et al., "A State Assessment of Private Lands Wildlife Conservation in the United States."; L. Wes Burger Jr. et al., "Private Lands Conservation: A Vision for the Future," *Wildlife Society Bulletin* 43, no. 3 (2019): 398–407.

6. U.S. Fish and Wildlife Service, "ESA Basics: 40 Years of Conserving Endangered Species," 2013, https://www.fws.gov/endangered/esa-library/pdf/ESA_basics.pdf; Megan E. Hansen et al., *Cooperative Conservation: Determinants of Landowner Engagement in Conserving Endangered Species* (Logan, UT: Center for Growth and Opportunity, Utah State University, November 2018).

7. These figures include only lands managed by the four public land management agencies (the Bureau of Land Management, Forest Service, Fish and Wildlife Service, and National Park Service) and the Department of Defense. They exclude federal lands managed by other agencies, such as the Bureau of Reclamation, and Indian lands, such as those held in trust by the United States. Carol Hardy Vincent, Laura A. Hanson, and Carla N. Argueta, *Federal Land Ownership: Overview and Data, Congressional Research Service Report R42346* (March 3, 2017).

8. Michael G. Brasher et al., "The History and Importance of Private Lands for North American Waterfowl Conservation," *Wildlife Society Bulletin* 43, no. 3 (2019): 338–54.

9. Bureau of Land Management, *Public Land Statistics 2017* (June 2018), 3–4.

10. Ibid., 5.

11. For example, Steve Jester, interview by Lowell E. Baier and Christopher E. Segal, July 19, 2019; Dave Smith and Ali Duvall, interview by Christopher E. Segal, August 12, 2019.

12. Bureau of Land Management, *Public Land Statistics 2017*, 5.

13. Bureau of Land Management, "Oil and Gas Statistics, Table 2: Acreage in Effect," 2018, https://www.blm.gov/sites/blm.gov/files/Table2_Number_of_Acres_Leased.xlsx; Bureau of Land Management, "Coal Data, National Coal Statistics Table," 2018, https://www.blm.gov/programs/energy-and-minerals/coal/coal-data.

14. Michael J. Bean and Melanie J. Rowland, *The Evolution of National Wildlife Law*, third edition (Westport, CT: Praeger, 1997), 427–28.

15. For example, Endangered Species Act Compensatory Mitigation Policy, 81 Fed. Reg. 95316 (December 27, 2016), 95346–47 (withdrawn).

16. Ibid., 95346.

17. U.S. Fish and Wildlife Service, Endangered Species Act of 1973, as amended through the 108th Congress (2005), 1. "Conserve" means "the use of all methods and procedures which are necessary to bring any endangered species or threatened species to the point at which the measures provided pursuant to [the ESA] are no longer necessary." Ibid., 2.

18. Under regulations adopted by FWS in 1981, habitat modification does not result in "harm" to a species unless individual members of that species are actually killed or injured. Final Redefinition of "Harm," 46 Fed. Reg. 54748 (November 4, 1981).

19. See chapter 12 for more discussion.

20. Aldo Leopold, "The Conservation Ethic," in *The River of the Mother of God and Other Essays by Aldo Leopold*, ed. Susan L. Flader and J. Baird Callicott (Madison, WI: University of Wisconsin Press, 1991), 185.

21. For example, Sunni L. Carr et al., "Influence of the Federal Landowner Incentive Program," *Wildlife Society Bulletin* 43, no. 3 (2019): 355–62.

Case Studies in Conservation
Red-Cockaded Woodpecker and Greater Sage-Grouse

2

Hundreds of species benefit from voluntary conservation, and new conservation actions are constantly bringing new species, new lands, and new landowners into the voluntary conservation movement. This book features many examples of species benefiting from conservation, but two species are used as case studies throughout: the red-cockaded woodpecker and the greater sage-grouse. These species have been selected for both their commonalities and differences.

Both species are birds, among the earliest and most heavily studied and conserved species in America.[1] Neither are migratory, meaning they are present in the same areas year-round, which makes it easy to identify populations and habitat and design targeted conservation programs. Each species is widespread, historically abundant, and closely associated with an iconic American landscape. Yet each suffered steep historic declines until serious efforts to conserve them were launched.

There are also critical differences between these species. The red-cockaded woodpecker lives in pine forests of twelve southeastern states, where land is overwhelmingly private, whereas the greater sage-grouse is found in the sagebrush sea that sprawls across eleven western public land states.[2] They also have different histories under the Endangered Species Act of 1973 (ESA). The red-cockaded woodpecker was listed as "endangered" in 1970.[3] The population hit its lowest point in 1990,[4] and since then a combination of conservation agreements, federal funding, and science-driven timber management practices, all under the aegis of the ESA, have set the species on a path toward recovery. Down- or delisting is a very real possibility in the near future, and as of 2019 the Fish and Wildlife Service (FWS) is in the process of conducting an assessment of the species.[5] The greater sage-grouse is not listed under the ESA but suffered population declines until 2013 despite a conservation program launched in 1995. Conservation efforts greatly accelerated after 2011 under threat of a possible ESA listing, ultimately stabilizing the species. These efforts included state governments, federal agencies, non-governmental organizations, private landowners, and local partnerships. Critical financial support came from the U.S. Department of Agriculture's Natural Resources Conservation Service and state governments. In 2015, the FWS concluded that it was not necessary to list the species under the ESA, and efforts to secure the species' recovery continue today.

Each of these species, the woodpecker and the grouse, is a conservation success story. Each required engagement by countless partners across millions of acres of land. Each benefited from a variety of regulatory and non-regulatory approaches, funding, and tools used by private landowners to engage with partnerships and implement conservation. And each showcases many of the conservation tools available to landowners. This chapter provides a brief introduction to these two species.

Red-Cockaded Woodpecker

The red-cockaded woodpecker (*Leuconotopicus borealis*) is a medium-sized (7.8 to 9.1 inches) woodpecker that lives exclusively in pine forests in the southeastern United States. Black and white (only mature males have the distinctive red patches, or cockades, for which the species is named), the species is identifiable by the ladder-like pattern of stripes on its wings and a distinctive white cheek patch. Red-cockaded woodpeckers breed in monogamous pairs but live in larger family groups: unattached males remain with their parents as "helpers" caring for subsequent clutches of eggs, whereas young females disperse in search of new territory. Population estimates for the species are based on the number of breeding groups, or "active clusters," rather than on a count of individual birds.[6]

Red-cockaded woodpeckers are highly reliant on mature pine forests. Trees generally need to be thirty years old before they can provide forage for woodpeckers and closer to one hundred years old before they can be used for nesting.[7] Red-cockaded woodpeckers nest in individual cavities that they excavate from pine trees over a period of one to three years. Unlike all other species of woodpecker, they excavate their cavities from living trees. As a result of this, the species has adapted to live in forests that experience frequent, low-intensity burning. These fires clear the woodlands of snags that host competing species and create an ecological niche for the red-cockaded woodpecker to occupy.[8] Historically, red-cockaded woodpeckers prefer mature longleaf pine trees, ideally those infected by red heart disease, a fungus

Figure 2.1. Red-cockaded woodpeckers building a nest.
Photograph by Doug Gochfeld.

that makes the interior of the tree soft and easier to excavate. For many years following European colonization of the southeast, the less valuable infected trees were often left standing. In more recent years, however, longleaf forests have been entirely converted to other pine species such as loblolly and slash. Although red-cockaded woodpeckers can live in these trees, they do not prefer them, in part because they are smaller. In addition, many of these trees are harvested for commercial lumber every twenty-five years, before red-cockaded woodpeckers can forage, let alone nest, in them. As a result, commercial pine forests are virtually devoid of red-cockaded woodpeckers.[9] Another distinctive characteristic of red-cockaded woodpecker nests is a series of small holes that cause sap to flow down the tree. This protects the nests from predators such as rodents and snakes. Twenty-seven other species of vertebrates have been documented using red-cockaded woodpecker cavities as dwellings. Other species of woodpecker, such as the pileated woodpecker, widen nesting cavity openings to the point that red-cockaded woodpeckers can no longer use the cavity.[10]

It is thought that at the time of European contact, the red-cockaded woodpecker numbered between 1 and 1.6 million active clusters, but by the time of its listing in 1970, the species had declined to approximately four thousand active clusters.[11] The primary threat to the species is a lack of suitable habitat. Frequent cutting of timber reduces the number of trees old enough

Figure 2.2. A red-cockaded woodpecker in front of a nesting cavity.
Photograph by Brady Beck.

to support woodpecker cavities, and fire suppression has made it easier for other species to outcompete red-cockaded woodpeckers. In addition, habitat fragmentation has made the species more susceptible to other threats such as natural disasters and lack of genetic variation.[12]

Conservation efforts began with development of a recovery plan in 1979, but at that time it was expected that most of the recovery effort would occur on public land.[13] Little progress was made in the first decade, and in 1990 the population reached a historic low of just 1,981 active clusters. One reason for this is that many landowners were hostile to red-cockaded woodpeckers. Fearing regulation under the ESA, landowners had an incentive to plant smaller, faster-growing species of pine and aggressively thin forests to prevent the maturating of suitable nesting trees for red-cockaded woodpeckers. Their concern was that if they were to cut down a tree occupied by red-cockaded woodpeckers, they could be held liable for taking an endangered species. In 1986, an outside review of the recovery program warned that the species was on a course toward being "restricted to state and national forests and other public lands."[14] This was the state of affairs despite the fact that there was no designated critical habitat for red-cockaded woodpecker—the threat of regulation of take under Section 9 of the ESA alone was enough to inspire fear among landowners.

In the 1990s, conservation incentives and ESA regulatory assurances began to yield conservation gains on private land and reverse the red-cockaded woodpecker's decline. In 1992, a private lands conservation strategy was developed.[15] A driving force behind this was the U.S. Department of Defense, which was unable to carry out its training mission on several major installations within red-cockaded woodpecker habitat, jeopardizing national defense. At Fort Bragg in the sandhills region of North Carolina, the Department of Defense, FWS, The Nature Conservancy, and other stakeholders collaborated to find a solution. By financing protection of woodpeckers on private land, they were able to protect woodpecker habitat, enable the military to meet its training objectives, and put money in landowners' pockets. As a result of this collaboration, in 1995 the FWS developed ESA regulatory assurances for landowners working to conserve woodpeckers, and this new tool spread rapidly across the woodpecker's range. Today, there are over four hundred landowners conserving 2.5 million acres of red-cockaded woodpecker habitat under ESA Safe Harbor Agreements alone.[16]

These programs were aided by the invention in the 1980s of artificial nesting cavities that red-cockaded woodpeckers would use, enabling active management of red-cockaded woodpecker habitat to accelerate colonization and establishment of new active clusters (see figures 2.3 and 2.4). Artificial cavities come in two varieties: drilled cavities and insert cavities. Drilled cavities were developed at North Carolina State University, and insert cavities were developed around the same time at Clemson University.[17] In a drilled cavity, a large drill bit, 1 to 1.5 inches in diameter, is pushed into a pine tree from two points, converging in the heartwood and forming a nesting cavity. Like natural cavities, these drilled cavities last an average of ten years, and they can only be placed in mature trees of approximately one hundred years old.[18] Inserted cavities are large rectangular blocks fashioned from woods such as western red cedar that are installed in existing pine trees. A space is mechanically excavated from the tree, the artificial cavity installed, and then the entire structure is sealed into the tree. Coating the lower trunk with pitch simulates the protective sap characteristic of red-cockaded woodpecker nests. Cavities feature metal face plates that prevent other woodpeckers from enlarging the holes, with small openings for red-cockaded woodpecker feeding and home-defense behavior.[19] Insert cavities are placed in sapwood rather than heartwood and can go in trees as small as fifteen inches in diameter. They are also significantly cheaper than drilled cavities, but do not last as long and must be replaced, or supplemented with more cavities in new trees, after an average of five years.[20]

Because the red-cockaded woodpecker is listed as an endangered species and artificial cavities pose a risk of harming woodpeckers if not drilled or installed properly, installation must be by trained professionals with permits from the FWS and the relevant state fish and wildlife agency. Installation services may be provided by retail outlets that sell insert cavities and other forest conservation products, biologists and environmental consultants that specialize in red-cockaded woodpecker conservation, government agencies, non-governmental organizations, and other partners.[21] These organizations, including non-governmental organizations such as The Nature Conservancy; Department of Defense installations such as Fort Bragg, North

Figure 2.3. Artificial nesting cavities have been key to red-cockaded woodpecker recovery. This model, made of western red cedar, features a stainless steel faceplate that prevents other species from enlarging the cavity. Holes in the faceplate allow red-cockaded woodpeckers to drill small holes around their cavity. In a natural pine tree, this defensive behavior would release sap that would dissuade terrestrial predators such as mice and snakes from reaching the nest. The artificial nesting cavity is shown uninstalled, left, and installed, right.
Courtesy of Raven Environmental Services, Inc.

Figure 2.4. Once installed, red-cockaded woodpecker artificial nest cavities are sufficiently similar to natural cavities for red-cockaded woodpeckers to occupy them.
Courtesy of the U.S. Fish and Wildlife Service.

Carolina; and private consultants such as Dr. J.H. Carter III & Associates, Inc., of Southern Pines, North Carolina, are often able to provide grant funding that allows small landowners to place artificial nesting cavities on their land at no expense.[22]

Today, the red-cockaded woodpecker population has greatly expanded from its historic low. In 2006 (the last year in which a five-year review was completed), the population had risen to 6,105 active clusters.[23] The 2003 revision to the red-cockaded woodpecker foresaw delisting between 2050 and 2075,[24] but recent process has accelerated recovery and delisting may now be possible by 2030 or sooner.[25] One challenge for recovery is that the species remains limited by the availability of cavities, and red-cockaded woodpecker habitat is management-intensive. The forest must maintain enough old, large trees for woodpecker cavities, controlled burning is needed to maintain woodpecker forage and prevent the development of forest too dense for woodpeckers, and dead snags must be left standing.[26] Our ability to continue to fund the management needed to maintain and expand red-cockaded woodpecker habitat will ultimately determine whether the species' recovery is a long-term success.

Figure 2.5. Although the red-cockaded woodpecker's range covers much of the Southeastern United States, habitat fragmentation and the decline of old-growth longleaf pine trees has greatly diminished the species' population.
Birds of North America, http://birdsna.org, published by the Cornell Lab of Ornithology.

Greater Sage-Grouse

Hundreds of miles to the west of the red-cockaded woodpecker lives the greater sage-grouse (*Centrocercus urophasianus*). The greater sage-grouse is a large (twenty-two to twenty-nine inches long), non-migratory, ground-nesting gallinaceous bird with a range that sprawls over 165 million acres in eleven states (California, Colorado, Idaho, Montana, Nevada, North Dakota, Oregon, South Dakota, Utah, Washington, and Wyoming). The species has declined 44 percent from its historic range of 297 million acres. The greater

Figure 2.6. Greater sage-grouse are known for their elaborate mating rituals. Here, a male spreads his tail feathers, puffs up his chest, and uses air sacs to emit a distinctive booming sound to attract mates. Each spring, breeding males congregate at the same locations, known as leks, to compete for mates. At most leks, a few or even a single male will mate with the vast majority of females.
Kenneth Rush/Shutterstock.

Figure 2.7. The sagebrush ecosystem sprawls over 120 million acres of the American west, supporting greater sage-grouse, mule deer, and more than 350 other species.
Wayne Broussard/Shutterstock.

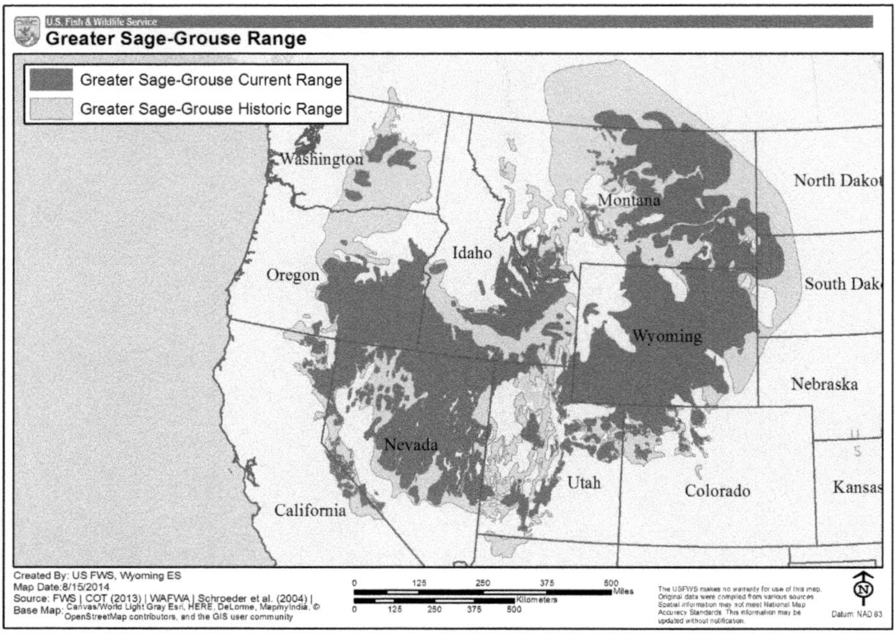

Figure 2.8. The range of the greater sage-grouse has declined 44 percent from its historic 296,646,000 acres to just 165,168,000 acres.
Courtesy of the U.S. Fish and Wildlife Service.

sage-grouse is a sagebrush-obligate species, feeding exclusively on sagebrush in winter and relying on sagebrush for cover from predators during nesting season. Widely recognized as the iconic species of the "sagebrush sea," the grouse is an ecological indicator species for 353 species that share the same ecosystem, including the pygmy rabbit, pronghorn, badger, loggerhead shrike, mule deer, golden eagle, western burrowing owl, greater short-horned lizard, and sage thrasher.

The greater sage-grouse was first described by Meriwether Lewis in a journal entry dated May 2, 1806, which noted that the species is "invariably found in the plains."[27] The population of greater sage-grouse is thought to have once been as high as sixteen million, but as settlers spread across the plains through the nineteenth century, the species declined, and by the early 1900s, when perhaps one million sage-grouse remained,[28] people began to take notice. Prominent conservationist Dr. William Temple Hornaday organized a campaign against overhunting of sage-grouse. In 1916, he published a pamphlet titled "Save the Sage Grouse from Extinction: A Demand from Civilization to the Western States," which was the first published prediction of the species' possible extinction.[29] Awareness of sage-grouse's decline gradually spread, and in 1954 the Western Association of State Fish and Wildlife

Agencies established the Sage and Columbian Sharp-tailed Grouse Technical Committee[30] to study sage-grouse and make recommendations. Although widespread hunting of greater sage-grouse was already on the decline, new threats were emerging in the form of habitat fragmentation due to land subdivision, exurban development, oil and gas development, wildfires, and the spread of invasive species and conifers.[31] In 1994, alarms were sounded regarding a possible ESA listing of the greater sage-grouse, "possibly within the next five years," and conservation efforts began in earnest. By 1995, every state in the sage-grouse's range had launched a conservation program to benefit the species.

The greater sage-grouse is best known for its elaborate mating ritual. Every spring, male sage-grouse gather in broad, open spaces called leks to perform elaborate displays. Males spread their feathers, force air through sacs on their chests to emit a distinctive booming noise, and strut in attempts to attract mates. These displays continue for several hours a day and for several weeks each spring. Typically, only one or two males at each lek will mate with the vast majority of females. After mating, females leave the lek to establish nests, usually under cover of sagebrush.

Accurately counting greater sage-grouse is difficult. Estimates are generally based on counting breeding males present on leks each spring. The range-wide

Figure 2.9. Male greater sage-grouse use the air sacs on their chests to emit a distinctive booming noise.
Photograph by Bert Filemyr.

Figure 2.10. After mating at leks, female greater sage-grouse travel short distances to establish nests under cover of prairie grasses. The grasses provide cover and food for the newborn chicks. Max Allen/Alamy Stock Photo.

population hit a low point in 2013, when surveys counted just 48,641 breeding males, a decline of 56 percent from 109,990 in 2007.[32] The most recent comprehensive, range-wide survey of greater sage-grouse was conducted in 2015, and it documented a sharp increase to 85,674 breeding males.[33] This has been extrapolated to suggest an overall population of 424,645.[34]

Over 50 percent of the greater sage-grouse's range is on public land managed by the Bureau of Land Management or the U.S. Forest Service. But throughout its range, there are opportunities for conservation work by private landowners. In the western part of the species' range, roughly encompassing the Great Basin, fire is the greatest threat to the species. Large areas of the range are caught in a fire/invasive species cycle, where cheatgrass and other invasive grasses create conditions ripe for devastating wildfires that kill sagebrush and open up the steppe for the growth of pinyon-juniper and other hardwood trees that provide hunting habitat for ravens and other sage-grouse predators. Wildfire prevention through controlled burns, mechanical treatments to remove pinyon-juniper woodlands and invasive species, and the banking and planting of native sagebrush seed have all helped reduce loss of sagebrush habitat due to this cycle.

In the eastern part of the sage-grouse's range, oil and gas development is the largest threat to the species. Roads, powerlines, and well pads all fragment habitat and provide perches for predators, which frightens sage-grouse and prevents hens from establishing nests. These threats can be reduced by protecting areas around known sage-grouse leks, using directional oil and gas drilling, and co-locating infrastructure in already disturbed habitat. Finally, throughout the range livestock grazing can be a threat to greater sage-grouse if overgrazing is allowed, destroying sagebrush and promoting cheatgrass. Properly managed grazing, on the other hand, can protect sage-grouse and improve overall habitat for wildlife and domestic livestock alike. Greater sage-grouse conservation is a wise investment because the species is far more easily conserved as opposed to restored. Sagebrush is a slow-growing plant that is easily outcompeted by invasive grasses, so recreating sage-grouse habitat is difficult. In addition, sage-grouse do not readily colonize new territory, instead returning to the same leks year after year. If an individual lek is destroyed or otherwise rendered unusable, its birds and their potential offspring will be lost.[35]

We have long known how to conserve sagebrush habitat for greater sage-grouse and other related species. But just as the possibility of an ESA listing

Figure 2.11. Oil and gas development near leks can provide habitat for greater sage-grouse predators such as ravens and can lead to lek abandonment.
Photograph by Joel Sartore/www.joelsartore.com.

stimulated interest in the species in the late 1990s, it took the imminent threat of an ESA listing to prompt widespread support for and investment in greater sage-grouse conservation programs. The species was formally petitioned for listing in 2002, but after a series of lawsuits FWS determined that listing the species was warranted but precluded by higher priorities. In 2011, the agency entered into a major legal settlement with two environmental litigation organizations and announced that it would make a final decision on whether to list the greater sage-grouse by September 2015. Beginning in 2007, state governments launched major investments in greater sage-grouse conservation. In 2010 the Natural Resources Conservation Service launched the Sage Grouse Initiative, which to date has invested over $760 million in sage-grouse conservation, counting both U.S. government appropriations and funds raised from partner organizations.[36] This money has enabled partnerships with eighteen hundred ranchers and conservation of over seven million acres of sage-grouse habitat.[37] Conservation actions funded by the Natural Resources Conservation Service include the use of grazing to eliminate invasive grasses and noxious weeds, mechanical treatments to remove invasive conifers, protection of land to reduce the impact of exurban development, protection of existing sagebrush habitat,[38] protection of moist habitat used by sage-grouse for feeding young, and removal of range fencing.[39]

When state government contributions are included, total investment in greater sage-grouse conservation between 2005 and 2020 will come to a total of $1.5 billion, by far the largest voluntary program ever established for the benefit of a single species. Because of this effort, in 2015 FWS announced that the greater sage-grouse did not warrant listing under the ESA. This decision relied in part on a database of sixty-two hundred projects carried out to benefit the species, with fully completed projects on three million acres of land.[40] This huge success for the species was also a huge success for landowners, and in the years since the decision conservation efforts for the greater sage-grouse have continued.

Notes

1. Much bird conservation has been driven by the 1918 Migratory Bird Treaty Act, which established federal jurisdiction over all migratory birds. However, neither the red-cockaded woodpecker nor the greater sage-grouse is migratory, meaning that federal versus state jurisdiction over the two species is based solely on their status under the ESA.

2. Although the majority of greater sage-grouse habitat is on public land, efforts by private landowners and leaseholders who graze livestock on public land have been critical to the conservation of the species.

3. This was under the authority of the Endangered Species Act of 1969, a predecessor to the modern ESA.

4. Caroline Roth, "Red-Cockaded Woodpecker Populations Begin Rebound," 2016. United States Department of Agriculture, https://www.usda.gov/media/blog/2016/03/28/red-cockaded-woodpecker-populations-begin-rebound.

5. Wendy Jackson, interview by Christopher E. Segal, January 28, 2019; Jay H. Carter, interview by Christopher E. Segal, September 30, 2019.

6. Cornell Lab of Ornithology, "All About Birds: Red-Cockaded Woodpecker Identification," last accessed April 25, 2019, https://www.allaboutbirds.org/guide/Red-cockaded_Woodpecker/id#; National Audubon Society, "Guide to North American Birds: Red-Cockaded Woodpecker," last accessed April 25, 2019, https://www.audubon.org/field-guide/bird/red-cockaded-woodpecker; National Wildlife Federation, "Red-Cockaded Woodpecker," last accessed April 25, 2019, https://www.nwf.org/Educational-Resources/Wildlife-Guide/Birds/Red-Cockaded-Woodpecker; Center for Biological Diversity, "Red-Cockaded Woodpecker," last accessed April 25, 2019, https://www.biologicaldiversity.org/campaigns/esa_works/profile_pages/RedcockadedWoodpecker.html.

7. Jay H. Carter, interview by Christopher E. Segal, September 30, 2019.

8. For example, Richard N. Conner, Daniel Saenz, and D. Craig Rudolph, "The Red-Cockaded Woodpecker: Interactions with Fire, Snags, Fungi, Rat Snakes, and Pileated Woodpeckers," *Texas Journal of Science* 56, no. 4 (2004): 415–26.

9. Jay H. Carter, interview by Christopher E. Segal, September 30, 2019.

10. U.S. Fish and Wildlife Service, "Red-Cockaded Woodpecker Recovery," December 19, 2016, https://www.fws.gov/rcwrecovery/rcw.html.

11. D. Craig Rudolph, Richard N. Conner, and Jeffrey R. Walters, "Red-Cockaded Woodpecker Recovery: An Integrated Strategy," in *Red-Cockaded Woodpecker: Road to Recovery*, ed. Ralph Costa and Susan J. Daniels (Blaine, WA: Hancock House Publishers, 2004), 70–76.

12. U.S. Fish and Wildlife Service, "Recovery Plan for the Red-Cockaded Woodpecker (*Picoides borealis*)," second revision (2003), 5–9.

13. David Drake and Edwin J. Jones, "Forest Management Decisions of North Carolina Landowners Relative to the Red-Cockaded Woodpecker," *Wildlife Society Bulletin* 30, no. 1 (2002): 122.

14. J. David Lzgon et al., "Report of the American Ornithologists' Union Committee for the Conservation of the Red-Cockaded Woodpecker," *The Auk* 103 (October 1986): 849.

15. U.S. Fish and Wildlife Service, "Recovery Plan for the Red-Cockaded Woodpecker (*Picoides borealis*)," second revision (2003), 121.

16. Michael J. Bean, *Endangered Species Act Safe Harbor Agreements: An Assessment, Working Paper* (Madison, WI: Sand County Foundation/Environmental Policy Innovation Center, 2017), 4. See chapter 13 for more information about this program.

17. Jay H. Carter, interview by Christopher E. Segal, September 30, 2019.

18. Ibid.

19. Wendy Jackson, interview by Christopher E. Segal, January 28, 2019; The Roost Stop, "Red-Cockaded Woodpecker Artificial Cavity Insert," last accessed April 25, 2019, https://www.therooststop.com/Red-cockaded-Woodpecker-insert-p/red-cockaded%20woodpecker%20insert.htm.

20. Jay H. Carter, interview by Christopher E. Segal, September 30, 2019.

21. Ibid.; The Roost Stop, "Red-Cockaded Woodpecker Artificial Cavity Insert"; Earth Tech Environmental LLC, "Red-Cockaded Woodpecker Cavity Installation," last accessed September 16, 2019, http://eteflorida.com/red-cockaded-woodpecker-cavity-installation/.

22. Jay H. Carter, interview by Christopher E. Segal, September 30, 2019.

23. Ralph Costa and U.S. Fish and Wildlife Service, "5-Year Review: Red-Cockaded Woodpecker (*Picoides borealis*)," (2006), 10. An updated review is in progress as of 2019.

24. U.S. Fish and Wildlife Service, "Recovery Plan for the Red-Cockaded Woodpecker (*Picoides borealis*)," second revision (2003), xvii.

25. Wendy Jackson, interview by Christopher E. Segal, January 28, 2019; Jay H. Carter, interview by Christopher E. Segal, September 30, 2019.

26. Jay H. Carter, interview by Christopher E. Segal, September 30, 2019.

27. "Journals of the Lewis & Clark Expedition, March 2, 1806," last accessed August 11, 2017, https://lewisandclarkjournals.unl.edu/item/lc.jrn.1806-03-02#n28030201.

28. The Natonal Wildlife Federation, "Saving the Greater Sage-Grouse," last accessed May 7, 2018, https://www.nwf.org/Our-Work/Wildlife-Conservation/Greater-Sage-Grouse.

29. William Temple Hornaday, *Save the Sage Grouse from Extinction: A Demand from Civilization to the Western States*, Permanent Wild Life Protection Fund Bulletin 5 (New York, NY: New York Zoological Park, 1916); Michael R. Conover and Anthony J. Roberts, "Declining Populations of Greater Sage-Grouse: Where and Why," *Human–Wildlife Interactions* 10, no. 2 (2016): 217.

30. U.S. Fish and Wildlife Service, "Who's on the Lek: A Guide to Players," last accessed August 11, 2017, https://www.fws.gov/mountain-prairie/species/birds/sagegrouse/Primer2SGWhoOnTheLek.pdf.

31. Today there is limited hunting of greater sage-grouse in those areas where the population is large enough to support it.

32. Rocky Barker, "Sage Grouse Numbers Plummeted after 2007, Study Shows," *Idaho Statesman*, April 24, 2015.

33. Western Association of Fish and Widlife Agencies, "WAFWA Report Documents Greater Sage-Grouse Population Rebound," August 17, 2015, https://www.wafwa.org/news/e_1606/News/2015/8/_WAFWA_Report_Documents_Greater_Sage-Grouse_Population_Rebound.htm.

34. Western Association of Fish and Wildlife Agencies, "Greater Sage-Grouse Population Trends: An Analysis of Lek Count Databases 1965–2015," August, 2015, http://www.wafwa.org/Documents%20and%20Settings/37/Site%20Documents/News/Lek%20Trend%20Analysis%20final%208-14-15.pdf; U.S. Fish and Wildlife Service, "The Greater Sage Grouse: Facts, Figures and Discussion," last accessed September 20, 2019, https://www.fws.gov/greatersagegrouse/factsheets/GreaterSageGrouseCanon_FINAL.pdf.

35. James R. Lyons, interview by Lowell E. Baier and Christopher E. Segal, October 9, 2018.

36. Natural Resources Conservation Service, "Sage Grouse Initiative 2.0 Investment Strategy, FY 2015-2018," August, 2015, http://www.sagegrouseinitiative.com/wp-content/uploads/2015/08/SGI2.0_Final_Report.pdf. See chapters 7 and 8 for more information.

37. Sage Grouse Initiative, "1,800 Ranchers and Counting!" December 18, 2018, https://www.sagegrouseinitiative.com/1800-ranchers-and-counting/.

38. This includes establishment of conservation easements on 450,000 acres of private land. Natural Resources Conservation Service, "Sage Grouse Initiative 2.0 Investment Strategy, FY 2015-2018," 11.

39. Ibid.

40. "12-Month Finding on a Petition to List Greater Sage-Grouse (Centrocercus Urophasianus) as an Endangered or Threatened Species," 80 Fed. Reg. 59858 (Ootober 2, 2015), 58871–72.

Getting Started with Conservation 3

PRIVATE LANDOWNERS OPERATE under competing demands and objectives. Foremost among these is earning a living and providing for their family. Other concerns may include maintaining their land to pass on to their descendants; being involved in community groups and trade associations; adopting practices that improve their business, including measures that improve the quality of their land and the biological community that it supports; and, in some cases, wildlife conservation. Wildlife conservation can be an end unto itself or can be the result of actions that are designed to benefit multiple resources such as soil quality, water quality, agricultural productivity, timber, and wildlife. Many wildlife conservation programs today seek to create win-win scenarios where wildlife and landowners themselves can both benefit.

Central to win-win conservation are efforts to eliminate regulatory and practical obstacles that discourage landowners from improving wildlife habitat and conserving wildlife. Most of this book is dedicated to profiling programs like these, which include financial and technical assistance under the federal Farm Bill (chapters 6–9), landowner assistance provided by a vast array of Federal, state, and NGO conservation partners (chapters 10–11), and Endangered Species Act programs run by the U.S. Fish and Wildlife Service and National Marine Fisheries Service (chapters 12–13). These programs provide financial assistance, technical support, and regulatory assurances to landowners engaged in voluntary conservation.

Planning is an important part of a well-designed, cost-effective conservation program, and is required to participate in many formal conservation programs. Landowners need to know what wildlife species and habitats are present on their land, must implement practices that benefit those resources, and must see a positive impact from those practices over time. In many cases, landowners

Figure 3.1. NRCS Texas range management specialists Lem Creswell (left) and Kent Ferguson (right) discuss ranch resource management goals with Montague County rancher Rooter Brite. Courtesy of USDA Natural Resources Conservation Service.

already have the knowledge they need to implement conservation practices, simply from living and working on the land. Examples of areas where landowners may lack knowledge are new farming or forestry techniques that could improve their operation, and information about conservation assistance programs that they may qualify for. Educating landowners about best farming, ranching, or silvicultural practices and conservation assistance programs is part of the mission of many NGOs, conservation districts, and local trade associations.

Conservation Planning

Formal conservation plans can be helpful to landowners, even landowners already practicing conservation. They are a requirement for participation in many of the programs discussed in this book, and may also be required by state law.[1] Fortunately, many partner organizations stand ready to assist landowners in preparing conservation plans, either in conjunction with participation in a conservation program or as a stand-alone product.

One of the ways that the U.S. Department of Agriculture's Natural Resources Conservation Service (NRCS) and its partners have worked to streamline participation in federal Farm Bill conservation programs (see chapters 6–9) is by making conservation planning part of the process of enrolling in its programs. When a producer (including farmers, ranchers, timber producers, and other agricultural producers) contacts a USDA Service Center to discuss enrolling in a Farm Bill conservation program (see chapter 6), NRCS employees, some of which are funded by partner organizations, will help them develop a conservation plan that meets the requirements of the program.[2] NRCS conservation planning is available through the agency's Conservation Technical Assistance program to all landowners free of charge, even those not intending to enroll in a Farm Bill conservation program. For more information, contact NRCS staff at your local USDA Service Center.[3] The NRCS conservation planning process applies to all of the resources identified in the conservation title of the Farm Bill, including soil quality and retention, water quality and conservation, and wildlife conservation. Conservation planning can vary on a case-by-case basis, but in general, NRCS uses a nine-step process.[4]

First, NRCS and the landowner identify problems and opportunities for conservation. This conversation is an opportunity for the landowner to make sure that NRCS understands all of their concerns. Second, NRCS and the landowner establish initial objectives. While in many cases the landowner will be focused on maximizing the efficiency of their operation, and NRCS will be working to ensure that ecological resources are considered, this is also an opportunity for the landowner to highlight any particular interest they have in wildlife conservation. This can be helpful because individual agency staff may be more or less focused on wildlife conservation.[5] Third, NRCS conducts an inventory of resources, including natural resources, economic information, and potential social impacts of the conservation plan. This ensures that the conservation plan will work for both the landowner and the natural resources. The scale of this inventory can vary, ranging from

Figure 3.2. NRCS field office in Bangor, Maine.
Photograph by Thomas Kielbasa, Natural Resources Conservation Service.

a single property to a watershed or conservation district. The same techniques are used by NRCS for state and even nation-wide inventories.

Fourth, resource data is analyzed and used to establish the baseline conditions of the property. This step is crucial for several reasons. One is that the goal of most conservation programs is to realize conservation benefits, and these will be measured against the baseline established in the conservation plan. Any adaptive management provisions in the conservation plan will also require comparisons to baseline. In addition, certain permits issued under the Endangered Species Act must include the ecological baseline of the covered land.[6] Fifth, the planners formulate alternatives. Alternatives may incorporate different conservation practices and financial assistance options, but the goal is always to "achieve the goals for the land, by solving all identified problems, taking advantage of opportunities, and meeting the social, economic, and environmental needs of the planning project."[7] Sixth, the alternatives are evaluated.

Seventh, the planning process comes to the point of making decisions. The landowner, and the landowner alone, must choose the conservation program, practices, and implementation plan that works best for them. A landowner's participation in this planning process is entirely voluntary throughout. NRCS staff may inform the landowner of what is possible and give advice on what might be best, but it is the landowner who makes the final decision. Eighth, the conservation plan must be implemented if the landowner so chooses—this is totally voluntary. If the planning process is being used independent of the Farm Bill, this may be the end of the conservationist's involvement. For Farm Bill conservation programs, NRCS remains involved and provides technical assistance in the installation of the agreed-upon conservation practices. Ninth, after a conservation plan has been implemented for an appropriate period of time it should be evaluated. This will show the landowner and any other stakeholders such as financial backers what effect the plan is having and how it could be made more effective. The baseline established in step four can be used at this point, as can adaptive management triggers built into the conservation plan.

Although this particular process is used by NRCS, it is an example of effective conservation planning in general. It establishes the needs of the landowner, conditions on the land, and opportunities for conservation. It facilitates informed decision making, adaptive management, and compliance with federal and state laws and regulations, including the Endangered Species Act. In addition to NRCS, landowners can turn to other partner organizations for assistance in preparing a conservation plan. These may include local conservation districts, agricultural extensions of local universities, and other NGOs.[8]

Resources for Landowners

There are a number of planning resources that landowners may wish to consult. Through NRCS's Conservation Technical Assistance program, USDA Service Center staff can advise landowners and even prepare conservation plans at no charge. In addition, the U.S. Fish and Wildlife Service's Partners for Fish and Wildlife program can be an invaluable resource. This program provides "technical assistance in the development, implementation, and evaluation" of Farm Bill and other conservation programs. Each FWS state office has an assigned Partners for Fish and Wildlife Coordinator, and their contact information is available online at https://www.fws.gov/partners/contactUs.html.[9]

For owners of private forest lands, the U.S. Forest Service's Forest Stewardship Program is available to provide technical assistance regardless of the landowner's individual objectives and needs. Landowners wishing to take advantage of this program should contact their respective state forestry agency. State forestry personnel will help landowners "discuss their management objectives, recommend management activities to enhance wildlife habitat, and develop a long-term forest management plan."[10] In addition, foresters can connect landowners to other organizations that provide conservation services for forest landowners. Finally, landowners who are interested in permanently protecting their land through a conservation easement, including the Forest Legacy Program, can also contact a local land trust.[11]

In addition to the Forest Stewardship Program, the U.S. Forest Service maintains web resources that can assist owners of private forest lands in a variety of ways. The Forest Service maintains a webpage on Forest Taxation (https://www.fs.fed.us/spf/coop/programs/loa/tax.shtml) and is one of the partners that support the National Timber Tax Website (https://www.timbertax.org/). These websites provide resources for forest landowners that include tips for tax planning, guides to the federal income tax and other tax issues that confront forest landowners, and resources such as webinars and in-depth articles on select tax topics.[12]

The National Agroforestry Center, a partnership between the Forest Service and NRCS, also maintains a website with many resources for private landowners (https://www.fs.usda.gov/nac/index.shtml).[13] These include information on agroforestry practices, demonstration sites for those practices, and publications, workshops, and trainings that promote those practices.[14] Publications include fact sheets, webinars, conference proceedings, and others. There are also planning tools, including design tools for riparian and general-purpose buffers, and economic tools, including cost/benefit calculators for buffers and windbreaks, and an income calculator for estimating income potential from conservation-oriented timber plantings.[15] New tools are constantly being developed, and the Center's website should become a more valuable resource over time.

At the state government level, both state fish and wildlife agencies and state forestry agencies can provide resources to private landowners for conservation. These programs are incredibly diverse, but examples can be found in chapter 11, and contact information for every state is included in appendix A. At the local level, a potential resource for any landowner is their local soil and water conservation district. There are 3,000 conservation districts in the United States, nearly one for every county.[16] Conservation districts assist landowners in determining their conservation needs, identifying resources and programs available to them, and developing conservation plans. The National Association of Conservation Districts maintains a directory that is available online at https://www.nacdnet.org/general-resources/conservation-district-directory/.[17]

Another resource for landowners are local agricultural extensions of state land-grant universities. Agricultural extensions were first established in the United States under the authority of the Smith–Lever Act of 1914, and today there is an agricultural extension serving nearly every county in the nation.[18] Agricultural extensions were created to provide education to farmers, and today they continue to do that while also providing information and services to landowners of all types. A directory of agricultural extensions, and other services provided by land-grant universities, can be found online at https://nifa.usda.gov/land-grant-colleges-and-universities-partner-website-directory.[19]

A new, privately developed resource for landowners across the nation is the Land Conservation Assistance Network (LandCAN), online at www.landcan.org.[20] LandCAN "is the only site in the country that hosts every Federal program that touches private landowners, all the equivalent state programs, all the conservation non-profits, every land trust in the country, every conservation district, and over 40,000 for-profit businesses serving rural America."[21] Altogether LandCAN hosts over 80,000 pages of private landowner assistance providers.[22] Its website is intuitively designed, with drop-down menus and clickable maps to allow easy navigation to specific types and locations of conservation programs. The website includes a Conservation Tax Center with an online law library of tax resources.[23]

Another private resource is the Conservation Finance Center, http://conservationfinancecenter.org/, which provides a financial calculator to help landowners understand opportunities for funding conservation work on their land. At this time the financial calculator is only designed for use in the Lower Mississippi River Valley, Arkansas, Kentucky, Louisiana, Mississippi, Missouri, and Tennessee, but other information on the site is applicable nationwide.[24] For landowners considering conservation easements, local land trusts (which can be found through the website https://www.findalandtrust.

org/), the Land Trust Alliance (https://www.landtrustalliance.org/what-you-can-do/conserve-your-land/getting-started), and the National Conservation Easement Database (https://www.conservationeasement.us/resources/) can all provide valuable information.[25] For those with particular concerns about threatened and endangered species, the U.S. Fish and Wildlife Service maintains a landowners portal that provides an orientation to the agency's programs (https://www.fws.gov/endangered/landowners/index.html).[26]

Conservation plans developed using the process described in this chapter are suitable for landowners who simply want to scientifically manage their property, or for those wishing to enroll in an established conservation program. These include federal Farm Bill programs, conservation programs run by federal or state agencies or NGOs, and U.S. Fish and Wildlife Service programs providing regulatory predictability. The next two chapters outline different types of conservation programs and their financial implications. The subsequent chapters describe in more detail many of the individual conservation programs available to landowners.

Notes

1. For example, Pennsylvania requires that all farming operations involving tilling and plowing develop and implement conservation plans detailing best practices and a timeline for their implementation. These are primarily focused on soil and water quality, but they can include wildlife conservation as well. Plans designed to comply with Farm Bill conservation programs will generally meet this requirement as well. Brandywine Conservancy, "Do You Need a New Conservation Plan?," last accessed June 18, 2019, https://www.brandywine.org/conservancy/landowner-resources/do-you-need-new-conservation-plan. Maryland, on the other hand, only requires conservation plans in select areas. Calvert Soil Conservation District, "Farm Planning," last accessed June 19, 2019, http://www.calvertsoil.org/c/262/farm-planning.

2. See chapters 6 and 9. In some cases, landowners may work with a Farm Service Agency, rather than NRCS, employee.

3. Natural Resources Conservation Service, "How to Ask for Conservation Technical Assistance," last accessed June 19, 2019, https://www.nrcs.usda.gov/wps/portal/nrcs/detail/national/programs/technical/cta/?cid=nrcs143_008232.

4. Natural Resources Conservation Service, "Nine Step Conservation Planning Process," last accessed June 18, 2019, https://www.nrcs.usda.gov/wps/portal/nrcs/detail/national/programs/financial/eqip/?cid=nrcs144p2_015695.

5. E.g. Joel Pedersen and Mark Hatfield, interview by Christopher E. Segal, June 18, 2019; Miles Moretti, interview by Lowell E. Baier and Christopher E. Segal, July 29, 2019.

6. See chapter 13.

7. Natural Resources Conservation Service, "Nine Step Conservation Planning Process."

8. E.g. Calvert Soil Conservation District, "Farm Planning"; University of New Hampshire Cooperative Extension Land & Water Conservation Program, "Preparing a Conservation Plan," March 2013, https://extension.unh.edu/resources/files/Resource001227_Rep1568.pdf.

9. U.S. Fish and Wildlife Service Partners for Fish and Wildlife Program, "Regional and State Contact Information," last accessed August 11, 2019, https://www.fws.gov/partners/contactUs.html. See chapter 10 for more information on this program.

10. Laurie Schoonhoven, email to Christopher E. Segal, July 2, 2019.

11. Ibid. For more information on these Forest Service programs, see chapter 10.

12. U.S. Forest Service, "Forest Taxation," last accessed August 11, 2019, https://www.fs.fed.us/spf/coop/programs/loa/tax.shtml; "National Timber Tax Website," last accessed August 11, 2019, https://www.timbertax.org.

13. U.S. Department of Agriculture National Agroforestry Center, "National Agroforestry Center," last accessed August 11, 2019, https://www.fs.usda.gov/nac/index.shtml. Agroforestry is an agricultural practice that uses trees and shrubs to address conservation needs. The classic example of agroforestry is using trees to establish a riparian buffer along the edge of a field, but agroforestry practices also include windbreaks, farming crops under or between trees (which may themselves be harvested for commercial purposes), or including trees in grazing areas (silvopasture). Agroforestry is particularly useful for producers wanting to conserve wildlife, increase wildlife habitat, and provide ecological connectivity across a landscape. USDA National Agroforestry Center, "What Is Agroforestry?," August 2012, https://www.fs.usda.gov/nac/assets/documents/workingtrees/infosheets/WhatIsAgroforestry07252014.pdf.

14. USDA National Agroforestry Center, "About the Center," last accessed July 3, 2019, https://www.fs.usda.gov/nac/about/index.php.

15. USDA National Agroforestry Center, "Tools," last accessed July 3, 2019, https://www.fs.usda.gov/nac/resources/tools/index.shtml.

16. National Association of Conservation Districts, "About NACD," last accessed May 23, 2019, https://www.nacdnet.org/about-nacd/. The history and functions of conservation districts are discussed more fully in chapter 6.

17. National Association of Conservation Districts, "Conservation District Directory," last accessed August 11, 2019, https://www.nacdnet.org/general-resources/conservation-district-directory/.

18. United States Department of Agriculture National Institute of Food and Agriculture, "Cooperative Extension History," last accessed July 28, 2019, https://nifa.usda.gov/cooperative-extension-history.

19. United States Department of Agriculture National Institute of Food and Agriculture, "Land-Grant University Website Directory," last accessed July 28, 2019, https://nifa.usda.gov/land-grant-colleges-and-universities-partner-website-directory.

20. Land Conservation Assistance Network, "Land Conservation Assistance," last accessed August 11, 2019, https://www.landcan.org/.

21. LandCAN Land Conservation Assistance Network, "Get to Know the Land Conservation Assistance Network (Landcan)," Septemper 22, 2017, https://www.youtube.com/watch?time_continue=237&v=OapyBl6tW4U. (Statement of Amos S. Eno.)

22. Ibid.

23. Land Conservation Assistance Network, "The Conservation Tax Center," last accessed July 22, 2019, https://www.landcan.org/conservation-tax-center/.

24. Conservation Finance Center, "Conservation Finance Center," last accessed June 19, 2019, http://conservationfinancecenter.org.

25. Land Trust Alliance, "Find a Land Trust," last accessed July 28, 2019, https://www.findalandtrust.org/; Land Trust Alliance, "What You Can Do," last accessed August 11, 2019, https://www.landtrustalliance.org/what-you-can-do/conserve-your-land/getting-started; National Conservation Easement Database, "Resources," last accessed August 11, 2019, https://www.conservationeasement.us/resources. The National Conservation Easement Database is a project of Ducks Unlimited and The Trust for Public Land, funded by the U.S. Endowment for Forestry and Community and the United States Geological Survey.

26. U.S. Fish and Wildlife Service, "For Landowners," last accessed August 11, 2019, https://www.fws.gov/endangered/landowners/index.html.

CONSERVATION TOOLS AND YOUR BUSINESS

II

"That land is a community is the basic concept of ecology, but that land is to be loved and respected is an extension of ethics."

–ALDO LEOPOLD, *A SAND COUNTY ALMANAC*

Common Mechanisms for Enrolling Land in Conservation Programs

4

LANDOWNERS NEED TO CONSIDER how adopting a conservation plan will fit into their business and/or household expenses. The vast majority of landowners may be able to make small-scale changes to their businesses, but significant habitat improvement projects can be major investments. And investments in conservation usually do not increase profits but instead remove land from cultivation, reduce grazing intensity, or reduce frequency of timber harvests.

Conservation assistance programs are designed to address this, but landowners need to find the program best suited for them. One consideration is the form in which financial assistance is received. This chapter describes major mechanisms through which conservation programs fund landowner conservation actions (grants, cost-sharing payments, incentive payments, rental contracts, and conservation easements) and common requirements and procedures for each mechanism. It concludes with a discussion of the tax consequences of the various conservation mechanisms.

Grants

The simplest of all conservation funding mechanisms is a grant. As in many areas, charitable organizations raise money for worthy causes and distribute it to individuals or organizations whose work advances those causes. In conservation, grant funding can be provided by government agencies, foundations, land trusts, and other non-governmental organizations. Conservation grants typically vary from a few thousand dollars to several hundred thousand dollars, and they are usually intended to fund a single project or conservation action to be completed within a few years, rather than an ongoing operation. Grant funding generally must be spent exclusively on specified conservation actions and may not be used for general operations or to subsidize other parts of an organization's program or a landowner's businesses.

In the overwhelming majority of cases, grants are awarded to organizations rather than directly to landowners. For example, the National Fish and Wildlife Foundation, a major grant funder, makes grants exclusively to federal, state, and local governments; educational institutions; and non-profit organizations.[1] Similarly, grant programs run by national organizations such as The Nature Conservancy and The Conservation Fund target organizations rather than small businesses and private landowners.[2]

Recipient organizations use grant funding to implement conservation actions, which include working with individual landowners. Details vary by program, including how much funding is available, how long it will take for funds to be distributed, and the application process. Landowners interested in conservation may want to reach out to local conservation organizations to see if they have funding available or might be interested in working with the landowner to develop a grant proposal. In some cases, landowners may be asked to contribute to a grant-funded conservation program either financially or through an in-kind contribution of labor.

From time to time, there are opportunities for landowners to apply directly for conservation grants. Information about these grants can be hard to find online because there are a limited number of central databases and some grants may not be included in them. However, there are places for landowners to start, including:

1. The Land Conservation Assistance Network's database of Federal & State Conservation Assistance Programs: https://www.landcan.org/Grant-and-Assistance-Programs/
2. Websites of federal agencies:
    a. The U.S. government's central grants database: https://www.grants.gov/
    b. Natural Resources Conservation Service: https://www.nrcs.usda.gov/wps/portal/nrcs/detail/national/programs/?&cid=stelprdb1048817
    c. U.S. Fish and Wildlife Service: https://www.fws.gov/grants/
    d. U.S. Forest Service: https://www.fs.fed.us/working-with-us/grants[3]

Cost-Sharing Payments

Many conservation programs under the federal Farm Bill incorporate cost-sharing payments as one of their components. For example, cost-shares may be included alongside incentive payments under the Environmental Quality

Incentives Program (EQIP), rental contract payments under the Conservation Reserve Program (CRP), or easement payments under the Agricultural Conservation Easement Program.[4] Cost-shares typically cover at least 50 percent of the costs landowners incur implementing conservation actions. The landowner is responsible for the remainder. Some programs permit a higher cost-share payment of 75 percent or even 90 percent, if the payment is being made to a landowner in a special category, such as a veteran, or in exchange for high-priority conservation work, such as actions to conserve Endangered Species Act–listed threatened or endangered species. Although cost-shares are helpful to landowners, cost-shares alone would rarely induce voluntary conservation, as landowners would still be shouldering part of the cost. Accordingly, most cost-shares are combined with other mechanisms described in the following pages.

Incentive Payments

An incentive payment is made to a landowner to encourage them to carry out conservation. Unlike a cost-share, an incentive payment is decoupled from the actual cost of the individual conservation activity, instead paying a predetermined, uniform rate based on an appropriate metric. For example, payments may be based on number of acres treated or number of feet of fence retrofitted to be wildlife-friendly. Incentive payments may be made at the start of conservation actions, after conservation actions are implemented, or over a period of years during the implementation of conservation actions. In select cases, including certain practices under EQIP, an incentive payment can be based on the income a landowner foregoes by removing land from production or reducing the intensity of farming or grazing. These incentives are usually more attractive.

Incentive payments are generally more valuable to landowners than cost-shares because it is possible for an incentive payment to fully cover the cost of a conservation practice. More often, however, whether receiving an incentive, cost-share, or both, the landowner will end up bearing some of the cost themselves. This means that conservation programs using these mechanisms are most attractive to landowners that already have a strong interest in conservation. Several Farm Bill programs utilize incentive payments, including EQIP, the Conservation Stewardship Program, and CRP. These programs are discussed in chapter 7.

Rental Contracts

Like incentive payments, rental contracts are decoupled from the individual cost of a conservation practice. The primary context in which landowners enter into rental contracts for conservation is through CRP, which is administered by the Farm Service Agency.[5] As implied by the term "rental," payments are made on an annual basis for the length of a contract, which is ten to fifteen years under CRP. In a CRP rental contract, rates are set on a case-by-case basis according to various criteria, but they commonly reflect the commercial rental rates of similar land. CRP rental payments are thus tied to actual land value in a way that cost-shares and inventive payments are not. This makes rental payments an attractive option for landowners, and they can actually profit from enrolling in CRP.

These profits, however, are small and are realized over the course of the contract. This means that rental contracts are most appealing when agricultural markets are weak or to landowners already interested in conservation. CRP rental contracts can include incentive payments, and like most incentive programs, they typically also include a cost-share component to provide technical assistance for implementing conservation actions.

Conservation Easements

The final mechanism for enrolling land in conservation is a conservation easement, which is radically different. Establishing a conservation easement is a real estate transaction in which a landowner sells (or donates) certain development and land use rights. In property law, an easement is "[a]n interest in land owned by another person, consisting in the right to use or control the land, or an area above or below it, for a specific limited purpose (such as to cross it for access to a public road)."[6] A conservation easement is defined by *Black's Law Dictionary* as follows:

> A real-estate covenant binding a parcel of land in a way that preserves a native plant or animal, a natural or physical feature of the land, or some aspect of the land that has some historical, cultural, or scientific significance. The easement is a recorded, perpetual, individually tailored agreement creating a nonpossessory interest in real property, the interest being held by a government entity or by a qualified nonprofit. It permanently restricts or imposes affirmative obligations on the property's owner or lessee to retain or protect natural, scenic, or open-space values of real property, ensure its availability for agricultural, forest, recreational, or open-space use, protect natural resources and habitat, maintain or enhance air or water quality, or preserve the historical, architectural, archeological, or cultural aspects of the real property while allowing the landowner to continue to own and use the land, sell it, or transfer it to heirs.[7]

In other words, a conservation easement is a property right held by an entity other than the landowner (usually a government or non-profit land trust) to use another's land to promote conservation of species or habitat. This usually takes the form of affirmatively declining to develop the land or a resource on it, which is made permanent by transferring the development right in question to the other party.

Conservation easements generally protect certain characteristics of land (e.g., open spaces, certain vegetation communities, etc.) and/or prohibit certain uses (e.g., oil and gas exploration, surface mining, urban development, etc.). However, they may or may not prohibit uses that are consistent with the protected characteristics (e.g., grazing, farming, hunting, tourism, in some cases even lumber harvesting or low-density housing[8]), provided that the protected characteristics are maintained. This makes conservation easements attractive to landowners who work on the land and want to protect their land and/or maintain their way of life. Reasons landowners place conservation easements on their land include love of the land, environmental protection, concern about development, societal values, open space, family legacy, preserving their culture, protecting their community, and, finally, financial and tax benefits.[9]

Conservation easements have existed since long before the Endangered Species Act. The first conservation easements were created over a century ago, in the 1880s to protect parkways designed by Frederick Law Olmstead in the Boston area.[10] However, conservation easements were uncommon until the passage of specific laws in each state that recognized conservation easements, the earliest of which were passed in the 1950s. This legislative process continued through the 1980s,[11] and the use of conservation easements is now widespread. A 2019 survey found that there were 158,168 conservation easements in the United States, protecting a total of 27,036,979 acres of land.[12]

When a landowner places a conservation easement on their property, they are surrendering a partial interest in that property. When a landowner sells a conservation easement, it is like any other real estate transaction: the easement is closed on and full payment made, with the payment taxed as capital gains. Easements may also be donated, which may entitle the landowner to a variety of income and/or property tax benefits.[13] For many landowners, especially those that are "land rich and cash poor," selling a conservation easement is attractive because it yields immediate payment in full. The value of a conservation easement is established by an assessment (usually paid for by the landowner) of the value of the land less the value of the land with the easement in place. For buyers of easements, which include government agencies and non-profit organizations, a conservation easement is generally cheaper to purchase than outright ownership of the same property, allowing protection of more land for the same amount of money.[14]

For landowners, the duration of a conservation easement is an important consideration. Most conservation easements are permanent and "run with the land," meaning that even if the land is transferred or sold to a new owner, the easement will still be in place.[15] A permanent commitment may not be right for everyone.[16] An alternative are term easements, which end after a specified period of time has elapsed,[17] commonly thirty years. Term easements are worth less than permanent easements, for example, thirty-year easements are typically worth 75 percent of the value of a permanent easement.

Conservation easements can be used in isolation to protect land. This commonly takes the form of creating a conservation easement and selling or donating it to a land trust, which then holds the easement. Land trusts are generally non-profit organizations, and they vary in size and structure from local volunteer organizations to regional and national organizations such as The Nature Conservancy or The Conservation Fund.[18] The land trust assumes responsibility for monitoring the land and ensuring compliance with the easement, whereas the landowner continues to exercise all direct control and management over the land. Other non-governmental organizations also own conservation easements, and in certain Farm Bill programs, such as Wetland Reserve Easements, the federal government owns conservation easements.[19] Conservation easements are also used as components of larger conservation programs. For example, easements are placed on land in conservation banks and other compensatory mitigation programs to ensure the permanent protection of wildlife habitat.[20] Conservation easements can include land management plans, but many do not.[21]

There are a number of resources for landowners interested in conservation easements. One place to start is by contacting a local land trust. National conservation organizations such as The Nature Conservancy and The Conservation Fund have offices in (or serving) every state, and landowners often reach out to them for advice.[22] Many conservation organizations are more interested in accepted donated easements, but some buy easements as well.[23] For Farm Bill easements, Natural Resources Conservation Service field offices are the best point of contact. Finally, a variety of local and state-based conservation organizations work with landowners and can connect them to easement-owning and -financing organizations.[24]

Tax Benefits and Consequences of Conservation

The mechanisms discussed in this chapter all represent sources of funding for landowners. Incentives, rental payments, and easement payments generate income that the landowner is free to spend as they see fit, whereas cost-shares offset expenditures. For each of these, the tax considerations form an important aspect of identifying the best conservation program. Landowners should consider consulting a tax professional before committing to any conservation program. This includes programs where the landowner receives outside funding—which may or may not be taxable—and also programs where the landowner expects to incur expenses—which may or may not be deductible. In addition, in some states the donation of a conservation easement may make the landowner eligible for a tax *credit*. Tax credits directly offset tax owed, whereas deductions merely offset income taxed, meaning that tax credits are far more valuable.[25]

Taxation of Income from Grants, Cost-Sharing Payments, Incentive Payments, Rental Payments, and Conservation Easements

As discussed previously, many conservation grants are designed for organizations rather than individuals. In those cases, recipient organizations are often tax-exempt organizations. But if a landowner receives a payment for a contribution to a grant-funded program, that payment is often taxed as ordinary income. For this reason, most grant-funded organizations and most landowners engaged in grant-funded conservation avoid any payments to the landowner.

Cost-sharing payments are treated differently because they offset conservation expenses that the landowner has incurred. Section 126 of the Internal Revenue Code (IRC) excludes from gross income cost-sharing payments made under any federal, state, or local government program "primarily for the purpose of conserving soil, protecting or restoring the environment, improving forests, or providing a habitat for wildlife."[26] This includes the cost-sharing components of Farm Bill incentive, rental, and easement programs as well as any other government conservation program. The IRC does not require any particular ratio of cost-sharing. However, payments that are not exclusively for qualifying conservation purposes or that increase the annual income derived from the property may not be fully deducted.[27]

Whereas the cost-sharing components of incentive and rental payments are tax free, the same is not true of the incentive and rental payments themselves. They are income for their recipients, typically taxed at the ordinary income rate, which is commonly in the 30 to 40 percent range once state and local taxes are included,[28] and such income is taxed in each year that it is received. This tax burden means that incentive and rental payments are financially much less attractive than they may appear at first glance.

The sale of a conservation easement, in contrast, is a real estate transaction taxed at the capital gains rate, which depending on income and filing status is usually either 15 or 20 percent. So when selling easements, not only do landowners receive the full appraised value of the easement immediately, they also pay a comparatively low tax rate on that income. This makes conservation easements by far the most attractive conservation mechanism for most landowners.[29]

Tax Benefits of Donated Conservation Easements

When a landowner places a conservation easement on their land, that permanent encumbrance reduces the fair market value of the property, and accordingly may reduce property taxes owed on that property.[30] This applies regardless of whether the easement was sold or donated. However, when easements are donated there are additional tax benefits available. Federal tax benefits include an income tax deduction, first enacted into law by Congress in 1976 and made permanent in 1980. As of 2019, qualifying farmers and ranchers (those earning more than half their income from agriculture) may deduct the value of a donated conservation easement up to 100 percent of their income over a sixteen-year period, and non-qualifying owners may deduct up to 50 percent of their income.[31] Since 1997, there has also been a federal estate tax benefit of 40 percent of the encumbered value of the land, with a cap of $500,000.[32] To qualify for these tax benefits, a conservation easement must be perpetual, must be donated to a non-profit organization, and must be exclusively for conservation purposes.[33] Note that an easement does not qualify as a donation if it was made to satisfy a legal obligation, such as compliance with a government regulation, as in a mitigation or species conservation bank.[34] If an easement is sold for less than its fair market value, the difference can be treated as a donation and qualify for tax benefits accordingly.[35]

Finally, there are also state income tax, property tax, and/or estate tax benefits associated with conservation easements in some states.[36] Unlike the federal government, some states also offer income tax *credits* for donating conservation easements.[37] Five states (Colorado, Georgia, New Mexico, South Carolina, and Virginia) offer *transferable* tax credits, which landowners are able to sell to other taxpayers if they don't pay enough tax to use the entire credit themselves. Another eight states (Arkansas, Connecticut, Delaware, Iowa, Maryland, Massachusetts, Mississippi, and New York) and Puerto Rico offer non-transferable tax credits.[38] The Land Trust Alliance maintains a regularly updated list of all state tax benefits available for conservation easements at https://www.landtrustalliance.org/topics/taxes/income-tax-incentives-land-conservation.[39]

Deductions of Conservation-Related Expenses for Farmers and Agricultural Producers

The IRC also allows farmers and agricultural producers to claim deductions for expenses incurred in soil and water conservation or endangered species recovery. This deduction is only available to farmers filing IRS Schedule F, Profit or Loss From Farming. Ordinarily these expenditures would be regarded as capital improvements that increase the landowner's basis in the land, reducing the eventual tax burden if they sell the land. For farmers with no long-term plans to sell, this is of limited value. IRC Section 175 allows these expenditures to be claimed instead as an income tax deduction. The deduction is limited in any single year to 25 percent of the farmer's gross income from farming, but unclaimed deductions may be carried over to subsequent years until the full amount has been claimed. The deduction can only be claimed for soil and water conservation actions carried out in accordance with a conservation plan approved by the Natural Resources Conservation Service, or for endangered species recovery actions carried out in accordance with a Fish and Wildlife Service–approved conservation plan.[40] Although the deduction cannot be taken for any expenditures offset by a cost-sharing payment, it is available for the remainder of such expenditures not covered by the cost-sharing payment. More information is available from the Fish and Wildlife Service and from knowledgeable tax professionals.[41]

Notes

1. For more information on the National Fish and Wildlife Foundation, see chapter 10.

2. These organizations work with small businesses and private landowners in a number of ways; see chapter 11 for more information.

3. Land Conservation Assistance Network, "Federal & State Conservation Assistance Programs," last accessed August 11, 2019, https://www.landcan.org/Grant-and-Assistance-Programs; "Grants.Gov," last accessed August 11, 2019, https://www.grants.gov/; Natural Resources Conservation Service, "NRCS Funding Opportunities," last accessed August 11, 2019, https://www.nrcs.usda.gov/wps/portal/nrcs/detail/national/programs/?&cid=stelprdb1048817; U.S. Fish and Wildlife Service, "Financial Assistance Home," July 8, 2015, https://www.fws.gov/grants/; U.S. Forest Service, "Grants," last accessed August 11, 2019, https://www.fs.fed.us/working-with-us/grants.

4. These programs are discussed in detail in chapters 7 and 8.

5. See chapter 7.

6. "Easement," in *Black's Law Dictionary*, ed. Bryan A. Garner (tenth edition, 2014).

7. "Conservation Easement," ibid.

8. See J.B. Ruhl, Alan Glen, and David Hartman, "Practical Guide to Habitat Conservation Banking Law and Policy," *Natural Resources and Environment* 20 (2005): 32 (discussing conservation banks).

9. James R. Farmer et al., "Motivations Influencing the Adoption of Conservation Easements," *Conservation Biology* 25 (2011): 827–34; see also Nancy A. McLaughlin, "Increasing the Tax Incentives for Conservation Easement Donations–A Responsible Approach," *Ecology Law Quarterly* 31 (2004): 41–50.

10. Julie Ann Gustanski and Roderick H. Squires, *Protecting the Land: Conservation Easements Past, Present, and Future* (Washington, DC: Island Press, 2000), xvii; Shea B. Airey, "Conservation Easements in Private Practice," *Real Property, Trust and Estate Law Journal* 44 (2010): 751.

11. Gustanski and Squires, *Protecting the Land: Conservation Easements Past, Present, and Future*, xvii–xviii; The Nature Conservancy, "Conservation Easements: The Evolution of

Conservation Easements," last accessed June 28, 2018, https://www.nature.org/about-us/private-lands-conservation/conservation-easements/evolution-of-conservation-easements.xml; Airey, "Conservation Easements in Private Practice," 751. See also Bean and Rowland, *The Evolution of National Wildlife Law*, 436.

12. National Conservation Easement Database, "National Conservation Easement Database," last accessed August 11, 2019, https://www.conservationeasement.us/.

13. Gregory M. Parkhurst and Jason F. Shogren, "An Economic Review of Incentive Mechanisms to Protect Species on Private Lands," in *Species at Risk: Using Economic Incentives to Shelter Endangered Species on Private Lands*, ed. Jason F. Shogren (Austin, TX: University of Texas Press, 2005), 114–21 (comparing purchased and donated conservation easements); Julia LeMense Huff, "Protecting Ecosystems Using Conservation Tax Incentives: How Much Bang Do We Get for Our Buck?" *Missouri Environmental Law & Policy Review* 11 (2004); Timothy Lindstrom, "State Tax Incentives for Conservation Easements Can Benefit Everyone," *Journal of Multistate Taxation and Incentives* 12-NOV (2002). See also discussion of tax consequences later in this chapter.

14. Bean and Rowland, *The Evolution of National Wildlife Law*, 437.

15. A small and shrinking number of states do not permit permanent conservation easements, instead specifying a maximum duration of years for a conservation easement. Ibid., 436.

16. Todd Gartner et al., "The Gopher Tortoise, Military Readiness, and Pre-Listing Conservation," in *Proactive Strategies for Protecting Species: Pre-Listing Conservation and the Endangered Species Act*, ed. C. Josh Donlan (Oakland, CA: University of California Press, 2015), 195; A.M. Merenlender et al., "Land Trusts and Conservation Easements: Who Is Conserving What for Whom?" *Conservation Biology* 18 (2004): 71.

17. Merenlender et al., "Land Trusts and Conservation Easements: Who Is Conserving What for Whom?" 67.

18. Ibid., 68. For more about land trusts, see chapter 11.

19. Wetland Reserve Easements are a continuation of the Wetlands Reserve Program. Healthy Forests Reserve Program easements are also owned by the federal government, whereas the Agricultural Land Easement Program provides for non-governmental organizations to purchase easements using Farm Bill dollars. All Farm Bill easements can be accompanied by additional cost-sharing payments. For more on these programs, see chapter 8.

20. See chapter 5 for information on compensatory mitigation.

21. Furthermore, a management plan for a conservation easement is more effective if it is in an addendum, rather than the easement itself, so that it can be updated as time passes. Moretti.

22. For example, Fred Annand, interview by Christopher E. Segal, February 22, 2019.

23. Airey, "Conservation Easements in Private Practice," 818–20.

24. These groups are discussed in more detail in chapter 11.

25. Timothy Lindstrom, *A Tax Guide to Conservation Easements*, second edition (Washington, DC: Land Trust Alliance, 2016), 204.

26. 26 U.S.C. § 126(a)(8).

27. 26 U.S.C. § 126.

28. James L. Cummins, interview by Christopher E. Segal, March 6, 2019.

29. Ibid.

30. Timothy Male and C. Josh Donlan, "The Future of Pre-Listing Conservation Programs for Wildlife Conservation," in *Proactive Strategies for Protecting Species: Pre-Listing Conservation and the Endangered Species Act*, ed. C. Josh Donlan (Oakland, CA: University of California Press, 2015), 226–27; The Nature Conservancy, "Conservation Easements: All About Conservation Easements," last accessed June 28, 2018, https://www.nature.org/about-us/private-lands-con

servation/conservation-easements/all-about-conservation-easements.xml; Lindstrom, "State Tax Incentives for Conservation Easements Can Benefit Everyone," 24.

31. John B. Wright and Anthony Anella, "Saving the Ranch: Fresh Eyes on Taxes, Development, and Conservation Easements," *Rangelands*, June 2007, 15–17; The Nature Conservancy, "Conservation Easements: The Evolution of Conservation Easements"; The Nature Conservancy, "Conservation Easements: All About Conservation Easements"; Land Trust Alliance, "Tax Incentives," last accessed June 29, 2018, http://www.landtrustalliance.org/issues-action/take-action/tax-incentives; Huff, "Protecting Ecosystems Using Conservation Tax Incentives: How Much Bang Do We Get for Our Buck?" 140–41; Airey, "Conservation Easements in Private Practice," 761–64. For a discussion of the mechanics and benefits of the tax deduction, see ibid., 764–90; McLaughlin, 'Increasing the Tax Incentives for Conservation Easement Donations–A Responsible Approach," 10–17; Lindstrom, *A Tax Guide to Conservation Easements*, 159–79.

32. Land Trust Alliance, "Estate Tax Incentives for Land Conservation: Keeping Land in the Family," last accessed June 28, 2018, https://www.landtrustalliance.org/topics/taxes/estate-tax-incentives-land-conservation; Huff, "Protecting Ecosystems Using Conservation Tax Incentives: How Much Bang Do We Get for Our Buck?" 141–42; Airey, "Conservation Easements in Private Practice," 790–800; Lindstrom, *A Tax Guide to Conservation Easements*, 303–10.

33. 26 U.S.C. § 170.

34. Timothy Lindstrom, "The Syndication of Conservation Easement Tax Deductions," *Saving Land*, Summer 2015.

35. Cummins.

36. The Nature Conservancy, "Conservation Easements: All About Conservation Easements"; Land Trust Alliance, "Income Tax Incentives for Land Conservation," last accessed May 9, 2019, https://www.landtrustalliance.org/topics/taxes/income-tax-incentives-land-conservation; Airey, "Conservation Easements in Private Practice," 800–12; Lindstrom, *A Tax Guide to Conservation Easements*; James L. Cummins, "Private Lands Conservation Programs," in *North American Wildlife Policy and Law*, ed. Bruce D. Leopold, Winifred B. Kessler, and James L. Cummins (Missoula, MT: Boone and Crockett Club, 2018), 243–44.

37. Cummins, "Private Lands Conservation Programs," 243–44; Land Trust Alliance, "Income Tax Incentives for Land Conservation."

38. Land Trust Alliance, "Income Tax Incentives for Land Conservation."

39. Ibid.

40. For example, Jordan Cooley, "Conservation Expense Federal Tax Deduction," April 10, 2015, https://valuewetlands.tamu.edu/2015/04/10/conservation-expense-federal-tax-deduction/; Roger A. McEowen, "Deductibility of Soil and Water Conservation Expenses," March 9, 2017, https://lawprofessors.typepad.com/agriculturallaw/2017/03/deductibility-of-soil-and-water-conservation-expenses.html.

41. U.S. Fish and Wildlife Service, "Farm Bill Endangered Species Tax Deduction," March 2010, https://www.fws.gov/endangered/esa-library/pdf/ES_TaxCredit2a.pdf.

Working with Regulated Entities and Investors

5

CHAPTER 4 DESCRIBED the mechanisms through which private landowners enroll land in conservation programs and receive payments, as well as the associated tax consequences. These programs were presented as two-party transactions involving the landowner and a government or non-profit organization providing funding and/or purchasing an easement. These relatively simple arrangements are limited by the availability of funding.

Additional resources for conservation are available from the private sector but involve a more complex regulatory landscape. Private entities are encouraged to invest in conservation due to regulatory pressure, primarily from the Endangered Species Act (ESA) and the Clean Water Act of 1972 (CWA). Funding comes from regulated entities such as industry that invest in compensatory mitigation to offset negative environmental impacts of their operations. In some cases, federal agencies or units of state or local governments are also required to fund compensatory mitigation, often as a result of road or bridge construction or other public works projects.

Introduction to Compensatory Mitigation

Human enterprise frequently has adverse effects upon the natural world. Whether converting prairie to farms, clearing forests for homes, extracting energy resources such as oil, gas, and coal from the earth, or even building highways, the expansion of civilization damages the natural world and creates conflict between humans and wildlife. One way this conflict can be reduced is through mitigation. In environmental law, "mitigation" means "[a] project or program designed to offset the effects of human activities on an existing historic or natural resource, such as a wetland, an endangered species, or an archaeological site."[1]

The modern concept of environmental mitigation dates back to the 1946 amendments to the 1934 Fish and Wildlife Coordination Act. The Fish and Wildlife Coordination Act authorizes federal "assistance to, and cooperat[ion] with, Federal, State, and public or private agencies and organizations in the development, protection, rearing, and stocking of all species of wildlife, resources thereof, and their habitat."[2] Like many environmental laws, it evolved from an initial concern about water resources: the original 1934 version focused on water pollution and impoundment.[3] An amendment in 1946 introduced the concept of mitigation.[4] It required agencies that "impounded, diverted, or otherwise controlled" waterways to consult with the Fish and Wildlife Service (FWS) regarding "means and measures that should be adopted to prevent loss of and damage to wildlife resources."[5] Further amendments in 1958, which remain law today, require consultation with the FWS "whenever the waters of any stream or other body of water are proposed or authorized to be impounded, diverted, the channel deepened, or the stream or other body of water otherwise controlled or modified for any purpose whatever. . . . Recommendations of the Secretary of the Interior . . . shall describe the damage to wildlife attributable to the project and the measures proposed for *mitigating* or *compensating* for these damages."[6]

Mitigation and the National Environmental Policy Act

The National Environmental Policy Act of 1969 (NEPA) and its implementing regulations created the modern definition of wildlife mitigation. NEPA is a linchpin of modern administrative and environmental law, requiring all federal agencies to evaluate the environmental impacts of their actions if they "significantly affect . . . the quality of the human environment,[7] which is the natural and physical environment and the relationship of people with that environment."[8] Like Section 7 of the ESA, this applies to actions carried out by federal agencies but also to private actions authorized or funded by federal agencies.[9] NEPA's implementing regulations establish a "mitigation hierarchy," which is used throughout U.S. environmental law, including under the ESA.

Mitigation includes:

1. Avoiding the impact altogether by not taking a certain action or parts of an action.
2. Minimizing impacts by limiting the degree or magnitude of the action and its implementation.
3. Rectifying the impact by repairing, rehabilitating, or restoring the affected environment.

4. Reducing or eliminating the impact over time by preservation and maintenance operations during the life of the action.
5. Compensating for the impact by replacing or providing substitute resources or environments.[10]

The Department of the Interior, including FWS, generally simplifies these five steps into three ("avoid, minimize, and compensate") and treats "rectify" and "reduce over time" as subcategories of "minimize."[11] This three-level mitigation hierarchy is then used to address potential impacts to natural resources. Impacts that cannot be avoided should be minimized. Impacts that remain after avoiding and minimizing must be compensated for. This final step, "compensatory mitigation," is sometimes referred to as just "mitigation," but this book uses the full term "compensatory mitigation" in order to avoid any confusion.

Compensatory Mitigation Under the Endangered Species Act

Compensatory mitigation is used under Sections 7 and 10 of the ESA. As discussed in chapter 1, Section 9 of the ESA makes taking a listed species illegal, and in some cases damaging the habitat of a listed species constitutes take, even on private land.[12] This means that developing land, extracting resources, and even engaging in wildlife-friendly agriculutural or silvicultural enterprises can violate the ESA. To address this, Congress passed Section 10 of the ESA in 1973, as well as later amendments. Section 10 provides mechanisms for regulated entities to receive permits granting exceptions to Section 9's prohibition on take.

These Section 10 conservation programs (discussed in detail in chapter 13) link the take of a listed species to other actions that offset the take through compensatory mitigation. This can allow land and resources to be developed while species and habitat are also conserved. It encourages advance compliance with the ESA, reducing the need for subsequent enforcement. Section 10 explicitly mentions compensatory mitigation. Section 10(a)(2) outlines requirements for Habitat Conservation Plans, which include "what steps the applicant will take to minimize and mitigate . . . impacts, and the funding that will be available to implement such steps."[13]

The Section 7 consultation process is another way to authorize take of a listed species if offset by compensatory mitigation. Section 7 describes compensatory mitigation as "measures, including, but not limited to, live propagation, transplantation, and habitat acquisition and improvement, as are necessary and appropriate to minimize the adverse effects of the agency action upon the endangered species, threatened species, or critical habitat."[14] Under Section 7, when FWS or the National Marine Fisheries Service finds that an agency action will jeopardize the continued existence of a listed species, or destroy or adversely modify critical habitat, it must provide "reasonable and prudent alternatives" to the proposed action, which may include compensatory mitigation. If the action will cause incidental take that does not rise to the level of jeopardy, FWS or the National Marine Fisheries Service provides "reasonable and prudent measures" to minimize incidental take, which may also include compensatory mitigation.[15] Compensatory mitigation under Section 7 is generally carried out by the agency involved. One example of this was when the Federal Highway Administration and Indiana Department of Transportation developed plans to extend Interstate 59 from Evansville to Indianapolis in 2006, potentially impacting the Indiana bat and bald eagle.[16] Compensatory mitigation can also be carried out by a private entity, such as when the Montanore Minerals Corporation developed plans for a copper and silver mine in the Kootenai National Forest in 2014, potentially impacting grizzly bears.[17] Compensatory mitigation may also play a role in Sections 4(d) and 5 of the ESA, but to date such applications of compensatory mitigation have been rare.[18]

Wetlands Compensatory Mitigation

Compensatory mitigation first matured under the CWA. Under Section 404 of the CWA, the Army Corps of Engineers uses compensatory mitigation for impacts from dredged or fill material on wetlands, streams, and other aquatic resources.[19] This is known as "wetlands compensatory mitigation." Under guidelines promulgated in 1980, Section 404 compensatory mitigation was initially on-site, but by 1995 the Corps determined that off-site compensatory mitigation was more effective.[20] CWA off-site compensatory mitigation is now governed by a regulatory rule issued in 2008 by the Corps and the Environmental Protection Agency.[21]

At present, wetlands compensatory mitigation is a mature regulatory regime. Compensation for impacts to wetlands must create equivalent functioning wetlands to replace wetlands that are damaged or lost, and regulatory agencies may prefer that mitigation be carried out within the same watershed. When and by whom the compensation is done can vary. In permittee-responsible compensatory mitigation, conservation is done by the same entity that causes the impacts. In in-lieu fee compensatory mitigation, the entity causing the impacts pays for a third party to carry out conservation. Finally, in compensatory mitigation banking, the entity causing the impacts purchases credits from a third party that has already carried out conservation when it created the bank.[22] The 2008

Figure 5.1. Habitat conservation work at the Lakes Creek Wetland Mitigation Bank. Courtesy of Amy S. Greene Environmental Consultants, Inc., Flemington, NJ.

regulations establish a preference for using credits from compensatory mitigation banks over other mechanisms, because "a mitigation bank must have an approved mitigation plan and other assurances in place before any of its credits can be used to offset permitted impacts, . . . which reduces some of the risks and uncertainties associated with compensatory mitigation."[23]

Wildlife Compensatory Mitigation

Wildlife compensatory mitigation is more complex than wetlands compensatory mitigation. Looking at two equivalent functioning wetlands within the same watershed is virtually an apples-to-apples comparison, and using the simple measure of acres of wetlands is often sufficient. There are many more variables between two parcels of land for wildlife compensatory mitigation. These include the presence of predator or prey species, elevation, climate, moisture, soil conditions, plant communities, and/or human disturbances, all of which affect whether a piece of land is suitable habitat for a species.[24]

One of the challenges of wildlife compensatory mitigation is that it is often difficult to create suitable habitat. For example, greater sage-grouse depend on slow-growing sagebrush, whereas red-cockaded woodpecker and northern spotted owl depend on old-growth forests. Once suitable habitat is created, occupation and use of that habitat is not guaranteed. For example, greater sage-grouse show a high fidelity to specific leks, returning to the same locations year after year. If a lek is destroyed and compensatory mitigation creates suitable land for a lek nearby, most birds from the old lek will not migrate to the new lek.[25]

Wildlife compensatory mitigation requires the preservation of complex ecological functions. Ecological functions must be carefully analyzed, and designers should not fall back on simple acre-to-acre comparisons.[26] It can be helpful to borrow terminology from the financial world, and think in terms of "debits" and "credits." Debits are "a quantification of the loss of ecological functions and/or services" and credits are "changes in . . . conditions that increase the value of . . . habitat relative to baseline."[27] For a species to survive, the credits generated by compensatory mitigation must be at least equal to the debits generated by negative impacts to the species. The language of credits and debits is also used in this way in the context of conservation banking. However, these are *ecological* credits and debits rather than dollars and cents. Although compensatory mitigation credits and debits can later be monetized by a conservation bank, it is ecology that defines what is needed to ensure that species survive.[28]

Despite the challenges of compensatory mitigation, and despite its position at the bottom of the mitigation hierarchy, compensatory mitigation is frequently used as a conservation tool because it creates flexibility. It allows development and other projects to go forward in sensitive habitat, even if they create impacts that cannot be eliminated through avoidance and minimization.[29]

The Concept of "Additionality"

One area in which compensatory mitigation is evolving is in its use on public land. Historically, compensatory mitigation has been limited to private land.[30] "Additionality," a concept borrowed from emissions trading systems such as carbon offsets, is helping to bring compensatory mitigation to public land, and it is also helpful for thinking about compensatory mitigation on private land.[31] As explained in a 2016 report relating to greater sage-grouse, "compensatory mitigation measures must provide conservation benefits that are . . . 'additional' to what would have occurred in the absence of the compensatory mitigation measure."[32] Private land already enrolled in government programs that provide financial incentives for landowner conservation, such as those under the Farm Bill, generally do not qualify as compensatory mitigation sites.[33] This avoids any counting of conservation benefits from the same land twice. In the context of public land, additionality must be demonstrated above the existing commitment of all federal agencies to protect wildlife on land they manage, which is mandated by the ESA.[34] Conservation activities on public land can generate compensatory mitigation credits provided that they are additional to activities already being carried out, including activities required as conditions for permits.[35]

One reason additionality is essential to effective compensatory mitigation is that compensatory mitigation exists in order to enable negative impacts on wildlife. Historically, compensatory mitigation has sought to achieve "no net loss" of ecological functions.[36] However, the no net loss standard lacks a proven track record of success, raising concerns that the historical use of compensatory mitigation may actually have allowed losses over time.[37] For example, replacing some number of acres of destroyed wetlands with equal acres of restored wetlands does not necessarily replace the ecological function of the destroyed wetlands. In the case of a wetland used by migratory birds, the ecological functions that must be considered include not just the presence of suitable vegetation, but also the regular utilization of the wetland by the birds in their seasonal migration patterns.[38] This is an area where our knowledge is still incomplete. Furthermore, achieving "no net loss" merely offsets negative impacts to resources and does not provide the cumulative environmental benefits needed to restore damaged ecosystems.[39]

As of 2019, the future direction of compensatory mitigation is uncertain, especially on public land. The Obama administration sought to strengthen compensatory mitigation by shifting from a goal of "no net loss" to "net gain" or "net conservation benefit"[40] and by issuing a Department of the Interior Secretarial Order, a presidential memorandum, a Department of the Interior Departmental Manual Chapter, and two FWS Policies to promote compensatory mitigation.[41] All of these were subsequently revoked or withdrawn by the Trump administration.[42] In addition, under the Trump administration the Forest Service reverted from net conservation benefit to no net loss, whereas the Bureau of Land Management concluded that it may not require compensatory mitigation at all.[43] These actions will discourage federal agencies and their permittees from investing in compensatory mitigation. In the long run, they will mean that more uncompensated environmental impacts will occur and it will be harder to identify, develop, and implement compensatory mitigation.[44] Hopefully, in the future government policy in this area will evolve in a more positive direction.

Types of Compensatory Mitigation

When it comes to implementing compensatory mitigation for at-risk species, there are a number of models to choose from. There are three different types of compensatory mitigation that have long been recognized by both the ESA and the CWA. The first, more basic, is permittee-responsible compensatory mitigation, in which the same party causing environmental impacts is also responsible for carrying out compensatory mitigation, either elsewhere on the project site or at another site. Permittee-responsible compensatory mitigation suffers from a major pitfall in that entities causing impacts on wildlife are, in most cases, not experienced with the conservation practices needed to mitigate those impacts. Accordingly, permittee-responsible compensatory mitigation is less popular than compensatory mitigation tools where the compensation is carried out by a different entity.

The simplest such compensatory mitigation programs are in-lieu fee programs. The entity causing negative environmental impacts pays a fee set by regulation in exchange for authorization to cause those impacts. The fee is subsequently spent on conservation carried out by a third party such as a conservation organization, a federal land management agency, or private landowners.[45] An example of a large-scale in-lieu fee program is the Western Association of Fish and Wildlife Agencies' 2013 Lesser Prairie Chicken Range-wide Conservation Plan. Under the Range-wide Plan, oil and gas companies pay per-acre enrollment fees and additional fees for impacting acres of lesser prairie chicken habitat. These fees are then spent on a combination of conservation easements and term contracts with willing landowners who carry out lesser prairie chicken conservation. Objective measures of habitat quality are used to weight relative impacts on and conservation of the species, ensuring a net conservation benefit to the species.[46]

Although in-lieu fee compensatory mitigation has many advantages over permittee-responsible compensatory mitigation, it still suffers from shortcomings. Because all the permittee must do is pay a fee, there is a danger of regulatory authorities underpricing the environmental impacts and authorizing a net loss of protected resources. This danger is avoided in a specialized type of in-lieu fee compensatory mitigation: a conservation bank.

Conservation Banking

A conservation bank is a piece of land that is managed under a conservation easement to benefit one or more species in perpetuity. These benefits are then quantified as credits and sold to third parties, which use the credits to offset negative impacts at other sites. Most conservation banks are operated by investors, called conservation bankers, who provide capital to pay landowners for the use of their land and fund conservation work, handle the regulatory process, and then sell credits for a profit. More rarely, a landowner might have enough land and capital to establish a bank using land they already own.

Like compensatory mitigation itself, conservation banking evolved out of wetlands and stream mitigation banking. Wetlands and stream mitigation banking was a small industry prior to 1995, when the Army Corps of Engineers, Environmental Protection Agency, FWS, and National Marine Fisheries Service developed regulations for the establishment and operation of mitigation banks. Since then, mitigation banking has grown into a $2 billion per year industry.[47] However, wildlife conservation banking is a much smaller part of this than wetlands mitigation banking, because conservation banks are more complex than wetlands banks. Although wetlands banks are meant to "replace the exact function and values of [a] specific wetland habitat," conservation banks offset impacts to a species. These can include loss of habitat, loss of food sources, harassment from increased human presence, and direct mortality, which can be mitigated through habitat preservation, habitat restoration, human-wildlife conflict reduction, and other actions.[48]

The first recognized conservation bank was the Carlsbad Highlands Bank in San Diego County, California, established in 1995 to provide coastal sage scrub habitat for the California gnatcatcher.[49] It was accompanied by publication of the state of California's "Official Policy on Conservation Banks."[50] In 2003, FWS published a nation-wide guidance for conservation banks.[51] Conservation banks are a small but growing part of wildlife conservation.[52] As of 2019 there are 182 approved conservation banks, according to the U.S. Army Corps of Engineers' Regulatory In-Lieu Fee and Bank Information Tracking System (RIBITS), which although mentioned by the Army Corps is also used by FWS to track credits and debits for conservation banks.[53]

Conservation banks are free market mechanisms that treat land providing habitat for listed species as a financial asset rather than a liability. Conservation banking, unlike other voluntary conservation mechanisms, pays landowners a market rate for the use of their property.[54] This allows conservation banks to compete with developers in the free market, and many conservation banks own land that would otherwise have been developed and lost all value as wildlife habitat.[55]

Conservation banks function by generating "credits," which are sold to third parties to meet compensatory mitigation obligations. The number of credits in a bank is a fixed number set when the bank is established, based on the quantity and quality of habitat available within the bank, determined by FWS. In some cases, credits awarded may be reduced by a fixed amount, ranging from 10 percent to 50 percent, in order to provide a margin for error in achieving no net loss.[56] As with any compensatory mitigation program, the exact calculation may include a number of different variables. The number of credits available in a given number of acres can be increased by locating those acres in higher-quality habitat or by improving the habitat present. Accordingly, when creating conservation banks there is an incentive for both landowners and investors to seek out land of the highest ecological value. Most costs of operating a bank are fixed, including staffing, ongoing habitat treatments, and surveys and monitoring. Thus finding better quality habitat and generating more saleable credits translates into larger profits,[57] and 91 percent of conservation banks are primarily profit-seeking ventures.[58] Because conservation banks are part of a regulatory system, conservation easements associated with banks are not charitable easements and are not eligible for income tax benefits.[59]

When a conservation bank is established by an entity other than a landowner, the conservation bank developer generally finds the land, negotiates a conservation easement, develops and implements the management plan, and handles negotiations with prospective credit purchasers.[60] Conservation banks come in many varieties, and in addition to commercial conservation banks, there are non-profit banks set up and owned by non-profit organizations. These function identically to commercial banks except that oftentimes there is no need to sell credits at a profit.[61] There are also banks owned by an entity that intends to use the credits itself to offset impacts caused by projects elsewhere. These "single-client" banks may also have financial motives in the form of mitigating impacts the bank owner is creating at another site, or they may be owned by a state or local government entity that needs to offset impacts of other government projects.[62]

Conservation banks are held to very vigorous standards. They must be established and conservation easements placed on the land before any credits are sold, meaning before any impacts to be mitigated occur. Conservation banks can generate credits by preserving existing habitat or by restoring habitat. As part of the bank establishment process, FWS informs prospective bank owners of how much of each type of habitat should be included, based on the habitat available and the needs of the species.[63] If restoration is required, that work must be done before any credits can be sold.[64]

One species that is benefiting from commercial conservation banking is the critically endangered Florida panther. Florida panthers, a subspecies of North American cougar,[65] once roamed the southeastern United States from Louisiana to South Carolina and as far north as Tennessee. After the population reached a low of twenty individuals in the 1970s, intense conservation efforts, including a "genetic rescue" by eight female cougars translocated from Texas

Figure 5.2. An endangered Florida panther (*Puma concolor coryi*).
RC Reid Photography/Shutterstock.

to Florida, restored the species to a population of 120 to 230 known adults, plus cubs and unknown adults.[66] Although male Florida panthers can range as far north as southern Georgia, the current range occupied by reproducing females is limited to the relatively open area between the Everglades and the Caloosahatchee River (west of Lake Okeechobee). This area has reached its carrying capacity for panthers, and panther recovery can only continue if the species can expand into new areas.

The species' official ESA recovery plan, last revised in 2008, calls for the eventual establishment of three distinct populations of at least 240 adults each. Limiting factors are habitat availability, prey availability, and the panther's lack of tolerance for humans.[67] Progress continues and small numbers of females and panther cubs have recently been spotted north of the Caloosahatchee River for the first time in forty years.[68] A large part of this is due to work by The Nature Conservancy to secure corridors to habitat outside the panther's current breeding range.[69] Seventy-five percent of the land within the panther's designated primary zone is now protected, and conservation efforts are focusing on the dispersal zone, a narrow band of twenty-eight thousand acres (forty-three square miles) that averages just three miles wide and connects the existing primary zone to the Caloosahatchee River and open land to the north.[70]

There are currently three private commercial conservation banks playing an important role in this by securing high-quality habitat within these corridors in exchange for the development of less vital habitat outside of them, and a fourth bank is in development. The three operational banks are the Florida Panther Conservation Bank and the Florida Panther Conservation Bank II (Panther II), both owned by Les and Jim Alderman, and the Panther Passage Conservation Bank, owned by the Wetlandsbank Company. The Florida Panther Conservation Bank is located in the species' primary habitat zone, whereas the other two banks are located in the dispersal zone. Each bank's credits are calculated based on the acres in the bank, modified by the zone where the bank is located and the quality of the habitat. Impacts on Florida panther, or debits, are calculated in the same way. Once modified, credits and debits are reduced to equivalent numerical values, with land in the primary zone and in higher-quality habitat providing more credits or debits.[71]

Conservation banks are durable and are usually made up of large contiguous parcels of land, allowing habitat restoration and other conservation actions to be highly effective. But establishing a bank is a time-consuming, expensive process. A conservation bank cannot sell any credits until it has in place its conservation plan, its conservation easement, and a permanent, non-wasting endowment sufficient to fund management of the land in perpetuity.[72] Expenses involved in a typical conservation bank include buying land or conservation easements from landowners, setting up the mitigation banking instrument, placing one or more conservation easements on the property, subsequent management of the property (often including habitat restoration), and annual monitoring and maintenance. Funding these is the main reason why outside investors are usually involved banks. In addition, the permitting process to establish a bank can be as long as three to five years.[73]

There are a number of ways to design conservation banks to minimize risk and maximize value for landowners. One is to develop a bank in phases, so that conservation easements only take effect when a credit sale is imminent.[74] Another is to provide increased benefits to both landowners and wildlife through credit stacking, which allows credits to be sold to benefit multiple species. The same land cannot be sold in more than one credit market, but including multiple species allows access to multiple credit markets. For example, the Panther II bank provides Florida panther credits and also credits for the threatened wood stork.[75] Species credits can also be stacked with credits for other resources, such as wetlands compensatory mitigation, stream compensatory mitigation, or regulatory carbon markets. The Panther II bank was also designed to provide carbon sequestration credits in a hypothetical future carbon market.[76]

Once a bank is permitted, its credits and their sales are tracked in a ledger attached to the banking instrument. A duplicate is maintained by FWS and entered into RIBITS, where it is publicly available via https://ribits.usace.army.mil/. Users can select the "Banks & ILF Sites" menu option; search for a conservation bank by name, location, or species; and view a summary of the current status of the credit ledger. FWS has the right to audit all components of a bank, including its books and land (with the landowner's permission) at any time. In addition, each conservation bank must submit an annual report documenting the status of its financial and biological commitments.[77] Buying credits from a conservation bank is simple. Anyone can search RIBITS by latitude and longitude and receive a list of all conservation banks with credits available for that location, including maps showing bank locations and service areas, which are set at the time the bank is created.[78] In addition, bank owners and investors are often actively looking for buyers within their service area.[79] Once a prospective buyer identifies a conservation bank with credits available, the buyer and the banker enter into a free market negotiation to establish a credit price.[80]

Establishing a conservation bank is a dynamic process that unfolds over a series of negotiations between investors and landowners, investor/landowners and FWS, and finally investor/landowners and prospective credit buyers. But the end result is always conservation reduced to a fixed number of dollars. This makes it easy for buyers to understand their obligations and for landowners to consider the benefits of a conservation bank as compared to farming, ranching, logging, or selling their land for development.

Variations on Conservation Banking

Conservation banking takes the in-lieu fee approach to compensatory mitigation and adapts it to work in a free market. This concept is being applied to other tools. One is a habitat exchange. Habitat exchanges are market-based, but unlike conservation banks they use an independently established central administrator to aggregate conservation efforts on a variety of lands, increasing flexibility and efficiency.[81] Pilot habitat exchanges are currently operating for lesser prairie chicken, greater sage-grouse, monarch butterfly, New England cottontail, and Swainson's hawk. Information about these habitat exchanges, including links to contact each individual exchange, is available from the Environmental Defense Fund through https://www.edf.org/ecosystems/habitat-exchanges-transform-conservation.[82] None of these species are listed under the ESA, but their habitat exchanges will continue to function if they become listed. Because habitat exchanges are operated by independent central clearinghouses, they can consolidate the benefits of numerous different conservation efforts, including traditional compensatory mitigation mechanisms such as conservation banks.[83] Compared to banks, habitat exchanges "allow for a more distributed base of suppliers—potentially hundreds or even thousands of different credit providers—so as to facilitate the achievement of conservation goals for wide-ranging, landscape-dependent species."[84] Habitat exchanges can also blend permanent and temporary protection of habitat. This may prove valuable if, for example, it allows for future relocation of protected land as species' ranges and migration corridors shift due to climate change.[85] Temporary protection may also allow limited conservation dollars to go further, or engage landowners unable to commit to permanent protection.[86]

In the future, there may be opportunities to build on habitat exchanges and establish more voluntary markets for tradable compensatory mitigation credits for unlisted species. A challenge to overcome is properly designing incentives for such investments.[87] One source of incentives is the emerging corporate sustainability movement, which encourages corporations to pursue social goals such as environmental protection, social justice, and economic development.[88] This movement is generating voluntary investments in carbon sequestration, sustainable forestry, soil health, air and water quality, and other resources that benefit humans and wildlife, for example through corporate donations to the National Fish and Wildlife Foundation.[89]

Compensatory Mitigation for Government Agencies

The ESA and CWA require government agencies to mitigate negative environmental impacts just as private actors must. One way to do this is through creation of a single-client conservation bank. Many state departments of transportation operate wetlands mitigation banks to offset impacts on protected wetlands caused by road and highway construction projects. Some agencies have also established conservation banks for compensatory mitigation for threatened and endangered species. For example, in 2008 the Oregon Department of Transportation established a conservation bank in the Rogue River Valley, which was actually the first conservation bank in the state.[90] An advantage of single-client conservation banks is that they eliminate both transaction costs and the need for investors to realize a profit.[91]

In addition to conservation banks, federal government agencies can take advantage of a FWS program called "recovery crediting," a hybrid of a conservation bank and a direct payment program. In a recovery crediting program, a government agency funds conservation on private land and creates a reserve of credits that the same agency can then draw upon for compensatory mitigation credits. Unlike conservation banks, recovery crediting conservation is not necessarily permanent. The original recovery crediting program was created to conserve golden-cheeked warbler on and around Fort Hood in Texas.[92] In 2008, FWS used that program as a model to create its Recovery Crediting Guidance.[93] A second recovery crediting pilot project was launched in 2017 to benefit the gopher tortoise in the eastern part of its range, where it is a candidate for listing under the ESA.[94] This project is a collaboration between FWS, the U.S. Department of Defense, and the states of Alabama, Florida, Georgia, and South Carolina. Like the red-cockaded woodpecker, the gopher tortoise is a keystone species within the longleaf pine ecosystem of the southeast, so conservation of the tortoise also benefits woodpeckers and many other species.[95] Assuming that recovery crediting proves effective for both tortoise conservation and military readiness, recovery crediting may become more widely adopted within the next five to ten years by the military and by other federal agencies.[96]

Notes

1. "Environmental Mitigation," in *Black's Law Dictionary*, ed. Bryan A. Garner (tenth edition, 2014).
2. Fish and Wildlife Coordination Act (as Amended), Pub. L. No. 73-121, 48 Stat. 401 (1934).
3. See An Act to Promote the Conservation of Wild Life, Fish, and Game, and for Other Purposes, Pub. L. No. 73-121, 48 Stat. 401 (1934).
4. "The [1946] act is mainly concerned with compensatory measures to mitigate the loss of or damage to fish and wildlife resources." Committee on Interstate and Foreign Commerce, Amending the Coordination Act, Report to Accompany H.R. 13138, S. Rep. No. 85-1981 (1958), 4.
5. An Act to Amend the Act of March 10, 1934 Entitled "an Act to Promote the Conservation of Wildlife, Fish, and Game, and for Other Purposes," Pub. L. No. 79-132, 60 Stat. 1080 (1946).
6. An Act to Amend the Act of March 10, 1934, to Provide for More Effective Integration of a Fish and Wildlife Conservation Program with Federal Water-Resource Developments, and for Other Purposes, Pub. L. No. 85-624, 72 Stat. 568 (1958), 569. (Emphasis added.)
7. National Environmental Policy Act of 1969, Pub. L. No. 91-190, 83 Stat. 852 (1970), 853.
8. Implementation of Procedural Provisions, 43 Fed. Reg. 55978 (November 29, 1978), 56004.
9. A NEPA evaluation has three possible outcomes. An action that does not individually or cumulatively have a significant effect is a "Categorical Exclusion" (CE) for which no further analysis is required. For actions that are not CEs, the agency prepares an "Environmental

Assessment" (EA). EAs look at the environmental impacts of the proposed actions, including any alternatives, and determines if the impacts are going to be significant. If not, the agency prepares a Finding of No Significant Impact (FONSI) and the NEPA process concludes. If there will be significant impacts, the agency proceeds to the third and final step in the NEPA analysis and prepares an Environmental Impact Statement (EIS). An EIS is a comprehensive analysis of the proposed action. Each EIS is subject to public comment and also to review by the Environmental Protection Agency, and the publication of a final EIS completes the NEPA process. Council on Environmental Quality, "A Citizen's Guide to the National Environmental Policy Act," last accessed November 5, 2018, https://www.fs.usda.gov/Internet/FSE_DOCUMENTS/stelprdb5310825.pdf. NEPA does not mandate specific outcomes, provided that the agency takes a "hard look" at the environmental consequences of its actions. *Robertson v. Methow Valley Citizens Council*, 490 U.S. 332, 352 (1989).

10. Implementation of Procedural Provisions, 43 Fed. Reg. 55978, 56005.

11. For example, U.S. Fish and Wildlife Service Mitigation Policy, 81 Fed. Reg. 83440 (November 21, 2016), 83457, 83477–78 (withdrawn).

12. *Babbitt v. Sweet Home Chapter of Communities for a Great Oregon*, 515 U.S. 687 (1995).

13. U.S. Fish and Wildlife Service, Endangered Species Act of 1973, as Amended through the 108th Congress, 28. See also Endangered Species Act Compensatory Mitigation Policy, 81 Fed. Reg. 95316, 95337–38 (withdrawn). For more on Section 10 conservation programs, see chapter 13.

14. U.S. Fish and Wildlife Service, Endangered Species Act of 1973, as Amended through the 108th Congress, 21 (Section 7(h)(1)(B)).

15. For example, U.S. Fish and Wildlife Service Mitigation Policy, 81 Fed. Reg. 83440, 83446 (withdrawn); Endangered Species Act Compensatory Mitigation Policy, 81 Fed. Reg. 95316, 95321, 95336–37 (withdrawn).

16. R. Andrew King, Revised Programmatic Biological Opinion on the Proposed Construction, Operation, and Maintenance of Alternative 3C of Interstate 69 (I-69) from Evansville to Indianapolis for the Federally Endangered Indiana Bat (*Myotis Sodalis*) and the Federally Threatened Bald Eagle (*Haliaeetus Leucocephalus*) Traversing Portions of Gibson, Warrick, Pike, Daviess, Greene, Monroe, Morgan, Johnson, and Marion Counties, Indiana Submitted to the Federal Highway Administration (August 24, 2006).

17. U.S. Fish and Wildlife Service, Final Biological Opinion on the Effects to Grizzly Bears from the Implementation of Proposed Actions Associated with Plan of Operations for the Montanore Minerals Corporation Copper/Silver Mine as Proposed by the U.S. Forest Service, Kootenai National Forest (March 31, 2014).

18. Endangered Species Act Compensatory Mitigation Policy, 81 Fed. Reg. 95316, 95338 (withdrawn). Section 4(d) authorizes special rules for threatened species, discussed in chapter 12. Section 5 authorizes land acquisitions.

19. U.S. Environmental Protection Agency, "Wetlands Compensatory Mitigation," last accessed October 4, 2019, https://www.epa.gov/sites/production/files/2015-08/documents/compensatory_mitigation_factsheet.pdf.

20. Ruhl, Glen, and Hartman, "Practical Guide to Habitat Conservation Banking Law and Policy," 26–27.

21. Compensatory Mitigation for Losses of Aquatic Resources, 73 Fed. Reg. 19594 (April 10, 2008). Earlier guidance was issued in 1995 and 2000. Federal Guidance for the Establishment, Use and Operation of Mitigation Banks, 60 Fed. Reg. 58605 (November 28, 1995); Federal Guidance on the Use of in Lieu-Fee Arrangements for Compensatory Mitigation

under Section 404 of the Clean Water Act and Section 10 of the Rivers and Harbors Act, 65 Fed. Reg. 66914 (November 7, 2000).

22. U.S. Environmental Protection Agency, "Wetlands Compensatory Mitigation."

23. Compensatory Mitigation for Losses of Aquatic Resources, 73 Fed. Reg. 19594, 15594. See also Wayne Walker, interview by Christopher E. Segal, October 5, 2018; U.S. Environmental Protection Agency, "Wetlands Compensatory Mitigation."

24. For example, Walker.

25. Lyons.

26. Ibid.

27. See Report to the Sage-Grouse Task Force, Greater Sage-Grouse Compensatory Mitigation (December, 2016), 20.

28. For example, Lyons.

29. For example, U.S. Fish and Wildlife Service Mitigation Policy, 81 Fed. Reg. 83440, 83455 (withdrawn); Lyons.

30. Sean Kiernan, "Perspective: Renewable Energy and Endangered Species," in *Proactive Strategies for Protecting Species: Pre-Listing Conservation and the Endangered Species Act*, ed. C. Josh Donlan (Oakland, CA: University of California Press, 2015), 42.

31. Michael Gillenwater, "What Is Additionality? Part 1: A Long Standing Problem," Discussion Paper No. 001 (Silver Spring, MD: Greenhouse Gas Management Institute, 2012), 4.

32. Report to the Sage-Grouse Task Force, Greater Sage-Grouse Compensatory Mitigation, 9.

33. Endangered Species Act Compensatory Mitigation Policy, 81 Fed. Reg. 95316, 95341 (withdrawn).

34. Specifically, Section 7(a)(1) states that "all . . . Federal agencies shall . . . utilize their authorities in furtherance of the purposes of this chapter by carrying out programs for the conservation of endangered species and threatened species."

35. Report to the Sage-Grouse Task Force, Greater Sage-Grouse Compensatory Mitigation, 9–10. See also Lyons.

36. For example, Compensatory Mitigation for Losses of Aquatic Resources, 73 Fed. Reg. 19594.

37. For example, Jennifer M. Germano et al., "Mitigation-Driven Translocations: Are We Moving Wildlife in the Right Direction?" *Frontiers in Ecology and the Environment* 13, no. 2 (2015): 100–05. See also Lyons.

38. Lyons.

39. Alex Echols, Alan Front, and James Cummins, "Broadening Conservation Funding," *Wildlife Society Bulletin* 43 (2019): 372–81.

40. For example, U.S. Fish and Wildlife Service Mitigation Policy, 81 Fed. Reg. 83440 (withdrawn); Endangered Species Act Compensatory Mitigation Policy, 81 Fed. Reg. 95316 (withdrawn); U.S. Department of the Interior Office of the Solicitor, Solicitor's Memorandum M-37037: The Bureau of Land Management's Authority to Address Impacts of Its Land Use Authorizations through Mitigation (December 21, 2016); Lyons; U.S. Forest Service, Greater Sage-Grouse Record of Decision: Idaho and Southwest Montana, Nevada, Utah (September, 2015).

41. U.S. Department of the Interior, Secretarial Order No. 3330: Improving Mitigation Policies and Practices of the Department of the Interior (October 31, 2013); U.S. Department of the Interior, Departmental Manual Release, Landscape-Scale Mitigation Policy (600 DM 6) (October 23, 2015); Presidential Memorandum: Mitigating Impacts on Natural Resources from Development and Encouraging Related Private Investment (November 3, 2015); U.S. Fish and Wildlife Service Mitigation Policy, 81 Fed. Reg. 83440; Endangered Species Act Compensatory Mitigation Policy, 81 Fed. Reg. 95316.

42. U.S. Department of the Interior, Secretarial Order No. 3349: American Energy Independence (March 29, 2017); U.S. Department of the Interior, Secretarial Order No. 3360: Rescinding Authorities Inconsistent with Secretary's Order 3349, "American Energy Independence" (December 22, 2017); Executive Order No. 13783: Promoting Energy Independence and Economic Growth (March 28, 2017); U.S. Fish and Wildlife Service Mitigation Policy, 83 Fed. Reg. 36472 (July 30, 2018); Endangered Species Act Compensatory Mitigation Policy, 83 Fed. Reg. 36469 (July 30, 2018).

43. Bureau of Land Management, Instruction Memorandum 2018-093: Compensatory Mitigation (July 24, 2018). For example, U.S. Forest Service, "Greater Sage-Grouse Proposed Land Management Plan Amendments (LMPA) and Draft Environmental Impact Statement (DEIS) for the Intermountain and Rocky Mountain Regions" (October 2018): ES-3 ("Net conservation gain changed to no net loss of habitat . . .").

44. Michael J. Bean, "Habitat Exchanges: A New Tool to Engage Landowners in Conservation," Working Paper (Madison, WI: Sand County Foundation/Environmental Policy Innovation Center, 2017), 6–7.

45. For example, Endangered Species Act Compensatory Mitigation Policy, 81 Fed. Reg. 95316, 95318 (withdrawn).

46. William E. Van Pelt et al., *The Lesser Prairie-Chicken Range-Wide Conservation Plan* (Cheyenne, WY: Western Association of Fish and Wildlife Agencies, 2013).

47. Echols, Front, and Cummins, "Broadening Conservation Funding"; Federal Guidance for the Establishment, Use and Operation of Mitigation Banks, 60 Fed. Reg. 58605.

48. U.S. Fish and Wildlife Service, Guidance for the Establishment, Use, and Operation of Conservation Banks (May 2, 2003), 3.

49. Marybeth Bauer, Jessica Fox, and Michael J. Bean, "Landowners Bank on Conservation: The U.S. Fish and Wildlife Service's Guidance on Conservation Banking," *Environmental Law Reporter News & Analysis* 34 (2004): 10719.

50. Ibid.

51. U.S. Fish and Wildlife Service, Guidance for the Establishment, Use, and Operation of Conservation Banks.

52. Jessica Fox and Anamaria Nino-Murcia, "Status of Species Conservation Banking in the United States," *Conservation Biology* 19, no. 4 (2005): 998.

53. U.S Army Corps of Engineers, "Regulatory in-Lieu Fee and Bank Information Tracking System," last accessed June 19, 2018, https://ribits.usace.army.mil; Walker.

54. Walker.

55. Fox and Nino-Murcia, "Status of Species Conservation Banking in the United States."

56. Echols, Front, and Cummins, "Broadening Conservation Funding."

57. Gregory M. Parkhurst and Jason F. Shogren, "Evaluating the Incentive Tools," in *Species at Risk: Using Economic Incentives to Shelter Endangered Species on Private Lands*, ed. Jason F. Shogren (Austin, TX: University of Texas Press, 2005), 247.

58. Fox and Nino-Murcia, "Status of Species Conservation Banking in the United States," 999.

59. Walker.

60. Ibid.

61. Fox and Nino-Murcia, "Status of Species Conservation Banking in the United States," 999–1001.

62. Ibid.

63. Walker.

64. When approving a conservation bank, FWS looks at whether conservation benefits are being measured in the same way as proposed impacts, whether the bank is sited such that

its habitat and connectivity are suitable, whether a bank's conservation plan includes suitable management activities to benefit the species and these activities are being carried out, whether a buffer area is included, whether credits are calculated appropriately, whether the bank has secured adequate funding, and whether the bank's land is permanently protected through a conservation easement or fee simple ownership. U.S. Fish and Wildlife Service, Guidance for the Establishment, Use, and Operation of Conservation Banks. Any restoration activities must be credited based upon demonstrated conservation outcomes and not just the potential future benefits. Bauer, Fox, and Bean, "Landowners Bank on Conservation: The U.S. FWS's Guidance on Conservation Banking," 10721. Banks with restoration activities in their management plans are awarded "potential" credits that are only released for sale as performance standards are met.

65. There is an argument within the scientific community that the Florida panther may not be a distinct subspecies, but at present it is listed as such under the ESA.

66. Florida Fish and Wildlife Conservation Commission, "Florida Panther," last accessed November 16, 2018, http://myfwc.com/wildlifehabitats/managed/panther/; U.S. Fish and Wildlife Service, Florida Panther (Puma Concolor Coryi) 5-Year Review (2009).

67. U.S. Fish and Wildlife Service, Florida Panther Recovery Plan: 3rd Revision (2008), ix–xi.

68. The Nature Conservancy, "Florida Panthers: Crossing the Caloosahatchee," last accessed November 16, 2018, https://www.nature.org/en-us/about-us/where-we-work/united-states/florida/stories-in-florida/florida-panther-kittens-north-of-caloosahatchee/.

69. Ibid.

70. Craig Pittman, "Saga of Florida Panther Is 'Sordid Story,'" *St. Petersburg Times*, April 19 2010; Nick Wiley and Paul Souza, "Conservation Banks Help Protect Panthers," ibid., March 1.

71. The Florida Panther Conservation Bank was awarded 15,541.4 credits (both potential and released) for its 1,930 acres. The Florida Panther Conservation Bank II was awarded 14,857.5 credits for its 472.4 acres, and the Panther Passage Conservation Bank was awarded 96,256 potential credits, of which 35,947 have been released, for its 4,009 acres. U.S Army Corps of Engineers, "Regulatory in-Lieu Fee and Bank Information Tracking System." Florida panther credits use "Panther Habitat Units" (PHUs) as a currency, calculated based on the Panther Habitat Assessment methodology, which is updated periodically. U.S. Fish and Wildlife Service, Florida Panther Recovery Plan: 3rd Revision, 62. The number of acres is first multiplied by a landscape compensation multiplier, weighted toward land within the primary zone with steep discounts for land in the secondary zone or "other" land outside of the panther's current range. U.S. Fish and Wildlife Service, Panther Habitat Assessment Methodology (September 24, 2012), 3. The resulting figure is multiplied by a habitat suitability value multiplier, with hardwood forests receiving the highest score, followed by shrub lands, barren or disturbed lands and areas overgrown with exotic plants, and finally open water and developed land, which receive a score of zero. Ibid., 10. Finally, land in the dispersal zone is awarded an additional multiplier of 3.5, in recognition of the fact that land in this area is more expensive. The Conservation Fund, "The Florida Panther Conservation Bank, II," last accessed September 20, 2019, https://www.conservationfund.org/images/programs/files/Florida-Panther.pdf. When setting the number of PHUs needed to mitigate a project, FWS further multiples the PHUs of the impacted habitat by a "base ratio" of 2.5, which reflects the fact that if all land within the primary zone at risk of development were to be developed, it would be necessary to secure 2.5 times the equivalent habitat in order to maintain a population of ninety panthers within the primary zone. U.S. Fish and Wildlife Service, Panther Habitat Assessment Methodology, 1–3.

72. Walker.

73. Ibid.

74. Ibid. However, in the interim the land must be maintained in suitable condition for the bank, including any required restoration work, and the funds needed to establish the endowment must be held in escrow.

75. Of the bank's 14,857.5 total credits, 11,175.15 are in habitat that support both panthers and wood storks. Habitat within the bank also supports snail kite, sandhill crane, and crested caracara, each of which may be added to the bank's credit system in the future. The Conservation Fund, "The Florida Panther Conservation Bank, II"; U.S Army Corps of Engineers, "Regulatory in-Lieu Fee and Bank Information Tracking System."

76. Because wetlands conservation banks also support wildlife and the two banking systems are closely related, wetlands banking credits cannot be unstacked from wildlife banking credits. Royal C. Gardner and Jessica Fox, "The Legal Status of Environmental Credit Stacking," *Ecology Law Quarterly* 4, no. 4 (2013): 744–45. However, carbon credits can be unstacked from wildlife banking credits, allowing each acre in the Panther II bank to potentially be sold for its wildlife value and again for its carbon value. The Conservation Fund, "The Florida Panther Conservation Bank, II"; Fox and Nino-Murcia, "Status of Species Conservation Banking in the United States," 745–46.

77. Walker.

78. U.S. Fish and Wildlife Service, Guidance for the Establishment, Use, and Operation of Conservation Banks, 8–9.

79. Walker.

80. Market prices are driven by many factors. These include supply of and demand for credits, local land values, the cost of capital, the risk being carried by investors, and of course the buyer's willingness to pay. Ibid.

81. Environmental Defense Fund, "Habitat Exchanges: How Do They Work?" last accessed July 12, 2018, https://www.edf.org/ecosystems/habitat-exchanges-how-do-they-work; Endangered Species Act Compensatory Mitigation Policy, 81 Fed. Reg. 95316, 95318 (withdrawn).

82. Environmental Defense Fund, "Habitat Exchanges Transform Conservation," last accessed November 20, 2018, https://www.edf.org/ecosystems/habitat-exchanges-transform-conservation.

83. For example, U.S. Fish and Wildlife Service Mitigation Policy, 81 Fed. Reg. 83440, 83458 (withdrawn).

84. Christopher S. Galik et al., "Improving Habitat Exchange Planning through Theory, Application, and Lessons from Other Fields," *Environmental Science & Policy* 73 (2017): 47.

85. Bean, "Habitat Exchanges: A New Tool to Engage Landowners in Conservation."

86. On the other hand, increased transaction costs such as from term contracts ending and being renewed or relocated may reduce the efficiency of habitat exchanges.

87. For example, in 2017 FWS adopted a Policy Regarding Voluntary Prelisting Conservation that sought to encourage state fish and wildlife agencies to develop voluntary markets for unlisted species. The idea was to provide flexibility for states, species, and landowners with different needs. Unfortunately, the policy lacked detail beyond the concept that benefits to a species from voluntary conservation prior to listing should be able to compensate for impacts occurring after listing, and as of 2019 it has seen no use. U.S. Fish and Wildlife Service, Policy Regarding Voluntary Prelisting Conservation Actions, Appendix 1, 735 FW 1; Timothy Male, *Headstarting Conservation: Reviewing New Policy to Incentivize Early Investments in Declining Species* (Madison, WI: Sand County Foundation/Mission: Wildlife, 2014).

88. Mel Wilson, "Corporate Sustainability: What Is It and Where Does It Come From?" *Ivey Business Journal*, 2003, https://iveybusinessjournal.com/publication/corporate-sustainability-what-is-it-and-where-does-it-come-from/.

89. Steve Jester, email to Christopher E. Segal, August 21, 2019. See chapter 10.

90. U.S. Fish and Wildlife Service Oregon Fish & Wildlife Office, "Oregon's First Conservation Bank," July 22, 2008, https://www.fws.gov/oregonfwo/LandAndWater/Conservation Planning/ConservationBank.asp; Oregon Department of Transportation, "Conservation Bank Helps ODOT," *Moving Ahead*, April 2015.

91. R. Neal Wilkins, "Improving the ESA's Performance on Private Land," in *Rebuilding the Ark: New Perspectives on Endangered Species Act Reform*, ed. Jonathan H. Adler (Washington, DC: American Enterprise Institute for Public Policy research, 2011), 70.

92. Recovery Crediting Guidance, 73 Fed. Reg. 44761 (July 31, 2008).

93. Ibid.

94. The eastern part of the range is east of the Mobile and Tombigbee Rivers in Alabama. West of the Mobile and Tombigbee Rivers the tortoise is listed as threatened. U.S. Fish and Wildlife Service Environmental Conservation Online System, "Gopher Tortoise (*Gopherus Polyphemus*)," last accessed July 15, 2019, https://ecos.fws.gov/ecp0/profile/speciesProfile?sId=6994.

95. U.S. Fish and Wildlife Service–Southeast Region et al., Department of Defense Gopher Tortoise (*Gopherus Polyphemus*) Conservation and Crediting Strategy (March 2017).

96. Bruce Beard, interview by Christopher E. Segal, July 10, 2019.

INCENTIVES FOR PRIVATE LAND CONSERVATION: THE FARM BILL

III

"There can be no doubt that a society rooted in the soil is more stable than one rooted in pavements."

–ALDO LEOPOLD, "ECOLOGY AND POLITICS"
(LECTURE, UNIVERSITY OF WISCONSIN, 1941)

Introduction to the Farm Bill 6

PREVIOUS CHAPTERS COVERED why and how to conserve wildlife on private land, the legal mechanisms to enroll in conservation programs, and financial considerations when doing so, including income from conservation programs and tax consequences. The remainder of this book describes individual conservation programs and the partner organizations that make them possible.

The most significant source of funding for private land conservation in the United States is the federal Farm Bill, which funds a variety of programs, most implemented by the U.S. Department of Agriculture's (USDA's) Natural Resources Conservation Service (NRCS). The Farm Bill is an omnibus bill passed by Congress roughly every five years that encompasses a vast array of programs relating to agriculture, goods, land, and energy. It is best known for agricultural subsidies, crop insurance, and food stamps.

The first predecessor of the modern Farm Bill, the Agricultural Adjustment Act of 1933, was focused on boosting agricultural prices during the Great Depression.[1] Farm policy began to consider conservation due to the Dust Bowl, which prompted passage of the Soil Conservation and Domestic Allotment Act of 1936, creating the Agricultural Conservation Program to conserve soil and prevent erosion.[2] This was followed in 1957 by the Great Plains Conservation Program. These two programs, as well as the Colorado River Basin Salinity Control Program and the Water Quality Incentives Program, were consolidated in 1996 into the Environmental Quality Incentives Program. The Environmental Quality Incentives Program provides funding to "enhance natural resources, particularly through improvements to soil health, conservation of water resources, improving air and water quality, enhancing habitat, and more."[3]

Congress passed Farm legislation irregularly until the Food and Agricultural Act of 1965, the first modern Farm Bill. Since then, Farm Bills have been passed roughly every five years, including in 1970, 1973, 1977, 1981, 1985, 1990, 1996, 2002, 2008, 2014, and 2018. The 1985 Farm Bill was the first Farm Bill to include the modern conservation title,[4] authorizing the Conservation Reserve Program and including the Highly Erodible Land and Wetland Conservation compliance requirements for anyone receiving federal Farm Bill program benefits. Another important milestone came in the 1996 Farm Bill, which established wildlife conservation as one of the goals of the Farm Bill's conservation programs, equal in importance to soil and water conservation.[5]

The Natural Resources Conservation Service

Most Farm Bill conservation programs are implemented by NRCS, an agency of the USDA. NRCS began in 1933 within the Department of the Interior as the Soil Erosion Service. The problem of soil erosion had been understood for some years, and solutions vigorously advocated for by Hugh Hammond Bennett, "the father of soil conservation," who became the first Chief of the Soil Erosion Service. But it was the Dust Bowl, which caused dust clouds to blow over Washington, DC, on three occasions in 1933 and 1934, that provided the impetus and public demand for a strong legislative response to the problem. In 1935, Congress passed the Soil Conservation Act, which transferred the young

Figure 6.1. Hugh Hammond Bennett, 1881–1960, known as "The Father of Soil Conservation," served as the first chief of the Soil Conservation Service from 1935 until 1951. Courtesy of Archives of Union County Agricultural Center, Union County, NC.

Figure 6.2. The Dust Bowl, a severe drought that gripped much of the American prairie in the 1930s, caused extensive loss of soil and dramatic dust storms that swept across the country. It inspired the creation of the Soil Conservation Service in 1935, which subsequently evolved into the Natural Resources Conservation Service. This iconic 1936 photo by Arthur Rothstein shows a farmer and his two sons caught in a dust storm in Cimarron County, Oklahoma.
Wikipedia.

Soil Erosion Service from the Department of the Interior to the USDA, renaming it the Soil Conservation Service.[6]

The Soil Conservation Service provided technical support for numerous New Deal programs, including the Civilian Conservation Corps, the Civil Works Administration, and the Works Progress Administration. In 1937, the agency led the push for the creation of soil conservation districts. Over the years it assumed responsibility for a number of water and flood-related programs, including the Flood Control Act of 1936, the USDA's drainage and irrigation assistance programs, the snow survey and water supply forecasting

Figure 6.3. The Jamie L. Whitten Building, the Washington D.C. headquarters of the U.S. Department of Agriculture, was built in 1908. Together with the larger U.S. Department of Agriculture South Building, constructed 1930–1936, it houses the Natural Resources Conservation Service, Farm Service Agency, and most other USDA agencies.
Archistoric/Alamy Stock Photo.

program, and the Water Facilities, Land Utilization, and Farm Forestry programs. Watershed conservation work followed as well as technical support for the Soil Bank Program, and the development of Ecological Site Descriptions to provide a stronger scientific basis for both agricultural and conservation activities,[7] rural development and recreation programs, and more. All of these programs expanded and deepened the agency's work with private landowners, ultimately transforming it into the premier agency for funding, promoting, and supporting conservation on private land. In 1994, Congress renamed the Soil Conservation Service the Natural Resources Conservation Service in recognition of the greatly broadened scope of the agency's mission.[8]

NRCS operations are structured to provide flexibility and decision-making at the local level. Conservation programs in each state are directed by a state conservationist. The NRCS also has a technical committee in each state that represents federal and state natural resource agencies, Indian tribes, agricultural and environmental organizations, and agricultural producers.[9] State technical committees provide information, analysis, and recommendations, and many Farm Bill conservation programs include an explicit role for them in identifying regional needs and objectives. State technical committees are an important part of the partnership model that NRCS and other USDA agencies use to deliver on-the-ground conservation.[10]

Other Farm Bill–Implementing Organizations

Although most Farm Bill conservation programs are administered by NRCS, a number of other entities also play a role in Farm Bill implementation. The most important is the Farm Service Agency (FSA). Like NRCS, the FSA is the end result of predecessor agencies that can be traced back to the early 1930s. The FSA's primary work is administration of the commodity pricing and financial assistance programs within the Farm Bill; it also plays a role in conservation as the primary implementing agency for the Conservation Reserve Program.[11] Another entity frequently mentioned in the Farm Bill and related legislation is the Commodity Credit Corporation. Chartered in 1933, the Commodity Credit Corporation is a government-owned and -operated corporation that is used to move funding from the U.S. Treasury to various Farm Bill programs, including many conservation programs.[12] The Commodity Credit Corporation is a financial entity with no staff of its own, and its operations are carried out by USDA agencies.[13]

From its earliest days, the Soil Conservation Service recognized a need for local leadership and expertise to effectively implement conservation. In 1935, the Secretary of Agriculture's Committee on Soil Conservation recommended that "Soil Conservation Associations" be created to assist the Soil Conservation Service. In 1937, a model Soil Conservation District Law was developed, and over the subsequent decades every state adopted legislation authorizing creation of soil conservation districts. Soil conservation districts are intended to empower volunteer local citizens to plan and install conservation practices based on local needs. Establishing the districts at the county level made it possible for each district to have personal contact with local landowners.[14]

Over the years, soil conservation districts have gradually broadened their focus to include water conservation, flood prevention, recreation, and other purposes, and many of them dropped the word "soil" from their names.[15] Today there are approximately three thousand conservation districts with seventeen thousand citizens serving on their governing boards.[16] Conservation districts are present in almost every county and provide resources for landowners interested in conservation. The National Association of Conservation Districts maintains a directory that is available online at https://www.nacdnet.org/general-resources/conservation-district-directory/.[17] Conservation districts also play a role in approving conservation plans under the Conservation Reserve Program.

Finally, many Farm Bill conservation programs rely on non-governmental organizations and other partnership organizations that assist in their implementation. The process of implementing a Farm Bill program "includes marketing projects to landowners, understanding program requirements, [partners] assisting USDA with administrative paperwork, ranking projects, obligating dollars, designing conservation practices, and guiding implementation."[18] Much of this work is done by NRCS, but some is done by third-party partners. Although funding for Farm Bill conservation programs has increased over the years, NRCS staff levels and technical support capacity have decreased, making support from third parties ever more critical.[19]

Congress recognized this need in the 2002 Farm Bill and created Technical Service Providers (TSPs).[20] TSPs are "individuals and entities [approved by the Secretary] to provide technical assistance to carry out programs under" the conservation title of the Farm Bill.[21] TSPs are hired by agricultural producers to provide services on behalf of NRCS, including the development of conservation plans and the design, installation, and monitoring of conservation practices.[22] TSPs are initially paid by landowners, but landowners are then reimbursed by NRCS according to set rates.[23]

The TSP model proved to be best suited to situations where landowners needed highly technical assistance, such as an engineering firm to address water drainage issues,[24] and less well suited to wildlife conservation.[25] Another approach, which gained popularity following the 2008 Farm Bill, is to use partnerships and volunteers to meet the public's demand for conservation planning and technical assistance.[26] Many organizations have created Farm Bill biologist programs.[27] For example, as of 2019 Pheasants Forever's Farm Bill Biologist program supports 188 biologists in thirty states. The organization shares the cost of these biologists with NRCS and other partners including other USDA agencies, state fish and wildlife agencies, Joint Ventures, the National Fish and Wildlife Foundation, other non-governmental organizations, and local organizations.[28] The exact makeup of funding for each biologist position varies. Farm Bill Biologists are located in USDA Service Centers. Since 2003, this program has provided information to over 231,500 landowners and impacted ten million acres of private land.[29] At the National Wild Turkey Federation, as of 2019, thirty staff are assigned to work with NRCS providing Farm Bill services to landowners. These positions are jointly funded by the organization, NRCS, and other partners.[30]

U.S. Department of Agriculture Service Centers

Although NRCS and FSA are both headquartered in Washington, DC, most of the agencies' work is done in the field. Each agency maintains a state office and a field office in virtually every county in the United States. These field offices are co-located in USDA Service Centers, which are designed to provide services from NRCS, FSA, and the Office of Rural Development at single locations.[31] This allows USDA staff to connect landowners with diverse needs to the correct agencies and programs with a minimum of additional effort.[32] USDA Service Center staff are often also the most trusted source of information for producers.[33] This is often where the private landowner starts their inquiry into Farm Bill funding opportunities for wildlife conservation programs. USDA maintains an easy-to-use, map-based online Service Center Locator at https://offices.sc.egov.usda.gov/locator/app.[34]

The Organization and Evolution of the Farm Bill

The 1985 Farm Bill, the Food Security Act of 1985, was the first to include the modern conservation title. The different subjects addressed by the Farm Bill are divided into distinct titles. The latest, 2018, Farm Bill contains twelve titles: I. Commodities; II. Conservation; III. Trade; IV. Nutrition; V. Credit; VI. Rural Development; VII. Research, Extension, and Related Matters; VIII. Forestry; IX. Energy; X. Horticulture; XI. Crop Insurance; and XII. Miscellaneous. Today, all but one of the conservation programs authorized and funded by the Farm Bill are located within the conservation title. The exception is the Healthy Forests Reserve Program, located in the Forestry Title.[35]

In order to understand the conservation programs within the Farm Bill and the way that Congress constantly adjusts them, it is helpful to briefly summarize the political process that takes place each time Congress negotiates and passes a new Farm Bill. The evolution of the Farm Bill over the years can be thought of as a dialogue between Congress, USDA, and the American people. Each time the Farm Bill is passed, Congress makes changes to numerous programs. USDA then creates or amends regulations to implement the Farm Bill. As programs are implemented, stakeholders develop new recommendations, which they take to Congress for consideration in the next Farm Bill. The constant changes to Farm Bill programs mean that many have complex histories.[36]

Some Farm Bill programs enjoy what is known as mandatory funding. The Farm Bill authorizes a certain amount of money to be appropriated each year, and for mandatory programs that money is then available in each year without any further act of Congress.[37] Other programs have discretionary funding. The Farm Bill may authorize appropriations at a certain level or may just say that appropriations are authorized in the amount necessary to carry out the program. But for the funds to actually be available, Congress must later include them in an annual appropriations bill. Some programs, for example the Healthy Forests Reserve Program, are discretionary and have not been funded by annual appropriations at anywhere near the levels authorized by the Farm Bill.[38]

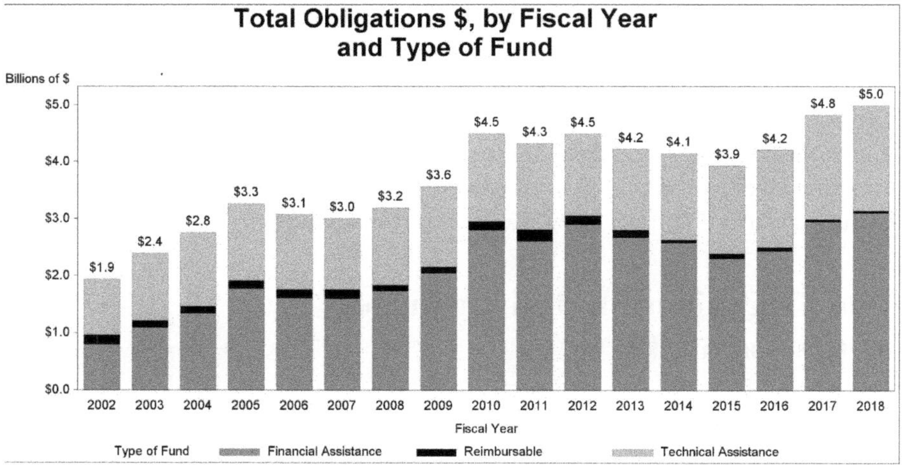

Figure 6.4. Natural Resources Conservation Service spending on conservation programs, 2002–2018. This includes most Farm Bill conservation programs, but excludes the Conservation Reserve Program, which is implemented by the Farm Service Agency. Total conservation spending under the 2018 Farm Bill is $5.97 billion per year.
Courtesy of USDA Natural Resources Conservation Service.

Even when a program has mandatory funding, it matters whether it is included in the Farm Bill's baseline. Programs in the baseline are automatically included in the next Farm Bill at the same funding levels at the start of negotiations.[39] This is an important advantage because when Congress is negotiating the cost of the Farm Bill, funding is first set at the per-title level and then further divided up between programs within each title. This means that conservation programs often compete with each other for funding. One of the most significant reversals was in 2014. Congress initially cut $400 million per year from the conservation title, and the across-the-board budget cuts known as sequestration then increased the cut to $600 million per year.[40] The 2018 Farm Bill provided $5.97 billion a year for the conservation title, a modest increase from $5.76 billion per year under the 2014 Farm Bill.[41] Demand for Farm Bill conservation programs, however, is estimated to be over $17 billion per year, meaning that much larger increases are needed in the future.[42]

Universal Compliance Requirements for Farm Bill Conservation Programs

In order to participate in most federal farm programs, there are certain requirements that every producer must meet. The first is compliance with the Highly Erodible Land Conservation provision, or "sodbuster," and the Wetland

Conservation provision, or "swampbuster," both of which originated with the 1985 Farm Bill.[43] Under these provisions, NRCS establishes requirements to protect highly erodible soils and wetlands, and producers who fail to comply are disqualified from Farm Bill conservation programs, Farm Bill loans and disaster assistance payments, Farm Bill commodity payments, federal crop insurance subsidies, and other programs.[44] Because of the many Farm Bill benefits impacted by sodbuster and swampbuster, compliance with the requirements among agricultural producers is strong, which has effectively slowed the loss of soil and wetlands to agriculture since 1985.[45] Unfortunately, these requirements do not provide any protection for native grasslands. In response, the 2008 Farm Bill introduced the "sodsaver" provision, strengthened in 2014, which imposes limits upon crop insurance subsidies to producers in six states (Iowa, Minnesota, Montana, Nebraska, North Dakota, and South Dakota) who convert native grasslands to agricultural production.[46]

All Farm Bill conservation programs are also subject to an adjusted gross income limit. In order to be eligible to participate, producers must have averaged *no more* than $900,000 in adjusted gross income per year over the three years prior to the tax year just concluded.[47] The 2018 Farm Bill modified this with a provision allowing the limit to be waived on a case-by-case basis for Farm Bill conservation programs if "environmentally sensitive land of special significance would be protected as a result of such waiver."[48] This waiver will make it easier for NRCS and its partner organizations to target the most ecologically valuable land for conservation, and is an important change in the 2018 Farm Bill.[49]

Notes

1. Jonathan Coppess, *The Fault Lines of Farm Policy: A Legislative and Political History of the Farm Bill* (Lincoln: University of Nebraska Press, 2018), 67.

2. North American Bird Conservation Initiative U.S. Committee, *2014 Farm Bill Field Guide to Fish and Wildlife Conservation* (2015), 4–5.

3. Ibid., 27.

4. For example, Coppess, *The Fault Lines of Farm Policy*, 293.

5. Morgan et al., "A State Assessment of Private Lands Wildlife Conservation in the United States."

6. NRCS, "More Than 80 Years Helping People Help the Land: A Brief History of NRCS," last accessed January 3, 2019, https://www.nrcs.usda.gov/wps/portal/nrcs/detail/national/about/history/?cid=nrcs143_021392; Coppess, *The Fault Lines of Farm Policy*, 47–48.

7. Ecological Site Descriptions are available online and can help both biologists and landowners identify ways to improve ecological conditions on a property. NRCS, "Ecological Site Descriptions," last accessed October 3, 2019, https://www.nrcs.usda.gov/wps/portal/nrcs/main/national/technical/ecoscience/desc/.

8. NRCS, "More Than 80 Years Helping People Help the Land: A Brief History of NRCS."

9. NRCS, "State Technical Committees," last accessed June 14, 2019, https://www.nrcs.usda.gov/wps/portal/nrcs/main/national/programs/technical/stc/.

10. Individuals or groups interested in joining their state technical committee should contact the NRCS state headquarters in their state. A directory of state headquarters offices is available online. NRCS, "Regional Boundaries, State Offices & Centers," last accessed August 12, 2019, https://www.nrcs.usda.gov/wps/portal/nrcs/detail/national/about/org/?&cid=nrcs143_021421#state.

11. See chapter 7.

12. Most conservation programs began to be funded through the Commodity Credit Corporation in 1996. Federal Agriculture Improvement and Reform Act of 1996, Pub. L. No. 104-127, 110 Stat. 888 (1996), 1016.

13. Farm Service Agency, "Commodity Credit Corporation," last accessed February 20, 2019, https://www.fsa.usda.gov/about-fsa/structure-and-organization/commodity-credit-corporation/index.

14. Huong N. Tran and Liu Chuang, "State Conservation District Laws Development and Variations" (NRCS Working Paper No. 3, 1996); St. Mary's Soil Conservation District, "History of Soil Conservation Districts," last accessed May 23, 2019, https://stmarysscd.com/about-us/history-of-soil-conservation-districts/.

15. Tran and Chuang, "State Conservation District Laws Development and Variations."

16. National Association of Conservation Districts, "About NACD."

17. National Association of Conservation Districts, "Conservation District Directory."

18. North American Bird Conservation Initiative U.S. Committee, *2014 Farm Bill Field Guide to Fish and Wildlife Conservation*, 11.

19. Ibid.

20. Committee on Conference, Farm Security and Rural Investment Act of 2002, Conference Report to Accompany H.R. 2646, H.R. Rep. No. 107-424 (2002), 499.

21. Farm Security and Rural Investment Act of 2002, Pub. L. No. 107-171, 116 Stat. 134 (2002), 279.

22. North American Bird Conservation Initiative U.S. Committee, *2014 Farm Bill Field Guide to Fish and Wildlife Conservation*, 12; Committee on Conference, Farm Security and Rural Investment Act of 2002, Conference Report to Accompany H.R. 2646, 499.

23. Natural Resources Conservation Service, "Technical Service Providers," last accessed February 6, 2019, https://www.nrcs.usda.gov/wps/portal/nrcs/main/national/programs/technical/tsp/.

24. Jim Inglis, interview by Christopher E. Segal, June 7 and October 2, 2019.

25. Pedersen and Hatfield.

26. Inglis.

27. These include state fish and wildlife agencies, Migratory Bird Joint Ventures, and nongovernmental organizations such as the Rocky Mountain Bird Observatory, Ducks Unlimited, the National Wild Turkey Federation, Ruffed Grouse Society, and Pheasants Forever. North American Bird Conservation Initiative Private Lands Subcommittee, *Delivering Fish and Wildlife Conservation: Building Farm Bill Capacity* (August 2009), 16.

28. See chapters 10 and 11 for information about many of these organizations.

29. Pheasants Forever, "Farm Bill and Conservation Technical Assistance Summary," May 2019; Pheasants Forever, "Farm Bill Biologists," last accessed June 14, 2019, https://pheasantsforever.org/getdoc/d68502f8-71ec-423c-a8a0-dbaf1e2851b2/Farm-Bill-Biologists.aspx.

30. Pedersen and Hatfield. Both Pheasants Forever and the National Wildlife Turkey Federation also have volunteers within local chapters who are able to assist landowners with in-kind

work or the loan of a piece of equipment. Inglis; Pheasants Forever, "Why Participate?" last accessed June 14, 2019, https://www.pheasantsforever.org/Participate/Why-Participate.aspx; Pedersen and Hatfield.

31. USDA, "USDA Local Offices," last accessed June 12, 2019, https://www.outreach.usda.gov/USDALocalOffices.htm. The Risk Management Agency, which operates federal crop insurance programs, is not co-located at USDA Service Centers but instead operates ten regional offices across the country.

32. In many cases, a producer will have one main contact at their local USDA Service Center whom they've worked with over the years, and that person will be their go-to contact for all of their Farm Bill needs and questions. Recently, the USDA has begun to use joint meetings between landowners and all of the co-located agencies to eliminate redundancies and improve customer service. Galon Hall and Tim Griffiths, interview by Christopher E. Segal, May 29, 2019.

33. Inglis.

34. USDA, "Service Center Locator," last accessed August 12, 2019, https://offices.sc.egov.usda.gov/locator/app.

35. Agricultural Improvement Act of 2018, Pub. L. No. 115-334 (2018).

36. Coppess, *The Fault Lines of Farm Policy*, 289–90, 302–08. This book focuses on programs authorized under the 2018 Farm Bill. Most former programs have been omitted.

37. All funding that is routed through the Commodity Credit Corporation is mandatory. Cummins.

38. Mark Rey, interview by Christopher E. Segal, January 30, 2019.

39. Jim Monke, *U.S. Library of Congress, Congressional Research Service, Farm Bill Programs without a Budget Baseline Beyond FY2018, CRS Report No. R44758* (2018), 1. These programs can still lose funding, but they enter budget negotiations with a considerable advantage.

40. National Sustainable Agriculture Coalition, "Path to the 2018 Farm Bill: Conservation," March 14, 2017, http://sustainableagriculture.net/blog/path-to-2018-farm-bill-conservation/.

41. National Sustainable Agriculture Coalition, "2018 Farm Bill by the Numbers," December 21, 2018, https://sustainableagriculture.net/blog/2018-farm-bill-by-the-numbers/; Megan Stubbs, *U.S. Library of Congress Congressional Research Service: Agricultural Conservation: A Guide to Programs, CRS Report No. R40763* (April 17, 2018), 3. The one Farm Bill program discussed in the following chapters that does not have mandatory funding is the Healthy Forests Reserve Program, which as previously stated is located in the forestry title of the Farm Bill. Rey.

42. See Echols, Front, and Cummins, "Broadening Conservation Funding."

43. National Sustainable Agriculture Coalition, "Enforcement of Conservation Compliance for Highly Erodible Lands," June 5, 2018, http://sustainableagriculture.net/wp-content/uploads/2018/06/CFRA-NSAC-Conservation-compliance-special-report.pdf.

44. NRCS, "Highly Erodible Land Conservation Compliance Provisions," last accessed February 4, 2019, https://www.nrcs.usda.gov/wps/portal/nrcs/detail/national/programs/alphabetical/camr/?cid=nrcs143_008440. The 1996 Farm Bill removed the sodbuster and swampbuster compliance requirements from crop insurance, but the requirements were reinstated by the 2014 Farm Bill. Sarah Munger, "Coupling Up: Are Crop Insurance and Conservation Compliance Mated for Life?" April 5, 2018, http://www.farmbilllaw.org/2018/04/05/coupling-crop-insurance-conservation-compliance-mated-life/; National Sustainable Agriculture Coalition, "Enforcement of Conservation Compliance for Highly Erodible Lands." Swampbuster serves as the primary protection for geographically isolated wetlands outside the jurisdiction of the Clean Water Act, which only regulates wetlands connected to navigable waterways. Brasher et al., "The History and Importance of Private Lands for North American Waterfowl Conservation."

45. Megan Stubbs, *U.S. Library of Congress Congressional Research Service: Conservation Compliance and U.S. Farm Policy, CRS Report No. R42459* (October 6, 2016); U.S. Sustainability Alliance, "Conservation Compliance," last accessed February 4, 2019, https://thesustainabilityalliance.us/conservation-compliance/.

46. Brasher et al., "The History and Importance of Private Lands for North American Waterfowl Conservation." As originally implemented under the 2008 Farm Bill, sodsaver applied in certain counties of Iowa, Minnesota, Montana, North Dakota, and South Dakota at the request of the state's governor, and prohibited crop insurance subsidies for the first five years of production. However, no governor requested sodsaver protections and the program was never implemented. The 2014 Farm Bill brought significant changes, making sodsaver protections mandatory throughout Iowa, Minnesota, Montana, Nebraska, North Dakota, and also South Dakota, but reduced non-compliance penalties to 50 percent of crop insurance subsidies for the first four years of production. The 2018 Farm Bill reauthorized sodsaver with technical amendments. Ruiqing Miao, David Hennessy, and Hongli Feng, "How Much Can the Sodsaver Provision in the 2014 Farm Bill Save Sod? It Depends on Crop Prices," *PolicyMatters*, January 19, 2017; National Sustainable Agriculture Coalition, "2018 Farm Bill Drilldown: Commodity Programs and Crop Insurance," December 14, 2018, https://sustainableagriculture.net/blog/2018-farm-bill-commodity-subsidies-crop-insurance/.

47. This limit has been in place since the 2014 Farm Bill, when it replaced previous adjusted gross income limits that were more complicated and had various special provisions and exceptions.

48. Food Security Act of 1985, as Amended through P.L. 115–334, Enacted December 20, 2018, 12–14.

49. Pedersen and Hatfield.

Farm Bill Incentives and Rental Contracts

7

SEVERAL FARM BILL PROGRAMS provide a combination of rental contracts, incentive payments, and cost-sharing payments. Over the years, these programs have been consolidated to three major programs in the 2018 Farm Bill. These are the Environmental Quality Incentives Program (EQIP), Conservation Stewardship Program (CSP), and Conservation Reserve Program (CRP). For each program, this chapter provides a brief history, followed by the purpose, required qualifications, and benefits of the program. Discussion of each program concludes with case studies and statistics for each program. Detailed tables showing annual expenditures and enrollment are available in appendix B.

Environmental Quality Incentives Program

EQIP, administered by the Natural Resources Conservation Service (NRCS), is a broad, rapidly growing incentive program promoted as the flagship conservation program of the Farm Bill.[1] The modern EQIP was created by the 1996 Farm Bill, which combined the Agricultural Conservation Program (1936), the Great Plains Conservation Program (1957), the Colorado River Basin Salinity Control Program (1974), and the Water Quality Incentives Program (1990) into a single program.

EQIP provides incentives for farmers and ranchers (referred to as "producers") to implement conservation practices on lands in active agricultural production. These practices span a broad range. In fact, any agricultural practice can qualify provided that it achieves the goal of improving environmental quality and the Secretary of Agriculture approves it. A non-exhaustive list is published in the NRCS Field Office Technical Guide, available at NRCS field offices and online at https://efotg.sc.egov.usda.gov/.[2] Approved practices include land management practices and structural practices, which means creating physical structures, either artificial or natural (e.g., planting trees).[3] The Farm Bill itself prioritizes some practices, including wildlife habitat improvement, water conservation, and organic farming. Additional competitive grants are available to promote innovative conservation approaches and air quality improvements.[4] A new type of EQIP contract introduced in the 2018 Farm Bill is a "conservation incentive contract" that authorizes five- to ten-year contracts for management practices to address "priority resource concerns." These priority resources concerns are identified for each watershed or other ecological region by NRCS in consultation with the state technical committee.[5]

The EQIP wildlife habitat improvement practice is unique in that the Farm Bill explicitly requires that 10 percent of all EQIP funds support "practices benefitting wildlife habitat."[6] This requirement first appeared in the 2014 Farm Bill as 5 percent of all EQIP funds, and subsequently increased to 10 percent in the 2018 Farm Bill.[7] Prior to 2014, there was no specific mandate to direct a percentage of EQIP funds to wildlife. Between 1996 and 2014, the Farm Bill included a separate Wildlife Habitat Incentives Program (WHIP). WHIP initially provided cost-sharing payments to landowners "to develop upland wildlife, wetland wildlife, threatened and endangered species, fish, and other types of wildlife habitat approved by the Secretary," with "habitat developed on pivot corners and irregular areas" added in 2008. This was the first Farm Bill program to specifically legislate cost-sharing payments for threatened and endangered species.[8] The 2014 Farm Bill consolidated WHIP into EQIP,[9] and the 2018 Farm Bill further expanded qualifying wildlife practices to include establishing temporary wetlands for migratory birds.[10] In addition to increasing EQIP's dedicated wildlife funding from 5 percent to 10 percent, the 2018 Farm Bill also decreased dedicated funding for practices relating to livestock production, including grazing management, from 60 percent to 50 percent. Although some grazing and water quality practices may benefit both livestock and wildlife, this shift reflects an increasing commitment to wildlife.[11]

EQIP funding can take the form of cost-sharing or incentive payments. EQIP provides up to 75 percent of costs associated with approved conservation practices and/or 100 percent of the income foregone by the producer in order to implement those practices. A higher cost-share of 90 percent is available to certain categories of producers (veterans, socially disadvantaged, and new farmers) and for certain categories of actions (each state government may identify up to ten "high-priority practices" to receive enhanced cost-sharing payments). Payment rates are not based on an evaluation of costs to individual producers but are set annually by state NRCS offices based on average costs for implementing a conservation practice. Current state payment schedules are available online through https://www.nrcs.usda.gov/wps/portal/nrcs/main/national/programs/financial/eqip/.[12]

There is a cap on total payments to a single recipient, regardless of the number of EQIP contracts with that recipient, of $450,000 in EQIP money within a five-year period. EQIP contracts are of variable length, with a maximum of ten years. Most contracts are shorter and run for one to three years.[13] In the 2018 Farm Bill, Congress recognized that it takes more time for wildlife habitat to develop and begin to be utilized. Accordingly, Congress stated that contracts focused solely on wildlife practices can be long-term contracts of up to ten years and prohibited NRCS from limiting such contracts to fewer than ten years. Such contracts may still be shorter at the request of the enrolling producer.[14]

To participate in EQIP, producers must "implement an environmental quality incentives program plan that describes conservation and environmental purposes to be achieved through 1 or more practices that are approved" by NRCS and must not conduct any practices on the enrolled land that would "tend to defeat the purposes of the program." They must also comply with any additional requirements imposed by NRCS.[15] NRCS, in turn, is directed by Congress to avoid duplication of regulatory paperwork and to streamline enrollment in EQIP for agricultural practices that meet the requirements of other regulatory programs.[16]

The directive to streamline EQIP enrollment for producers enrolled in other programs combined with the ability of state governments to identify high-priority practices makes EQIP a good source of funding for public/private partnerships. Since the 2008 Farm Bill, NRCS has used Landscape Conservation Initiatives to focus its conservation spending around priority resources, species, and regions. Examples include the Migratory Bird Habitat Initiative, Monarch Butterfly Initiative, and National Water Quality Initiative.[17] Two species-focused initiatives largely funded through EQIP are the Sage Grouse Initiative (SGI) and the Lesser Prairie Chicken Initiative (LPCI).[18]

At their launches in 2010, both SGI and LPCI represented a new approach for NRCS. The initiatives emphasize partnerships and local leadership, with NRCS providing funding and support. Both use emerging science to deploy the correct interventions in the most important habitat. SGI utilizes partnerships with dozens of organizations, eighty-two as of 2019, including other federal agencies; state fish and wildlife agencies; conservation districts; universities; non-

Figure 7.1. Like greater sage-grouse, lesser prairie chickens (*Tympanuchus pallidicinctus*) reproduce by congregating on leks each spring. The species was listed as threated in 2014 but the listing was overturned by a court in 2015.
D. Longenbaugh/Shutterstock.

profits such as the Mule Deer Foundation, National Audubon Society, National Wild Turkey Federation, The Nature Conservancy, Nevada Wildlife Federation (an affiliate of the National Wildlife Federation), North American Grouse Partnership, Pheasants Forever, Rocky Mountain Elk Foundation, Ruffed Grouse Society, Safari Club International's Utah Chapter, Theodore Roosevelt Conservation Partnership, World Wildlife Fund, and many others; corporations such as ConocoPhillips; and trade associations such as the National Cattlemen's Beef Association.[19] LPCI lists a similarly impressive twenty-eight partners, including many of the same organizations.[20] These partners provide additional financial support, connect landowners with NRCS, and help carry out conservation actions.[21] The breadth of these coalitions is a reflection of the interconnectedness of life. For example, large populations of both mule deer and elk rely on the same habitat as greater sage-grouse, drawing conservation organizations dedicated to these species into the SGI coalition.[22] NRCS funding, partnerships, and the enthusiastic participation of landowners have driven both SGI and LPCI, building relationships and gaining momentum over time.

Between 2010 and 2015, NRCS invested more than $296.5 million into SGI, with partner organizations contributing another $128 million, for a total of $424.5 million.[23] This money was used to engage more than "1,300

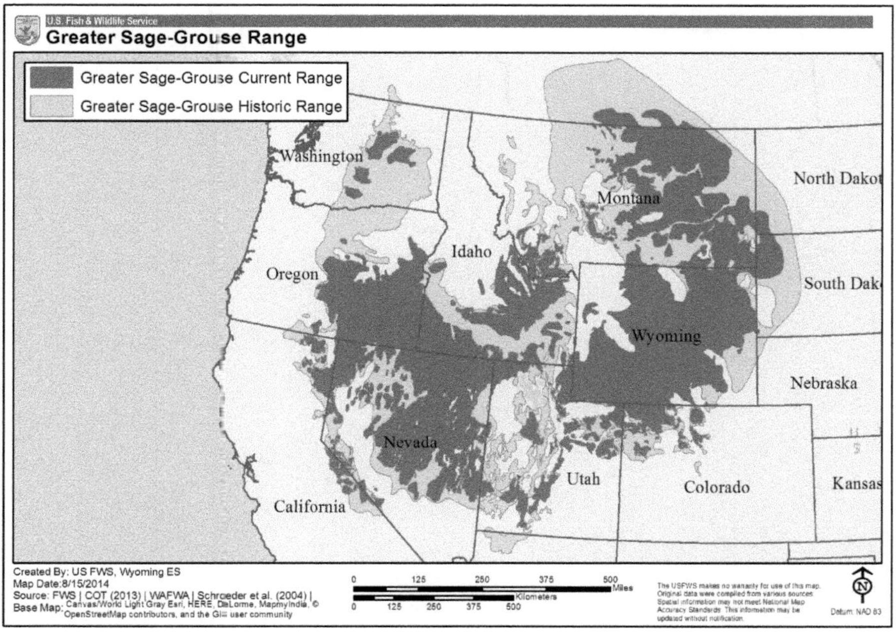

Figure 7.2. Greater sage-grouse habitat range map.
Courtesy of the U.S. Fish and Wildlife Service.

ranchers to conserve over 5 million acres of sagebrush habitat."[24] The years 2016 to 2018 brought another $134 million in NRCS funding,[25] joined with $125 million in partner funding, bringing the total investment in SGI, from 2010 to 2018, to $760 million for sage-grouse conservation. In the period 2010 to 2018, this money enabled partnerships with eighteen hundred ranchers and conservation of over seven million acres of sage-grouse habitat.[26] NRCS' investment has been split between EQIP and the Agricultural Conservation Easement program.[27] Conservation actions funded include grazing to eliminate invasive grasses and noxious weeds, mechanical treatments to remove invasive conifers, protection of land to reduce the impact of exurban development, protection of existing sagebrush habitat, protection of mesic (moist) habitat used by sage-grouse for feeding young, and removal of range fencing.[28]

Although just under 50 percent of the greater sage-grouse's range is on private land, fully 95 percent of the lesser prairie chicken's range is on private land, making Farm Bill conservation even more critical for that species. Between 2010 and 2015, LPCI invested $24 million in lesser prairie chicken conservation, with partners contributing another $10 million.[29] Conservation actions funded include prescribed grazing to improve rangeland health, treat-

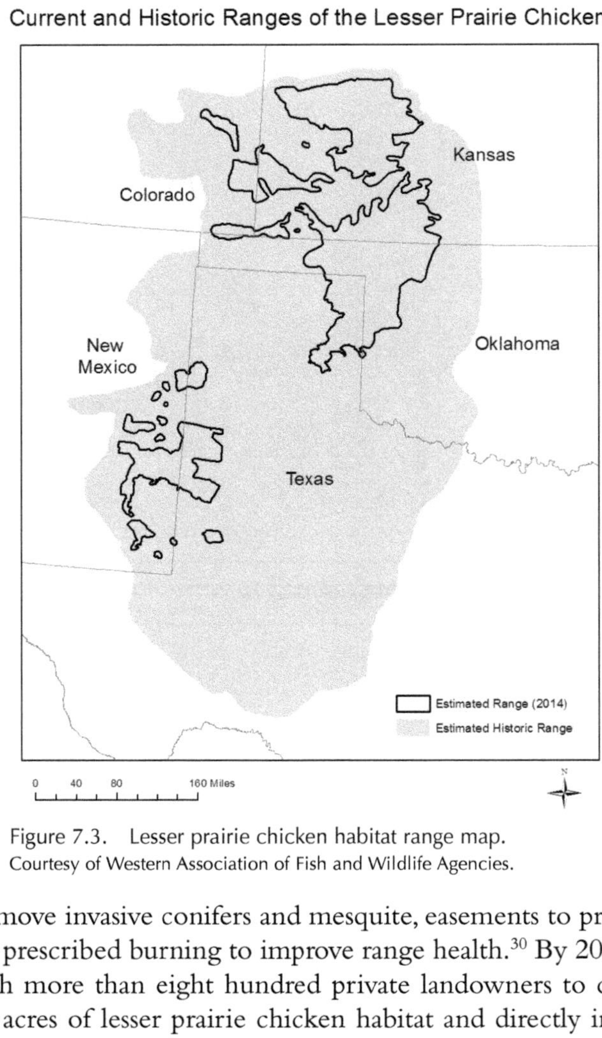

Figure 7.3. Lesser prairie chicken habitat range map.
Courtesy of Western Association of Fish and Wildlife Agencies.

ments to remove invasive conifers and mesquite, easements to protect existing habitat, and prescribed burning to improve range health.[30] By 2018, LPCI had worked with more than eight hundred private landowners to conserve over 1.6 million acres of lesser prairie chicken habitat and directly invested $41.6 million of NRCS funding into lesser prairie chicken conservation.[31]

EQIP represents a massive source of funding for private land conservation. Between 2005 and 2018, the only years for which data is available, the program as a whole invested $18.25 billion in conservation on 286.5 million acres of land.[32] Landowners interested in applying for EQIP should contact their local NRCS field office for more information. Applications are accepted year-round, although individual state offices rank applications and make funding decisions at different times. The application form is available on NRCS' website.[33] Each application must include an individual EQIP Plan of Operations, listing conservation objectives, quantifiable goals to measure progress toward objectives, the NRCS-approved conservation practices to be used to achieve these objectives, an implementation schedule, and any additional information needed for NRCS to evaluate the effectiveness of the plan.[34] U.S. Department of Agriculture Service Center staff can help landowners prepare these plans.[35]

Conservation Stewardship Program

CSP is a working lands program similar to EQIP in many ways. In 2018, the House of Representatives proposed to eliminate CSP as a standalone program and roll it into EQIP, similar to the way the 2014 Farm Bill combined WHIP and EQIP. The final version of the 2018 Farm Bill, however, retrained CSP as its own program. The modern version of CSP dates only to the 2008 Farm Bill, when it was created as a replacement for an older program, the Conservation Security Program, which itself originated with the 2002 Farm Bill. The Conservation Security Program and CSP share the same purpose of rewarding and incentivizing agricultural producers to improve soil, water, and air quality; promote wildlife habitat; and reduce water and energy consumption.[36] The 2018 Farm Bill added a goal of adapting to or mitigating against increasing weather volatility, promoted rotational grazing, and provided one-time incentive payments for planting cover crops to improve soil quality.[37]

CSP is designed to provide an incentive for producers to improve conservation practices over time. Under the program, NRCS designates five "priority resources of concern" for a watershed or region. Examples of priority resources of concern include soil erosion, water quality, air quality, or fish and wildlife habitat.[38] Participating landowners must meet or exceed the "stewardship threshold," the management required to improve a natural resource, for two priority resources of concern in order to enter the program, and must meet the stewardship threshold for a third priority resource of concern within five years. CSP contracts run for five years and can be renewed once for another five years on the condition that during the course of the second contract, the landowner much meet the stewardship threshold for a fourth priority resource of concern.

Payments under CSP vary but may not exceed 75 percent of the cost of implementing and maintaining conservation practices to meet the stewardship threshold requirements of the contract. Supplemental payments are available for certain high-priority activities. Producers that implement qualifying resource-conserving crop rotations or advanced grazing management may receive payments equal to 150 percent of that they would otherwise receive.[39] Activities involving planting and maintaining cover crops are eligible for payments of 125 percent the normal value.[40] In addition, the Farm Bill provides for one-time incentive payments to CSP enrollees for preparation of a comprehensive conservation plan that meets the stewardship thresholds for all five priority resource concerns applicable to their area.[41] No landowner may receive a total of more than $200,000 in payments over a five-year period, regardless of how many CSP contracts they enter into. And unlike other Farm Bill conservation programs, a CSP contract must include all eligible land within the property enrolled. This makes CSP a "whole farm" program that rewards past investments in conservation while simultaneously incentivizing further investments, all aligned with the specific ecological needs of the watershed or region.

CSP applications are accepted at any time, although most NRCS offices conduct evaluations once per year. Before applying, producers may find it helpful to fill out the agency's online Conservation Stewardship Self-Screening Checklist.[42] Also available online are lists of approved "CSP Enhancements" designed to meet or exceed stewardship thresholds, and a list of supporting activities that make up these enhancements. These resources can be found on the CSP homepage at https://www.nrcs.usda.gov/wps/portal/nrcs/main/national/programs/financial/csp/.[43] NRCS staff evaluate each application to identify eligible land and resources requiring conservation to meet stewardship thresholds. Finally, NRCS presents the landowner with activities to consider implementing in order to participate.[44] The end result is a conservation plan tailored to the specific property.[45]

CSP priority resources of concern tend to be broad categories such as "animals," "erosion," or "water quality." As a result, the benefits of the program are wide-ranging and impact producers, livestock and crops, and a variety of wildlife. Some landowners will find that CSP is the right program for them, whereas others may prefer EQIP, due to variations in contract length, payment amounts, whole versus partial farm enrollment, and targeted practices. The NRCS website highlights farmers and ranchers using CSP to advance conservation, and the diversity of their objectives and techniques illustrates the breadth of the program. For example, California wine grape grower Matt Terry enrolled his vineyard in CSP in 2010 and uses the program to support cover crops that improve soil quality and water retention. Iowa corn and soybean farmer Adam Boge enrolled in 2014 and uses CSP to widen stream buffers and improve nutrient and pesticide application techniques. South

Figure 7.4. Riparian buffers are important conservation tools for controlling erosion and maintaining stream quality. This example filters drainage from agricultural fields before it reaches a tributary of the South Skunk River in central Iowa.
Courtesy of Dan Jayes, USDA Agricultural Research Service.

Dakota rancher Fanny Brewer enrolled in 2010 and uses the financial security provided by the program to implement rotational grazing and improve the natural forage on her ranch.[46]

CSP is one of the largest Farm Bill conservation programs, with $10.5 billion invested in conservation from 2004 to 2018. Over the life of the program, over forty-six thousand producers have enrolled over seventy million acres of land in CSP.[47] Like EQIP, CSP annually receives mandatory funding under the Farm Bill.[48] CSP is also used as a vehicle for a variety of sub-programs, which may only exist during the life of a single Farm Bill. For example, the 2018 Farm Bill established a new, temporary Grasslands Conservation Initiative within CSP.[49]

Conservation Reserve Program

CRP is the only Farm Bill conservation program administered by the Farm Service Agency (FSA) rather than NRCS. The modern version of CRP was created by the Food Security Act of 1985 to conserve "highly erodible lands."[50] The 1990 Farm Bill greatly expanded the mandate of CRP, authorizing the enrollment of lands in order to protect water quality, wetlands, and wildlife habitat.[51] CRP is a term rental contract program into which landowners enroll land for ten to fifteen years. Enrollees agree to remove land from crop production and plant grass and other conservation cover (cover crops) to improve soil retention, water quality, wildlife habitat, and other environmental values. There are a number of established conservation practices within CRP, and information about each can be found on the FSA website at https://www.fsa.usda.gov/programs-and-services/conservation-programs/crp-practices-library/index.[52] In exchange for adopting these conservation practices, landowners receive annual rental payments and cost-share assistance.[53] The historical importance of CRP is immense. It was the first modern Farm Bill conservation program, and when it was introduced it made conservation profitable for private landowners, yielding immediate benefits for grassland-reliant birds and other species.[54] CRP has been shown to reduce soil loss, mitigate pesticide loss, improve water and air quality, sequester carbon, increase populations of pollinators and other beneficial insects, and provide wildlife habitat.[55] CRP has had its greatest benefits in the waterfowl breeding grounds of the Prairie Pothole region. CRP-enrolled lands in Montana, North Dakota, and South Dakota alone are estimated to have increased waterfowl recruitment by 1.5 million birds per year and over 12.3 million birds from 2005 to 2011.[56]

Lands eligible for CRP include highly erodible cropland that was either planted in four of the last six years or already enrolled in CRP; marginal pastureland suitable for use as a riparian buffer, wetland, or to improve water quality or wildlife habitat; grasslands that contain forbs or shrubs for grazing and are ecologically significant; or any wetland or riparian buffer that could be farmed. As of the 2018 Farm Bill, land subject to an Agricultural Land Easement may also be enrolled in CRP.[57] In addition, FSA may enroll otherwise ineligible land under a number of special circumstances, including if keeping the land in production would degrade soil, water, or air quality, or enrolling the land would improve water conservation. Otherwise ineligible land may also be enrolled in CRP if it includes one or more of a grassed waterway, counter strip, snow fence, wildlife habitat, windbreak, shelterbelt, salt-tolerant vegetation, field borders, or practices that benefit a legally identified wellhead protection area.[58] In addition, a producer must have owned or operated the land for at least twelve months prior to the first year of the CRP contract, unless the land was acquired due to the death of the previous owner or due to foreclosure, or if FSA determines that the land was not acquired specifically for the purpose of enrolling it in CRP. Finally, since 2008 the Farm Bill has limited CRP to 25 percent of the cropland in any single county, although there are provisions for exceptions in certain situations.[59]

CRP enrollment occurs annually, and the Farm Bill authorizes enrollment up to a statutory acreage cap, rather than a fixed budget. This cap has declined over the years, from an initial high of forty-five million acres in 1986 to a historic low of twenty-four million acres at present. Under the 2018 Farm Bill, the cap will gradually increase to twenty-seven million acres.[60] Demand to enroll in CRP usually exceeds the statutory cap, making the program highly competitive. Current demand is estimated to be as much as forty million acres.[61] Enrolled land remains in the program for ten to fifteen years. Because in any given year some land remains in the program while other land leaves the program, and land that has been in the program before may be enrolled again, it is impossible to calculate the total number of acres that have benefited from CRP over the years.

When landowners enroll in CRP, they receive a rental payment for the use of their land as well as a cost-sharing payment of 50 percent of the cost of establishing appropriate cover on their land, and some CRP practices qualify for additional payments. Cost-sharing payments under CRP cannot, when combined with other funding sources available to the producer, exceed 100 percent of the actual cost of the conservation practice. Rental payments are determined on a case-by-case basis according to criteria in the Farm Bill and implementing regulations. FSA bases its offers on the average prevailing dryland (non-irrigated) cash rental rates for each county, which can be adjusted by the amount necessary to encourage producers to enroll in the program. However, to save money, the 2018 Farm Bill limited CRP payments to 85 percent of estimated rental rates for the general CRP program and 90 percent for continuous signup contracts (see following discussion).[62] In addition, landowners have the option of offering to join the program at a lower price, although the only benefit of this is making their application more competitive. Because demand for CRP has exceeded the acreage cap in recent years, Farm Bill proponents hope that the new rental rate reductions will not reduce the size of the program, but as of 2019 it is too early to say.

A beneficial change in the 2018 Farm Bill is that rental rates will no longer be reduced due to certain economic uses if the area impacted is suffering from drought or natural disaster. Covered practices include emergency haying, emergency grazing, and other uses of forage.[63] The Farm Bill also authorizes additional incentive payments "in an amount sufficient to encourage proper thinning and other practices to improve the condition of resources, promote forest management, or enhance wildlife habitat on the land."[64] Under the

2014 Farm Bill these payments provided 150 percent of the cost of implementing such approved practices, whereas other activities qualified for 10 or 20 percent bonuses to rental rates.[65] The 2018 Farm Bill limits these payments to 100 percent of cost, and additional implementing changes based on the 2018 Farm Bill are pending as this book goes to press.[66]

Within CRP there are a number of sub-programs for landowners to consider.[67] The first is the Conservation Reserve Enhancement Program (CREP), which targets high-priority conservation concerns identified by a state.[68] CREP offers landowners enhanced cost-sharing payments for certain practices, the possibility of cost-sharing payments over 100 percent of their costs, and other additional incentives.[69] A second sub-program is Continuous CRP, which is an enrollment mechanism within CRP that allows enrollment at any time for specific high-priority conservation actions.[70] Like CREP, Continuous CRP may offer heightened payments and other incentives.[71] Under the 2018 Farm Bill, 40 percent of the acres in Continuous CRP must be in certain practices that will have a positive impact on water quality and/or be in riparian or wetland areas.[72]

A third sub-program is the Farmable Wetlands Program, which sets aside a number of CRP acres in each state to restore small tracts of previously farmed wetlands and wetland buffers.[73] Finally, under CRP the 2018 Farm Bill established two pilot programs that modify the length of contracts. Normally all CRP programs require ten- to fifteen-year contracts. The CLEAR 30 program allows for certain lands to be enrolled in CRP for thirty years,[74] whereas the Soil Health and Income Protection Pilot Program is the first CRP program ever to offer contracts of fewer than ten years.[75]

Although CRP is administered by FSA, NRCS provides assistance in the form of technical land eligibility determinations, conservation planning, and practice implementation, and other partners including local conservation districts, state agencies, and non-governmental organizations provide support as well.[76] Landowners interested in CRP, including sub-programs such as Continuous CRP or CREP, should contact their local FSA field office for information, including the time of the next enrollment window. To enroll, landowners will have to develop a conservation plan approved by a local conservation district.[77]

CRP encompasses a wide range of locations, conditions, and conservation practices, and as a result CRP's value for wildlife conservation can vary. Landowners enrolling in CRP can help ensure that their enrollment is as beneficial to wildlife as possible by indicating that habitat and wildlife are priorities for them.[78] One CRP program that has been helpful for wildlife conservation is FSA's State Acres for Wildlife Enhancement (SAFE) initiative, established in 2008.[79] This agency-led initiative is a strategy that FSA uses to address state and regional high-priority wildlife objectives. Under SAFE, projects are designed by government agencies, local and regional conservation groups, and producers themselves in order to leverage local knowledge and prioritize CRP practices most beneficial to wildlife.[80] As of 2019, almost 10 percent of CRP contracts included SAFE practices.[81] Participation in SAFE does not impose any requirements on landowners that are different from those of CRP or the relevant sub-program.

Once species that has benefited from CRP is the lesser prairie chicken, especially in Kansas, which provides an example of how specific CRP practices can benefit wildlife.[82] CRP Practice CP-25 is used to establish groundcover that conserves rare and declining wildlife habitat. Land under CP-25 cannot be harvested or grazed and must be seeded with native grasses, forbs, legumes, and shrubs; noxious weeds must be controlled.[83] The long-term maintenance of native grasses under CP-25 without grazing leads to deep root systems that allow those grasses to be more productive through droughts.[84] CP-25 has been widely used in Kansas, and since the mid-1990s the state's SAFE practices have enrolled close to thirty thousand acres of lesser prairie chicken habitat.[85] As a result, the lesser prairie chicken's Kansas population is the healthiest across the species' range.[86] In addition, EQIP contracts through NRCS' LPCI include contracts specifically designed to extend lesser prairie chicken conservation on land coming out of CRP, prolonging the benefits of the habitat established by the initial CRP enrollment.[87]

Notes

1. For example, *To Review USDA Farm Bill Conservation Programs: Hearing before the Subcommittee on Conservation and Forestry of the Committee on Agriculture, U.S. House of Representatives*, 116th Cong. (2019). (Testimony of Matthew Lohr.)
2. Natural Resources Conservation Service, "NRCS Field Office Technical Guide," last accessed August 13, 2019, https://efotg.sc.egov.usda.gov.
3. Farmers' Legal Action Group Inc., "Is Your Farm 'EQIPed' for Conservation?" 2007, http://www.flaginc.org/wp-content/uploads/2013/03/FGtoEQIP.pdf.
4. Agricultural Improvement Act of 2018, 70–77.
5. Food Security Act of 1985, as Amended through P.L. 115–334, 72–74.
6. Agricultural Improvement Act of 2018, 70–71.
7. Agricultural Act of 2014, Pub. L. No. 113-79, 128 Stat. 649 (2014), 730.
8. Federal Agriculture Improvement and Reform Act of 1996, 1020; Food, Conservation, and Energy Act of 2008, Pub. L. No. 110-234, 122 Stat. 923 (2008), 1068.
9. Agricultural Act of 2014, 730.

10. Agricultural Improvement Act of 2018, 71. This was based on a successful pilot program launched in 2015 by Delta Waterfowl, a research-oriented conservation organization started in 1938 that focuses on ducks and other migratory game species. Delta Waterfowl, "Delta Waterfowl Applauds Inclusion of 'Working Wetlands' in 2018 Farm Bill," press release, https://deltawaterfowl.org/delta-waterfowl-applauds-inclusion-of-working-wetlands-in-2018-farm-bill/; John Devney, interview by Christopher E. Segal, July 2, 2019.

11. Food Security Act of 1985, as Amended through P.L. 115–334, 84; Pedersen and Hatfield.

12. Natural Resources Conservation Service, "Environmental Quality Incentives Program," last accessed August 13, 2019, https://www.nrcs.usda.gov/wps/portal/nrcs/main/national/programs/financial/eqip/.

13. National Sustainable Agriculture Coalition, "Environmental Quality Incentives Program," last accessed May 28, 2019, http://sustainableagriculture.net/publications/grassroots guide/conservation-environment/environmental-quality-incentives-program/.

14. Committee on Conference, Agriculture Improvement Act of 2018, Conference Report to Accompany H.R. 2, H.R. Rep. No. 115-1072 (2018), 575; Pedersen and Hatfield.

15. Food Security Act of 1985, as Amended through P.L. 115–334, 88–89.

16. Ibid., 89–90.

17. Natural Resources Conservation Service, "NRCS Landscape Conservation Initiatives," last accessed October 1, 2019, https://www.nrcs.usda.gov/wps/portal/nrcs/detailfull/national/programs/initiatives/?cid=stelprdb1042113.

18. Both initiatives also received WHIP funding before WHIP was consolidated with EQIP by the 2014 Farm Bill. In this chapter, WHIP figures are included within EQIP figures.

19. Sage Grouse Initiative, "Partner List," last accessed January 14, 2019, https://www.sagegrouseinitiative.com/about/partners/.

20. Lesser Prairie Chicken Initiative, "Partners," last accessed January 14, 2019, http://lpcinitiative.org/about/partners/.

21. Sage Grouse Initiative, "New Paradigm," last accessed January 14, 2019, https://www.sagegrouseinitiative.com/about/new-paradigm/.

22. Blake Henning, interview by Christopher E. Segal, July 17, 2019; Moretti.

23. Natural Resources Conservation Service, "Sage Grouse Initiative 2.0 Investment Strategy, FY 2015-2018," 6.

24. Sage Grouse Initiative, "Collaborative Conservation is Paying Off for the Sagebrush Ecosystem," August, 2015, https://www.sagegrouseinitiative.com/collaborative-conservation-paying-off-sagebrush-ecosystem/.

25. Natural Resources Conservation Service, "Working Lands for Wildlife Obligations, All Years All Species" (2019).

26. Sage Grouse Initiative, "1,800 Ranchers and Counting!"

27. Natural Resources Conservation Service, "Sage Grouse Initiative 2.0 Investment Strategy, FY 2015-2018," 6. For discussion of the Agricultural Conservation Easement program, see chapter 8.

28. Ibid.

29. Natural Resources Conservation Service, "Lesser Prairie-Chicken Initiative FY16-18 Conservation Strategy," 2015, https://www.nrcs.usda.gov/wps/PA_NRCSConsumption/download?cid=nrcseprd967006&ext=pdf.

30. Ibid.

31. Natural Resources Conservation Service, "Working Lands for Wildlife Obligations, All Years All Species."

32. Natural Resources Conservation Service, "Environmental Quality Incentives Program (EQIP)," last accessed June 21, 2019, https://www.nrcs.usda.gov/Internet/NRCS_RCA/

reports/fb08_cp_eqip.html; Natural Resources Conservation Service, "NRCS Conservation Programs," last accessed August 9, 2019, https://www.nrcs.usda.gov/Internet/NRCS_RCA/reports/cp_nat.html.

33. Natural Resources Conservation Service, "How to Apply for USDA-NRCS Conservation Programs," last accessed August 13, 2019, https://www.nrcs.usda.gov/wps/portal/nrcs/detail/national/technical/nra/?cid=nrcs144p2_027058.

34. 7 CFR § 1465.7(C).

35. Farmers' Legal Action Group Inc., "Is Your Farm 'EQIPed' for Conservation?" 10.

36. North American Bird Conservation Initiative U.S. Committee, *2014 Farm Bill Field Guide to Fish and Wildlife Conservation*, 32–33.

37. Agricultural Improvement Act of 2018, 78–70.

38. Natural Resources Conservation Service, "Nine Major Resource Concern Categories and Their Subcategories," last accessed October 4, 2019, https://www.nrcs.usda.gov/Internet/FSE_DOCUMENTS/nrcseprd1140206.pdf.

39. "Resource-conserving crop rotations" must include at least one resource-conserving crop identified in regulations that "reduces erosion, improves soil fertility and tilth [tillage], interrupts pest cycles, builds soil organic matter, and (where relevant) reduces depletion of soil moisture or otherwise reduces the need for irrigation." "Advanced grazing management" means grazing practices identified by regulation that provide for improved soil health and carbon sequestration, drought resilience, wildlife habitat, wildfire mitigation, control of invasive plants, and water quality improvement. Food Security Act of 1985, as Amended through P.L. 115–334, 101–02.

40. Ibid., 101.

41. Ibid., 102.

42. Natural Resources Conservation Service, "Conservation Stewardship Program Producer Self-Screening Checklist," last accessed January 16, 2019, https://www.nrcs.usda.gov/wps/PA_NRCSConsumption/download?cid=stelprdb1269861&ext=pdf.

43. Natural Resources Conservation Service, "Conservation Stewardship Program," last accessed August 13, 2019, https://www.nrcs.usda.gov/wps/portal/nrcs/main/national/programs/financial/csp.

44. Natural Resources Conservation Service, "Apply for CSP," last accessed January 16, 2019, https://www.nrcs.usda.gov/wps/portal/nrcs/detailfull/national/programs/financial/csp/?cid=nrcseprd1288620.

45. A more complete discussion of the application process can be found in the National Sustainable Agriculture Coalition's Farmer's Guide to the Conservation Stewardship Program, available at http://sustainableagriculture.net/wp-content/uploads/2015/02/CSP-Farmers-Guide-2015-final.pdf.

46. Natural Resources Conservation Service and Dave Sanden, "Building Sustainability on a California Vineyard, Ranch," April 4, 2017, https://www.nrcs.usda.gov/wps/portal/nrcs/detail/national/newsroom/features/?cid=nrcseprd1323428; Natural Resources Conservation Service and Jason Johnson, "USDA Helps Iraqi War Veteran Enhance Conservation on Farm," September 2016, https://www.nrcs.usda.gov/wps/portal/nrcs/detail/national/programs/financial/csp/?cid=nrcseprd1290732; Natural Resources Conservation Service and Colette Kessler, "We Are the Brewers. We Are Conservationists," September 2016, https://www.nrcs.usda.gov/wps/portal/nrcs/detail/national/programs/financial/csp/?cid=nrcseprd1292065.

47. Natural Resources Conservation Service, "NRCS Conservation Programs"; Natural Resources Conservation Service, "Conservation Stewardship Program (CSP)," last accessed June 21, 2019, https://www.nrcs.usda.gov/Internet/NRCS_RCA/reports/fb08_cp_cstp.html; Wild Farm Alliance, "Farm Bill 2018 Issue Brief: Conservation Stewardship Program," 2018, https://d3n8a8pro7vhmx.cloudfront.net/wildfarmalliance/pages/302/attachments/original/1531441423/csp-issue-brief_FinalWeb.pdf?1531441423.

48. This approach began with the 2018 Farm Bill. Agricultural Improvement Act of 2018, 90. When the Conservation Security Program began in 2002, it was simply funded out of leftover Commodity Credit Corporation funds, with no annual limit. Farm Security and Rural Investment Act of 2002, 278; Food, Conservation, and Energy Act of 2008, 1072. The 2014 Farm Bill adopted a limit of ten million new acres of CSP land each year at an average payment of $18 per acre. Agricultural Act of 2014, 726. As a result of this history, the expense of the Conservation Security Program/CSP program has fluctuated over the years.

49. The need for this program arose because to save money, the 2018 Farm Bill removed from the commodity program some acres that had been planted in grass for the past ten years. In order to encourage producers to maintain those lands in grass, the Grassland Conservation Initiative was created. This new program is part of CSP but has some special conditions. First, producers need to meet the stewardship threshold for just one priority resource of concern within five years, as opposed to two priority resources to enroll and a third within five years. Second, producers may only enroll for one five-year term with no opportunity to renew. Finally, all qualifying applicants with be automatically accepted, but payments are statutorily limited to $18 per acre per year. These special conditions reflect the program's status as a temporary replacement for suspended commodity payments for these lands. Food Security Act of 1985, as Amended through P.L. 115–334, 103–05; National Association of Conservation Districts, "2018 Farm Bill Breakdown: Conservation Stewardship Program," Janaury 15, 2019, https://www.nacdnet.org/2019/01/15/2018-farm-bill-breakdown-conservation-stewardship-program/; National Ssutainable Agriculture Coalition, "A Closer Look at the 2018 Farm Bill: Lands Conservation Programs," January 14, 2019, http://sustainableagriculture.net/blog/a-closer-look-at-the-2018-farm-bill-working-lands-conservation-programs/.

50. Food Security Act of 1985, Pub. L. No. 99-198, 99 Stat. 1354 (1985), 1509. CRP's roots were in the Agricultural Act of 1956, which established the Acreage Reserve Program, dedicated to reducing production of wheat, corn, cotton, tobacco, rice, and peanuts, and CRP, which rewarded farmers with rental and cost-share payments for removing land from production entirely. Both programs proved ineffective and expensive, and the Acreage Reserve Program was terminated in 1958 with CRP following in 1960. Coppess, *The Fault Lines of Farm Policy*, 108–12, 127, 131; Natural Resources Conservation Service and J. Douglas Helms, "Brief History of the USDA Soil Bank Program," January 1985, https://www.nrcs.usda.gov/Internet/FSE_DOCUMENTS/stelprdb1045666.pdf.

51. Food, Agriculture, Conservation, and Trade Act of 1990, Pub. L. No. 101-624, 104 Stat. 3359 (1990), 3577–78; Natural Resources Conservation Service and Helms, "Brief History of the USDA Soil Bank Program." Wildlife benefits from improvements to water quality and wetlands as well as wildlife habitat.

52. Farm Service Agency, "CRP Practices Library," last accessed October 3, 2019, https://www.fsa.usda.gov/programs-and-services/conservation-programs/crp-practices-library/index.

53. Farm Service Agency, "Conservation Reserve Program," last accessed January 7, 2019, https://www.fsa.usda.gov/programs-and-services/conservation-programs/conservation-reserve-program/index.

54. Morgan et al., "A State Assessment of Private Lands Wildlife Conservation in the United States."

55. Mark D. McConnell, "Bridging the Gap between Conservation Delivery and Economics with Precision Agriculture," *Wildlife Society Bulletin 43*, no. 3 (2019): 391–97.

56. Brasher et al., "The History and Importance of Private Lands for North American Waterfowl Conservation." Recruitment is the number of young added to a population by reproduction from adults each year.

57. Agricultural Improvement Act of 2018, 106. For information on the Agricultural Land Easement program, see chapter 8.

58. Food Security Act of 1985, as Amended through P.L. 115–334, 47–48.

59. Food, Conservation, and Energy Act of 2008, 1080; Food Security Act of 1985, as Amended through P.L. 115–334, 120–21.

60. Food Security Act of 1985, Pub. L. No. 99-198, 1509; Food, Agriculture, Conservation, and Trade Act of 1990, 3577; Federal Agriculture Improvement and Reform Act of 1996, 994; Farm Security and Rural Investment Act of 2002, 239; Food, Conservation, and Energy Act of 2008, 1028-29; American Taxpayer Relief Act of 2012, Pub. L. No. 112-240, 126 Stat. 2313 (2012), 2363; Agricultural Act of 2014, 714; Agricultural Improvement Act of 2018, 44.

61. Echols, Front, and Cummins, "Broadening Conservation Funding."

62. Food Security Act of 1985, as Amended through P.L. 115–334, 73–74.

63. Ibid., 67–68.

64. Ibid., 71.

65. National Sustainable Agriculture Coalition, "Conservation Reserve Program," October, 2016, http://sustainableagriculture.net/publications/grassrootsguide/conservation-environment/conservation-reserve-program/.

66. Food Security Act of 1985, as Amended through P.L. 115–334, 71.

67. Farm Service Agency, "Conservation Reserve Enhancement Program," last accessed September 6, 2018, https://www.fsa.usda.gov/programs-and-services/conservation-programs/conservation-reserve-enhancement/index; Farm Service Agency, "Conservation Reserve Program State Acres for Wildlife Enhancement (SAFE)," January 2008, https://www.fsa.usda.gov/Internet/FSA_File/safe08.pdf; Farm Service Agency, "Farmable Wetlands Program," last accessed January 30, 2019, https://www.fsa.usda.gov/programs-and-services/conservation-programs/farmable-wetlands/index; Cummins, "Private Lands Conservation Programs," 237–40. For case studies for many of these programs, see North American Bird Conservation Initiative U.S. Committee, *2014 Farm Bill Field Guide to Fish and Wildlife Conservation*, 38–40.

68. State governments work with FSA to identify state or nationally significant conservation needs, set quantifiable goals to meet these needs, locate suitable land, set payment amounts, and identify appropriate CRP conservation practices. State governments and other partners also provide matching funds for CREP enrollees, which may include cash, in-kind contributions, or technical assistance. Food Security Act of 1985, as Amended through P.L. 115–334, 52–54.

69. CREP pays 100 percent of the fair market value of installing stream fencing, crossings, and alternative water development on marginal pastureland. Additional benefits include 100 percent cost-sharing payments to manage riparian buffers and authorization to plant and harvest food-producing woody plants on forested riparian buffers. CREP enrollments that address regional drought concerns may include land otherwise ineligible for CRP, can authorize certain dryland agricultural uses of the land, and may feature higher rental prices based on the price of similar drought and water conservation agreements in the same region. Ibid., 54–55, 72.

70. Lands eligible for Continuous CRP include marginal pastureland in or near riparian areas or otherwise enrolled in order to promote water quality, other riparian and wetland areas, and land enrolled in CREP. The 2018 Farm Bill established a minimum number of acres to be enrolled in Continuous CRP, starting at eight million acres in 2019 and rising to 8.6 million acres by 2023. Ibid., 50–51.

71. Incentives include a 100 percent cost-share for establishing CRP practices on lands enrolled in Continuous CRP, an increase in the limit on rental payments from 85 percent to 90 percent, and one-time payments of at the time of enrollment equal to 32.5 percent of the first year's rental payment. Ibid., 71–73.

72. This is part of the Clean Lakes, Estuaries and Rivers Initiative, which directs FSA to prioritize conservation practices aimed at reducing sediment and nutrient loading and algae blooms. Ibid., 49–50.

73. Enrolled land is limited to forty acres per contract for wetlands and constructed wetlands and twenty acres for flooded farmland; FSA and its state partners are authorized to set a limit for buffers. The program is limited to one hundred thousand acres in any one state, which FSA may increase to two hundred thousand acres, and 750,000 acres nationwide. Eligible lands include wetlands that were farmed in at least three of the previous ten years, constructed wetlands that capture nitrogen running off of adjacent fields, certain lands previously used for aquaculture, certain lands previously used for prairie wetland overflow purposes, and buffers around otherwise qualifying land. Enrollees are required to restore the hydrology of the land to the maximum extent practicable, establish cover crops, and refrain from commercial use of enrolled lands. Ibid., 55–58.

74. This option is limited to lands exiting the Clean Lakes, Estuaries and Rivers Initiative, which may reenroll for thirty years. Ibid., 58–61.

75. The program is limited to the Prairie Pothole region, and to fifty thousand total acres. It allows less-productive land that was cropped in each of the last three years to be enrolled in CRP for three, four, or five years in order to conserve and improve soil, water, and wildlife resources. Payments are limited to 50 percent of the average rental rate for CRP lands within the same county, and no more than 15 percent of the eligible land on a property may be enrolled. Producers are required to plant the lowest-cost cover crop at their own expense, although they may harvest that crop outside of the nesting season or accept a payment from a non-governmental organization in exchange for not harvesting the crop. These conditions are loosened for certain classes of producers (beginning, limited resource, racially disadvantaged, or veterans), which may also receive a 50 percent cost-sharing payment for the cover crop and may receive payments of up to 75 percent of the applicable average rental rate. Ibid., 61–63; Ed Maixner, "On Tap: First Ever under-10-Years CRP Contracts," *Agri-Pulse*, February 20 2019.

76. Natural Resources Conservation Service, "Conservation Reserve Program," last accessed May 22, 2019, https://www.nrcs.usda.gov/wps/portal/nrcs/detail/national/programs/?cid=stelprdb1041269.

77. Where no conservation district is present, the Farm Bill assigns this responsibility to the U.S. Department of Agriculture.

78. Inglis.

79. Ibid.

80. North American Bird Conservation Initiative U.S. Committee, *2014 Farm Bill Field Guide to Fish and Wildlife Conservation*, 37.

81. Inglis.

82. Natural Resources Conservation Service Kansas State Office, "Lesser Prairie-Chicken Habitat in Jeopardy–A Conservation Plan Can Help," last accessed June 17, 2019, https://www.nrcs.usda.gov/wps/portal/nrcs/detail/ks/technical/cp/?cid=nrcs142p2_033117.

83. Inglis.

84. Ibid.; Natural Resources Conservation Service, "Conservation Reserve Program CP-25," last accessed September 20, 2019, https://www.fsa.usda.gov/Assets/USDA-FSA-Public/usdafiles/FactSheets/2015/CRPProgramsandInitiatives/Practice_CP25_Rare_and_Declining_Habitat.pdf; Natural Resources Conservation Service, "CRP Practice CP25 (Idaho)," June 2010, https://www.nrcs.usda.gov/Internet/FSE_DOCUMENTS/nrcs144p2_042378.pdf.

85. Natural Resources Conservation Service Kansas State Office, "Lesser Prairie-Chicken Habitat in Jeopardy–a Conservation Plan Can Help"; Farm Service Agency, "Conservation Reserve Program State Acres for Wildlife Enhancement (SAFE)."

86. Natural Resources Conservation Service Kansas State Office, "Lesser Prairie-Chicken Habitat in Jeopardy–a Conservation Plan Can Help."

87. Jim Pitman, Jon Ungerer, and Rod Winkler, "Lesser Prairie-Chicken Conservation Plan & Mitigation Framework (Slideshow)."

Farm Bill Easements 8

IN ADDITION TO THE INCENTIVE and rental contract programs discussed in chapter 7, the Farm Bill also funds a number of programs dedicated to financing conservation easements on private lands. As discussed in chapter 4, easements are often attractive options for landowners because a landowner can place on easement on their property and immediately receive a fair market payment for the rights being purchased under that easement, taxed as capital gains. In many cases, the landowner subsequently may continue to carry out agricultural and recreational practices on that land. Most Farm Bill easements have been consolidated into the Agricultural Conservation Easement Program (ACEP) under the conservation title. ACEP is divided into two subprograms: Agricultural Land Easements (ACEP-ALE) and Wetland Reserve Easements (ACEP-WRE). A third Farm Bill easement program, the Healthy Forests Reserve Program, established in 2003, is under the forestry title and is considerably smaller. All Farm Bill easement programs are run by the Natural Resources Conservation Service (NRCS). For detailed statistics on the annual impact of these programs, see appendix B.

Agricultural Conservation Easement Program–Agricultural Land Easements

ACEP-ALE was created by the 2014 Farm Bill as a consolidation of the Farm and Ranch Lands Protection Program and the Grassland Reserve Program.[1] The Farm and Ranch Lands Protection Program protected land with "prime, unique, or other productive soil" from conversion to non-agricultural use.[2] The Grassland Reserve Program protected "grassland, land that contains forbs, or shrubland," and land that historically contained those vegetative communities and "has potential to serve as habitat for animal or plant populations of significant ecological value" if restored.[3] These two programs were virtually identical in their operation, and the 2014 Farm Bill consolidated them while automatically transferring land enrolled in either into the new program.[4]

ACEP-ALE is designed to facilitate and purchase conservation easements on private farm and ranch lands, protecting both producers and wildlife from the threat of development. In order to qualify, there must be a pending offer by an eligible entity to purchase a conservation easement on the land. Land trusts are eligible entities, as are state and local governments, Indian tribes, and non-profit conservation organizations such as The Conservation Fund, The Nature Conservancy, and the Rocky Mountain Elk Foundation. The land must exhibit "prime, unique, or other productive soil"; historical or archeological resources; the potential to protect grazing "and related conservation values"; or characteristics that further a state or local policy consistent with ACEP-ALE. Eligible lands include cropland; rangeland; "grassland or land that contains forbs, or shrubland," and land that historically contained those vegetative communities; pastureland; and nonindustrial private forest land that contributes to the economic value of or buffers land that otherwise qualifies.[5] ACEP-ALE easements are always permanent.[6]

Under ACEP-ALE, the Farm Bill contributes up to 50 percent of the fair market value of the easement, provided that the entity purchasing the easement contributes an equal amount. Farm Bill funds may be used for the purchase of the easement as well as for technical assistance for conservation on the covered land. In the case of grasslands of special environmental significance, the Farm Bill may contribute up to 75 percent of the fair market value of the easement.[7] Many conservation organizations focus on protecting these grasslands, and in the last five years 55 percent of all acres enrolled in ACEP-ALE have included grasslands of special environmental significance.[8] ACEP-ALE is not subject to an annual acreage cap and is limited only by the Farm Bill's authorization of mandatory funding. Because ACEP-ALE easements are sold rather than donated by the landowner, the income and estate tax benefits discussed in chapter 4 are not available to participating landowners.

ACEP-ALE provides for multi-year agreements with eligible entities that govern the terms and conditions under which that entity may receive funds for purchasing easements. Any eligible entity may enter into a three- to five-year contract with NRCS and may choose to undergo an additional certification process to qualify for a contract of five years or longer. The Farm Bill establishes minimum terms and conditions for conservation easements to qualify. The easement deed must be consistent with the purposes of the program, permit effective enforcement of the conservation purposes of the easement, include a right of enforcement for the government as well as for the easement holder, include a limit on impervious surfaces, and include a conservation plan for any highly erodible cropland.[9]

The 2018 Farm Bill made some significant changes to ACEP-ALE. First, it removed a requirement from the 2014 Farm Bill that all ACEP-ALE easement deeds include a conservation plan "where appropriate." This change was a priority for the Land Trust Alliance, the American Farmland Trust, and other groups that found that ACEP-ALE easement plans were a disincentive to participation in the program.[10] This makes ACEP-ALE unique among Farm Bill conservation programs in that landowners may participate without preparing a conservation plan, although conservation plans are still required for highly erodible cropland, and the Farm Bill provides for technical assistance in the preparation of such conservation plans.[11]

A second change in the 2018 Farm Bill was the introduction of "buy-protect-sell" transactions. For most of ACEP-ALE's history, eligible entities have been required to purchase easements from existing landowners. In "buy-protect-sell" transactions, however, the eligible entity owns or intends to purchase land outright, place an easement on the land, and then transfer the land to a farmer or rancher while retaining the easement. Under the 2018 Farm Bill, eligible entities planning buy-protect-sell transactions may receive 50 percent of the easement value from ACEP-ALE, but must return the money if the final sale does not take place within three years, subject to an extension if NRCS finds it justified.[12] Buy-protect-sell projects are "intended to help farmers and ranchers gain access to affordable farmland."[13] Finally, the 2018 Farm Bill provides that land subject to an ALE may also be enrolled in the Conservation Reserve Program.[14]

To enroll in ACEP-ALE, landowners need to be matched up with an eligible entity willing to purchase an ACEP-ALE easement on their land. Landowners can find eligible entities through local NRCS field offices or directly. An online resource they can turn to is the Farmland Information Center, https://www.farmlandinfo.org/, jointly run by the U.S. Department of Agriculture and the American Farmland Trust. This website has many resources for landowners interested in conservation easements. In particular, the Farmland Directory, https://www.farmlandinfo.org/directory, allows landowners to search by state and zip code for easement programs in their area.[15] Further inquiries can be made to individual land trusts and other organizations to see if they participate in ACEP-ALE.

Agricultural Conservation Easement Program–Wetland Reserve Easements

In addition to ACEP-ALE, ACEP also includes ACEP-WRE. ACEP-WRE differs from ACEP-ALE in several ways. ACEP-WRE is a continuation of the Wetlands Reserve Program (WRP), created by the 1990 Farm Bill and consolidated into ACEP by the 2014 Farm Bill. WRP/ACEP-WRE has been particularly effective at promoting wildlife conservation: since WRP began in 1991, populations of wetland species across the United States have increased 51 percent.[16] Whereas ACEP-ALE applies to agricultural lands, ACEP-WRE applies to wetlands. Like ACEP-ALE, ACEP-WRE finances conservation easements on land where the owner then retains title, but unlike ACEP-ALE, under ACEP-WRE NRCS purchases easements directly.

ACEP-WRE easements restore, protect, and enhance wetlands. Through the easement, NRCS takes possession of the rights to farm, graze, harvest timber, subdivide, build, and drain wetlands on the property, none of which the agency will ever do. The landowner retains the title to the land and the right to hunt, fish, enjoy quiet recreation, manage water resources, and, in limited cases, graze the land and thin timber.[17] Easements may be permanent, or they may run for the maximum duration allowed by state law or for thirty years.[18]

The amount NRCS pays for easements is variable. Applications to sell easements are ranked based on the value of the easement for protecting and enhancing habitat for migratory birds and other wildlife and for improving water quality. Additional considerations include the cost-effectiveness of the easement, the availability of matching funds from the landowner or a third party, and other factors.[19] Payments are further limited to the lower of the fair market value of the easement, based on a per-acre Geographic Area Rate Cap for the area, or an offer made by the landowner.[20] Payments for thirty-year easements and contracts are limited to 75 percent of the payment for a permanent easement.[21] For easements valued at $500,000 or less, payments may be spread out over ten years. For easements valued at more than $500,000, payments *must* be spread out over at least five but not more than ten years.[22]

To be eligible for ACEP-WRE, land may not be a successful forested wetland created under the Conservation Reserve Program, may not have been put into agricultural production or converted from wetlands after passage of the 1985 Farm Bill, and may not have changed ownership within the previous two years, with exceptions for the death of the previous landowner, a foreclosure, and situations where it can be shown that the land was not acquired for the purpose of entering the ACEP-WRE program.[23] There is no acreage cap for the ACEP-WRE program, although no more than 15 percent of the cropland in any county may be enrolled in the program,[24] and like ACEP-ALE the program is limited by mandatory funding provided by the Farm Bill.

In addition to purchase of easements, ACEP-WRE also provides financial assistance payments to landowners in order to support actions that restore and improve the function of enrolled wetlands. Under the Farm Bill, NRCS may provide between 75 percent and 100 percent of the cost of wetlands management for permanent and maximum-term easements, and between 50 percent and 75 percent of the cost of wetlands management for thirty-year easements and contracts. These improvements must be carried out under an established wetland reserve easement plan developed by NRCS.[25] In addition, NRCS may enter into partnership agreements with states, non-governmental organizations, and Indian tribes to allow those entities to assist in the execution of wetland reserve easement plans.[26] WRP/WRE has been responsible for restoring over 700,000 acres of bottomland hardwoods in Arkansas, Louisiana, and Mississippi, resulting in the delisting of the Louisiana black bear. Because ACEP-WRE easements are owned by NRCS, landowners interested in placing an ACEP-WRE easement on their property should contact their local NRCS field office.

Agricultural Conservation Easement Program: Impact

Both ACEP-ALE and ACEP-WRE are high-impact programs because they place land into permanent conservation status, unlike the term-contract programs discussed in chapter 7. They are correspondingly expensive, however, so the amount of land placed under easements in a given year is limited. Because ACEP represents a consolidation of several programs, its impact can only be understood by totaling up the impacts of all of those programs over the years. The figures are staggering: an investment of $9.3 billion in federal funds and the protection of 4.85 million acres of land.[27] Going forward under the 2018 Farm Bill, ACEP will provide more annual funding for conservation than the entire Land and Water Conservation Fund.[28]

The Sage Grouse Initiative (SGI) was introduced in chapter 7 under the Environmental Quality Incentives Program. Roughly half of SGI's funding, however, comes from ACEP and is used to acquire conservation easements.[29] Easements are SGI's preferred tool for protecting land against exurban development, cultivation of grasslands, and mesic area (habitat with a moderate amount of moisture) degradation and loss,[30] and ACEP easements have been used to protect habitat in every state within the sage-grouse's range.[31] Between 2010 and 2018, SGI funded 227 conservation easements that protected 620,000 acres of sage-grouse habitat at a cost of $249 million.[32] The vast majority of this land was threatened by exurban development.[33]

Healthy Forests Reserve Program

In addition to ACEP, the Farm Bill also funds conservation easements through the Healthy Forests Reserve Program (HFRP). HFRP was not originally part of the Farm Bill, but was separately established by the Healthy Forests Restoration Act of 2003.[34] This law primarily focused on improving fire management in national forests, but provided an opportunity to expand conservation practices on private forests as well.[35] Starting with the 2008 Farm Bill, Congress used the Farm Bill as a vehicle to reauthorize and amend the program. Unlike other conservation programs, HFRP falls under the forestry title of the Farm Bill rather than the conservation title.[36]

Unlike ACEP and other Farm Bill conservation programs, HFRP was established with the primary purpose of protecting land in order to "restore, enhance, or otherwise measurably increase the likelihood of recovery" of species listed under the Endangered Species Act, and to increase the well-being of candidate species, state-listed species, species of special concern, and species of greatest conservation need under State Wildlife Action Plans.[37] This focus on at-risk species is unique among Farm Bill conservation programs.[38] HFRP also has a secondary purpose of promoting biodiversity and carbon sequestration.[39]

Land may be enrolled in HFRP with a permanent easement, a maximum duration easement under state law, a thirty-year easement, or a ten-year cost-share agreement. HFRP easements are owned by NRCS. Property enrolled in HFRP must be subject to a restoration plan to benefit threatened, endangered, and at-risk wildlife. The 2018 Farm Bill provided, for the first time, legislative guidance on the content of restoration plans, which may include "land management practices, vegetative treatments, structural practices and measures, practices to increase carbon sequestration, practices to improve biological diversity, and other practices and measures."[40]

Funding under HFRP is based on the fair market values of easements, with additional percentage caps based on the length of the easement. Permanent or maximum duration easements are eligible for payments of 75 percent to 100 percent of the value of the easement and the cost of planned conservation actions. Thirty-year easements are eligible for payments of no more than 75 percent of the value of the easement and the cost of conservation. Ten-year cost-share agreements are eligible only for payments of 50 percent of the cost of conservation.[41] HFRP is also subject to a nationwide enrollment cap of two million acres.[42] HFRP is subject to discretionary funding, which has led to large fluctuations in funding and limited its growth. From 2006 to 2018, HFRP invested $53.5 million in federal funds for the protection of 676,000 acres.[43]

HFRP has funded easements in thirteen states: Arkansas, California, Georgia, Indiana, Kentucky, Maine, Michigan, Mississippi, Ohio, Oklahoma,

Figure 8.1. The gopher tortoise (*Gopherus polyphemus*) is listed as threatened in the western portion of its range, including throughout Mississippi.
psmphotography/Shutterstock.

Oregon, Pennsylvania, and South Carolina.[44] Landowners in these states that are interested in the program should contact their local NRCS field office. As of 2019, the program is extremely limited. For example, in Mississippi HFRP easements are only available in the area around Camp Shelby Joint Forces Training Center on the edge of the De Soto National Forest, where easements are available to conserve the endangered dusky gopher frog, the threatened gopher tortoise, and the threatened black pine snake. There are also red-cockaded woodpeckers in the De Soto National Forest, which may colonize HFRP easements in the future.[45]

Notes

1. North American Bird Conservation Initiative U.S. Committee, *2014 Farm Bill Field Guide to Fish and Wildlife Conservation*, 42–46. The Farm and Ranch Lands Protection Program was established by the 1996 Farm Bill as the Farmland Protection Program and renamed by the 2002 Farm Bill. The Grassland Reserve Program was established by the 2002 Farm Bill.
2. Federal Agriculture Improvement and Reform Act of 1996, 1020–21.
3. Farm Security and Rural Investment Act of 2002, 259.
4. Agricultural Act of 2014, 731–32.
5. Food Security Act of 1985, as Amended through P.L. 115–334, 132–33.

6. Lori Faeth, interview by Christopher E. Segal, January 28, 2019.

7. Food Security Act of 1985, as Amended through P.L. 115–334, 134–35.

8. When an easement includes any grasslands of special environmental significance, NRCS pays 75 percent of the value of the entire easement, meaning that conservation organizations and other partners need contribute only 25 percent rather than 50 percent. Jerome Faulkner and Jessica Groves, Presentation at the Association of Fish and Wildlife Agencies 109th Annual Meeting (September 24, 2019).

9. The Farm Bill also lists additional deed provisions that eligible entities may elect to adopt, including provisions to keep land under the control of a working farmer or rancher, and provisions relating to subsurface mineral development. Food Security Act of 1985, as Amended through P.L. 115–334, 135–37.

10. Andrew Bowman, "Pass the Farm Bill," *Land Trust Alliance Blog*, December 11, 2018; Madeline Bodin, "The Farm Bill: Celebrating Passage of a Key Land Conservation Resource," *Saving Land*, Spring 2019; American Farmland Trust, "Farm Policy Update: The 2018 Farm Bill," December 21, 2018, https://farmland.org/farm-policy-update-the-2018-farm-bill/.

11. Food Security Act of 1985, as Amended through P.L. 115–334, 134.

12. Agricultural Improvement Act of 2018, 99–100.

13. Ibid., at Joint Explanatory Statement of the Committee of Conference, 596.

14. Ibid., 106.

15. Farmland Information Center, "Farmland Information Center," last accessed August 13, 2019, https://www.farmlandinfo.org; Farmland Information Center, "Farmland Directory," last accessed August 13, 2019, https://www.farmlandinfo.org/directory.

16. Cummins.

17. Water management was added to this list by the 2018 Farm Bill. North American Bird Conservation Initiative U.S. Committee, *2014 Farm Bill Field Guide to Fish and Wildlife Conservation*, 44–45.

18. Indian tribes may enter into thirty-year management agreements rather than surrendering any land rights through easements, but the program treats such thirty-year contracts in the same way as thirty-year easements. Food Security Act of 1985, as Amended through P.L. 115–334, 139.

19. Ibid., 139–40; Agricultural Improvement Act of 2018, 103.

20. Food Security Act of 1985, as Amended through P.L. 115–334, 142. Landowners generally have little incentive to offer an easement for less than the Geographic Area Rate Cap. See Pete Hall, "Wetland Reserve Easements (WRE) and Benefits to Landowners," last accessed January 22, 2019, https://www.johnhallco.com/wetland-reserve-easements-and-the-practical-benefits-to-landowners/.

21. Food Security Act of 1985, as Amended through P.L. 115–334, 142.

22. Ibid., 142–43.

23. Ibid., 139. There is an exception for forested wetlands created under CRP that proved unsuccessful, and these may be enrolled in ACEP-WRE.

24. This limit was first established for WRP by the 2008 Farm Bill, which set it at 10 percent, but the 2018 Farm Bill increased it to 15 percent. Like the Conservation Reserve Program, ACEP-WRE is also limited to 25 percent of the total land in a county. Ibid., 120–21. Prior to the transition to ACEP-WRE in the 2014 Farm Bill, WRP had also been subject to an acreage cap. The cap was set at one million acres under the 1990 Farm Bill, adjusted to 975,000 acres in 1996, and then increased to 2,275,000 acres in 2002 and 3,041,200 acres in 2008. Food, Agriculture, Conservation, and Trade Act of 1990, 3585; Federal Agriculture Improvement and Reform Act of 1996, 995; Farm Security and Rural Investment Act of

2002, 252; Food, Conservation, and Energy Act of 2008, 1037, 1080; Agricultural Improvement Act of 2018, 94.

25. Unlike ACEP-ALE, the 2018 Farm Bill did not modify this requirement for ACEP-WRE. If the landowner retains existing grazing rights, grazing must comply with the wetland reserve easement plan or a separate grazing management plan.

26. Food Security Act of 1985, as Amended through P.L. 115–334, 143; Agricultural Improvement Act of 2018, 104–05.

27. These figures represent a summation of the impact of the Farmland Protection Program, Farm and Ranch Lands Protection Program, Grassland Reserve Program, WRP, and ACEP from 1991 to 2018. Note that the Farm Bill authorizes expenditures for ACEP as a whole, and NRCS subsequently divides the ACEP budget into ACEP-ALE and ACEP-WRE funding based on demand. Natural Resources Conservation Service, "NRCS Conservation Programs: Farm and Ranch Lands Protection Program (FRPP)," last accessed June 23, 2019, https://www.nrcs.usda.gov/Internet/NRCS_RCA/reports/fb08_cp_frpp.html; Natural Resources Conservation Service, "NRCS Conservation Programs: Grassland Reserve Program (GRP)," last accessed June 23, 2019, https://www.nrcs.usda.gov/Internet/NRCS_RCA/reports/fb08_cp_grp.html; Natural Resources Conservation Service, "NRCS Conservation Programs: Wetlands Reserve Program (WRP)," last accessed June 23, 2019, https://www.nrcs.usda.gov/Internet/NRCS_RCA/reports/fb08_cp_wrp.html; Natural Resources Conservation Service, "NRCS Conservation Programs: Agricultural Conservation Easement Program (ACEP)," last accessed June 23, 2019, https://www.nrcs.usda.gov/Internet/NRCS_RCA/reports/srpt_cp_acep.html; Office of Management and Budget, *The Budget for Fiscal Year 1998: Department of Agriculture* (1997), 168; Office of Management and Budget, *The Budget for Fiscal Year 1999: Department of Agriculture* (1998), 99; Office of Management and Budget, *The Budget for Fiscal Year 2000: Department of Agriculture* (1999), 104; Office of Management and Budget, *The Budget for Fiscal Year 2001: Department of Agriculture* (2000), 110; Office of Management and Budget, *The Budget for Fiscal Year 2002: Department of Agriculture* (2001), 108; Office of Management and Budget, *The Budget for Fiscal Year 2003: Department of Agriculture* (2002), 111; U.S. Department of Agriculture, *FY 2004 Budget Summary* (2003), 24; U.S. Department of Agriculture, *FY 2005 Budget Summary* (2004), 34; U.S. Department of Agriculture, *FY 2006 Budget Summary and Annual Performance Plan* (2005), 29; Natural Resources Conservation Service, "NRCS Conservation Programs"; United States Department of Agriculture Economic Research Service, "Wetland Reserve Program Expenditures," January 15, 2009, https://web.archive.org/web/20090115231303/http://www.ers.usda.gov/Briefing/ConservationPolicy/images/wrpexpenditures.htm.

28. Smith and Duvall.

29. See Natural Resources Conservation Service, "Sage Grouse Initiative 2.0 Investment Strategy, FY 2015-2018."

30. Ibid. Mesic areas are lands adjacent to water.

31. Sage Grouse Initiative, "#Fridaysonthefarm for Future Generations of Ranchers and Sage Grouse," November 2, 2018, https://www.sagegrouseinitiative.com/fridaysonthefarm-for-future-generations-of-ranchers-and-sage-grouse/; Sage Grouse Initiative, "#Habitathero Mike Fenn: Leaving a Legacy for Wildlife in Wyoming," September 25, 2017, https://www.sagegrouseinitiative.com/habitat-hero-mike-fenn-wyoming/; Sage Grouse Initiative, "Round-up of Ranching Success Stories for Sage Grouse," May 20, 2015, https://www.sagegrouseinitiative.com/round-up-of-ranching-success-stories-for-sage-grouse/; Natural Resources Conservation Service, "Sage Grouse Initiative 2.0 Investment Strategy, FY 2015-2018."

32. Natural Resources Conservation Service, "Working Lands for Wildlife Obligations, All Years All Species."

33. Natural Resources Conservation Service, "Sage Grouse Initiative 2.0 Investment Strategy, FY 2015-2018," 11.

34. Healthy Forests Restoration Act of 2003, Pub. L. No. 108-148, 117 Stat. 1887 (2003).

35. Rey.

36. Because HFRP is funded under the forestry title, income and conservation compliance limits applicable to the conservation title do not apply. For example, Texas Parks & Wildlife, "2014 Farm Bill Programs: Healthy Forest Reserve Program (HFRP)," last accessed June 7, 2019, https://tpwd.texas.gov/landwater/land/private/farmbill/hfrp/.

37. Species of greatest conservation need were added by the 2018 Farm Bill. For more information on State Wildlife Action Plans, see chapter 10.

38. Rey.

39. Healthy Forests Restoration Act of 2003, 1911–12; Agricultural Improvement Act of 2018, 364.

40. Agricultural Improvement Act of 2018, 365.

41. Healthy Forests Restoration Act of 2003, 1913.

42. Ibid., 1912.

43. Natural Resources Conservation Service, "NRCS Conservation Programs: Healthy Forests Reserve Program (HFRP)," last accessed June 23, 2019, https://www.nrcs.usda.gov/Internet/NRCS_RCA/reports/fb08_cp_hfrp.html; U.S. Department of Agriculture, *FY 2008 Budget Summary and Annual Performance Plan* (2007), 63; U.S. Department of Agriculture, *FY 2009 Budget Summary and Annual Performance Plan* (2008), 56; U.S. Department of Agriculture, *FY 2010 Budget Summary and Annual Performance Plan* (2009), 72.

44. Natural Resources Conservation Service, "NRCS Conservation Programs: Healthy Forests Reserve Program (HFRP)."

45. Natural Resources Conservation Service Mississippi, "Healthy Forests Reserve Program," last accessed August 13, 2019, https://www.nrcs.usda.gov/wps/portal/nrcs/main/ms/programs/easements/forests/.

9

Partnership Programs Under the Farm Bill

MANY ASPECTS OF FARM BILL CONSERVATION are delivered utilizing partnerships. Moving from the Farm Bill to on-the-ground conservation is a multi-step process, from program design and implementation to marketing programs to landowners, preparing conservation plans, awarding contracts, and guiding implementation.[1] The process engages the Department of Agriculture and its agencies, other federal agencies, state fish and wildlife agencies, state technical committees, conservation districts, land trusts, other non-governmental organizaions (NGOs), other partners, and of course landowners.

Partnerships are even critical to staffing Farm Bill programs. Chapter 6 discussed Farm Bill biologist programs where NGOs, state fish and wildlife agencies, other federal agencies, and other organizations help staff Natural Resources Conservation Service (NRCS) offices. Similar groups of partners, including NRCS, may fund staff positions at state fish and wildlife agencies or NGOs. The guiding principle for these partnerships is to get the staff and funding in the right place to achieve on-the-ground conservation. This can lead to complicated arrangements, such as a state agency paying for an NGO's employee to work in an NRCS office.[2] Many partnerships also help identify priorities and provide other input for Farm Bill conservation programs.[3]

In addition to partnerships as tools to deliver conservation in all Farm Bill programs, there are specific programs designed from the ground up to promote partnerships. Two of these, the Regional Conservation Partnership Program and the Voluntary Public Access and Habitat Incentive Program (VPA-HIP), are Farm Bill programs. A third, Working Lands for Wildlife (WLFW), is not a formal Farm Bill program but a strategy developed by NRCS and the Fish and Wildlife Service (FWS) for implementing Farm Bill programs.

Regional Conservation Partnership Program

The centerpiece of the Farm Bill's partnership programs is the Regional Conservation Partnership Program (RCPP), established by the 2014 Farm Bill.[4] RCPP allows partner organizations to design conservation projects. Partners are empowered to define the benefits sought, activities to achieve these benefits, and the timeframe and geographic area of the project. Partners then use a combination of Farm Bill funding and other resources to enroll landowners in the project, implement the project, and monitor compliance with and progress of the project.[5] Eligible partners include Indian tribes, non-profit organizations, state and local governments, private industry, conservation districts, universities, and many other types of organizations. Between 2014 and 2018, over two thousand organizations worked with NRCS to create 362 different projects, with impacts in all fifty states and Puerto Rico.[6] Partners who wish to participate in RCPP and implement their own Farm Bill project submit proposals to NRCS, and proposals are assessed and ranked in a competitive process.[7] In general, successful projects share four traits: they are innovative, they leverage additional non-federal contributions, they offer impactful solutions to real-world problems, and they engage more participants than NRCS would otherwise be able to reach.[8]

When first established in 2014, RCPP was funded with $100 million each year plus 7 percent of the funding and acres allocated to each of the following programs: the Environmental Quality Incentives Program (EQIP), the Conservation Stewardship Program, the Agricultural Conservation Easement Program, and the Healthy Forests Reserve Program.[9] This was an easy way to make substantial funds available without greatly increasing the total cost of Farm Bill conservation, but it led to criticism because it diverted funds from existing, oversubscribed programs.[10] The 2018 Farm Bill removed diversion of other programs' funds and acres to RCPP and increased RCPP's mandatory funding from $100 million to $300 million per year.[11]

The actual dollars available to landowners through RCPP are much greater. One of the requirements of the program is for non-federal funds to be utilized, and project proposals are rated in part on how efficiently they leverage federal dollars with other sources of funding. Partners, including NGOs, state government agencies, and others, raise these funds from a variety of sources, including grants, corporate donations, and their own budgets. As of 2016, the last time RCPP published a detailed report, the program had invested $590 million in partner-led conservation efforts, with partners contributing $900 million in financial and technical assistance. NRCS projected that through 2018, RCPP investments by NRCS and partners would total at least $2.4 billion.[12] The $300 million per year in dedicated funding under the 2018 Farm Bill is projected to generate $3 billion in matching non-federal investments in RCPP projects over the life of the 2018 Farm Bill, for a total investment by NRCS and partners of $4.5 billion over the five years from 2019 through 2023.[13]

Past RCPP projects have addressed diverse wildlife conservation needs, including monarch butterfly habitat conservation, fish passage structure installation, and early successional forest habitat development. RCPP projects may utilize grants, cost-shares, rental agreements, and/or conservation easements. One example of an RCCP easement project is the Coastal Headwaters Longleaf Forest Conservation and Restoration Project, which benefits gopher tortoise, flatwoods salamander, and other at-risk and threatened and endangered species in the forests of southwestern Alabama and western Florida. The project began in 2014 as a primary partnership between Resource Management Service, LLC (RMS), and The Conservation Fund. The vision for the Coastal Headwaters Longleaf Forest is to permanently conserve approximately two hundred thousand acres of working longleaf forests and functional longleaf ecosystem habitat to support approximately forty-five at-risk, threatened and endangered, and NatureServe G1/G2 species. Additionally the project area encompasses 755 miles of flowing streams in four major river basins that empty into the Gulf of Mexico. Currently, one-quarter of the land is made up of streams, wetlands, and riparian areas, whereas the remaining three-quarters is forested. Of the current forestland, approximately 10 percent remains in longleaf pine forest, whereas approximately 90 percent has been converted to the faster-growing loblolly pine. RMS manages this land on behalf of institutional investor owners, such as pension funds and educational endowments, and it has a fiduciary responsibility to maximize the return on their investments. Although some landowners and businesses might choose to reduce their profitability in order to practice conservation, RMS is legally obligated to find ways to practice conservation without making any compromise on its fiduciary responsibility. This has led to an innovative approach to funding conservation with the Coastal Headwaters Longleaf Forest.

The Coastal Headwaters Longleaf Forest project proposes to restore approximately 150,000 upland acres to longleaf pine, grow the forest to at least forty-five years of age before final harvest, and burn the forest every four years to promote tree health and maintain the ground cover that was historically present in the region. Periodic thinning throughout the rotation will help

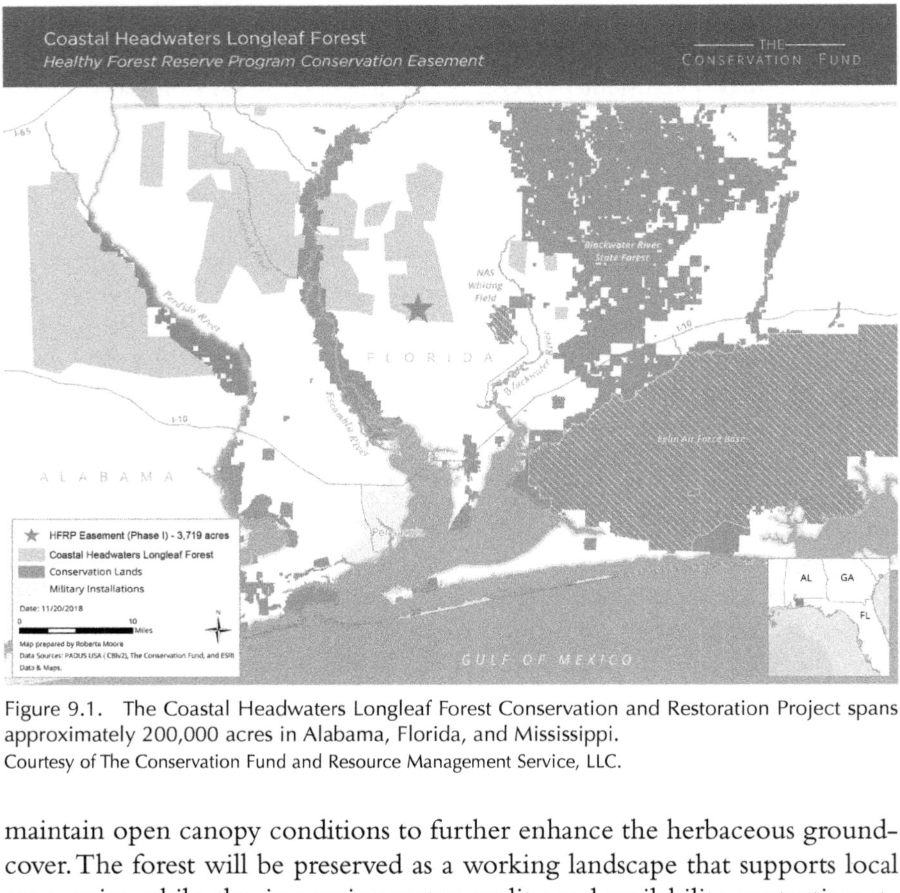

Figure 9.1. The Coastal Headwaters Longleaf Forest Conservation and Restoration Project spans approximately 200,000 acres in Alabama, Florida, and Mississippi.
Courtesy of The Conservation Fund and Resource Management Service, LLC.

maintain open canopy conditions to further enhance the herbaceous groundcover. The forest will be preserved as a working landscape that supports local economies while also improving water quality and availability, protecting at-risk and threatened and endangered species, and maintaining a healthy, native longleaf ecosystem. To accomplish this, RMS and The Conservation Fund have engaged over thirty partners and funders, including federal agencies, state agencies, environmental groups, sportsmen's groups, and private companies, all of whom are contributing funding or in-kind support to the effort. As of early 2019 collectively they had raised approximately $40 million of the projected total cost of $200 million.[14]

Grant funding has come from a variety of sources, including from the Florida Forest Service, Florida Department of Environmental Protection, Walmart, U.S. Forest Service, and four RCPP grants.[15] RMS places conservation easements on parcels of land as funding is available and brings land into the planned management practices as easements are placed. The Conservation Fund has the lead responsibility for assembling funding, and the easements are ultimately owned by NRCS.[16] The first RCPP grant of

Figure 9.2. As part of the Coastal Headwaters Longleaf Forest Conservation and Restoration Project, Resource Management Service, Inc., plans to plant 150,000 of longleaf pine.
Photograph by Jimmy Bullock, Resource Management Service, LLC.

$5 million was awarded in 2016 and activated in 2018 to secure the first 3,719-acre easement for the project. This land, now protected as a working longleaf forest and functional longleaf ecosystem in perpetuity, also supports approximately two thousand forest management, harvesting, and manufacturing jobs in local communities.[17]

Securing this easement was a major milestone because it required a new approach to easement appraisals. In order for the easement to be financially viable, the appraisal needed to reflect not only the value of the surrendered development rights but also both the additional management costs committed to for restoring and maintaining a functional longleaf ecosystem and the revenue foregone by harvesting longleaf pine every forty-five years instead of loblolly pine every twenty-five years. All were incorporated into the valuation of the easement and the payment RMS received when the easement was sold.[18]

Since that first 2016 RCPP award, the project has received three additional RCPP awards that will soon finance additional conservation easements. According to Jimmy Bullock, RMS Senior Vice President of Forest Sustainability, sufficient funding has been secured to place approximately twenty thousand acres under conservation easements by the end of 2019 and at least thirty thousand acres under conservation easements by the end of 2021. Getting from there to 150,000 upland acres will be a matter of time, money, and persistence. One possibility for future easements that may make the project more cost-effective is incorporating compensatory mitigation credits for carbon, wetlands, or endangered species.[19]

The Coastal Headwaters Longleaf Forest project embodies all of the characteristics of a successful Farm Bill–funded conservation program. It engages a wide variety of stakeholders, brings together diverse funding sources including non-federal funding and in-kind support, achieves conservation at scale while protecting a working landscape, and shows creativity and ambition.[20] Especially significant is the project's pioneering approach to easement valuation and the opportunity through the RCPP program for the Farm Bill's Adjusted Gross Income limits to be waived, thus enabling meaningful conservation through partnerships with large forest landowners. In this project, RCPP funding truly brings about conservation at scale. If this approach is adopted widely, easements will become more feasible for large landowners and put dollars to work for conservation in areas where it will have the most benefit. The Coastal Headwaters Longleaf Forest Project is proof economic and ecologic benefits can be achieved at scale, and it will be a model for future conservation success. The Coastal Headwaters Longleaf Forest project also builds on many other programs dedicated to conservation of the longleaf pine ecosystem, including the Longleaf Alliance, NRCS's Longleaf Pine Initiative, the National Fish and Wildlife Foundation's Longleaf Stewardship Fund, and the multi-stakeholder America's Longleaf Restoration Initiative, made up of twenty-two state and federal agencies and other organizations.[21]

Voluntary Public Access and Habitat Incentive Program

A comparatively small Farm Bill partnership program is VPA-HIP, established by the 2008 Farm Bill. VPA-HIP provides grant funding for state and tribal governments to secure access to and improve conditions on private land for wildlife-dependent recreation, including hunting and fishing. Projects designed to strengthen wildlife habitat improvement projects on lands already enrolled in CRP and Agricultural Conservation Easement Program–Wetland Reserve Easements are given priority, as are projects with non-federal matching funds available. From 2009 through 2018, VPA-HIP received $90 million in mandatory funding, funded projects in thirty states, and generated ninety-nine dollars in recreational spending for every one dollar in federal funding. Grants made between 2014 and 2018 opened more than 950,000 acres of land to public access.[22] Although many VPA-HIP grants emphasize hunting and fishing, wildlife viewing and other non-consumptive uses of wildlife are also part of the program, and public access for these purposes advances conservation of at-risk species as well. This can happen directly. In 2015 the Montana Department of Fish, Wildlife and Parks received a VPA-HIP grant to enable public viewing of greater sage-grouse.[23]

Working Lands for Wildlife

WLFW is an innovative partnership between NRCS and FWS. Rather than a Farm Bill program, WLFW is a strategy for the deployment of Farm Bill funds. It began in 2012 on an ad hoc basis. Most WLFW funding comes from EQIP and Agricultural Conservation Easement Program–Agricultural Land Easements, although any Farm Bill working lands program can use the WLFW strategy. Under WLFW, NRCS and FWS target priority species and land of high ecological value for conservation. NRCS then "provides technical and financial assistance to participants who voluntarily make improvements to their working lands," and FWS "provides participants with regulatory predictability" under the Endangered Species Act of 1973 (ESA) through tools such as Candidate Conservation Agreements with Assurances for up to thirty years.[24]

WLFW launched in 2012 with seven initial species, including both the greater sage-grouse and the lesser prairie chicken. Eleven more species were added in 2016. Both the Sage Grouse Initiative and the Lesser Prairie Chicken Initiative began in 2010 but were brought into WLFW in 2012.[25] Across all of its species, between 2012 and 2018 WLFW invested $542.5 million Farm Bill dollars in work with 6,451 producers on more than 9.3 million acres of land in twenty-four different types of landscapes in all fifty states.[26]

The greatest innovation of WLFW is its use of the ESA. Landowners that enroll in a WLFW Farm Bill program are automatically compliant with ESA regulations, without any action or even knowledge on their part. When NRCS and FWS select a species and associated conservation practices to include in WLFW, they engage in a programmatic Section 7 consultation for each species. This means that additional individual Section 7 consultations are not required to enroll landowners in WLFW programs.[27] In the case of the Sage Grouse Initiative, for example, a single Section 7 consultation covered two thousand landowners and two million acres of land.[28] Although WLFW largely focuses on individual species, each species is an indicator for an entire ecosystem, and both landowner enrollments and regulatory predictability are designed to work for the entire ecosystem. For example, the WLFW ESA Section 7 Biological Opinion for the southwestern willow flycatcher covered all seventy-five listed species within that ecosystem.[29] Thus any landowner participating in that WLFW program was fully protected from *any* additional regulations under the ESA.

Figure 9.3. The southwestern willow flycatcher (*Empidonax traillii extimus*) is listed as endangered. Scarlett Howell/USGS.

The 2018 Farm Bill provided Congressional recognition of WLFW and directed the two agencies to continue the collaboration and expand it to include the Farm Service Agency. Congress further provided that although WLFW calls for ESA permits to provide thirty years of regulatory predictability for participating landowners, the agencies could offer landowners regulatory predictability beyond thirty years.[30] The value of WLFW to at-risk species is immense. Because of the regulatory assurances provided by FWS, under WLFW landowners can work on listed and at-risk species without any fear of ESA regulations. Landowners further value WLFW because participation results in more efficient, lower-impact agricultural operations.[31] Landowners interested in benefiting from ESA regulatory assurances under WLFW should contact their local NRCS field office to see if their proposed conservation activities might be covered by WLFW.

Notes

1. For example, North American Bird Conservation Initiative U.S. Committee, *2014 Farm Bill Field Guide to Fish and Wildlife Conservation*, 11.

2. Ibid., 12; NRCS Washington, "Partnerships Support Wildlife Programs for NRCS-WA," last accessed February 7, 2019, https://www.nrcs.usda.gov/wps/portal/nrcs/detail/wa/newsroom/factsheets/?cid=nrcs144p2_036546; Inglis.

3. These include NGOs, state governments, state technical committees, local working groups, and conservation districts. North American Bird Conservation Initiative U.S. Committee, *2014 Farm Bill Field Guide to Fish and Wildlife Conservation*, 8–13.

4. When it established RCPP, Congress consolidated four similar programs. The first program was the Cooperative Conservation Partnership Initiative, which dated to the 2008 Farm Bill and provided for partner organization to propose specific projects under EQIP, the Conservation Stewardship Program, and the Wildlife Habitat Incentives Program in order to "better fit the local circumstances and goals and objectives of the . . . project." Funding was drawn from covered programs at a rate of 6 percent of the available money and acres. Committee on Conference, Food, Conservation, and Energy Act of 2008, Conference Report to Accompany H.R. 2419, H.R. Rep. No. 110-627 (2008), 740–41; Natural Resources Conservation Service, "Cooperative Conservation Partnership Initiative," last accessed February 8, 2019, https://www.nrcs.usda.gov/wps/portal/nrcs/main/national/programs/financial/ccpi/; Food, Conservation, and Energy Act of 2008, 1079.

The second program was the Agricultural Water Enhancement Program, which also dated to the 2008 Farm Bill and was a contract program within EQIP that provided funding to conserve surface and groundwater and improve water quality. Partner organizations identified suitable areas for the program and entered into multi-year agreements with NRCS to implement it. Between 2009 and 2017, AWEP spent $356 million and enrolled 1.4 million acres. Natural Resources Conservation Service, "Agricultural Water Enhancement Program," last accessed February 8, 2019, https://www.nrcs.usda.gov/wps/portal/nrcs/main/national/programs/financial/awep/; Natural Resources Conservation Service, "Agricultural Water Enhancement Program (AWEP)," last accessed February 8, 2019, https://www.nrcs.usda.gov/Internet/NRCS_RCA/reports/fb08_cp_awep.html.

The third program was the Chesapeake Bay Program for Nutrient Reduction and Sediment Control, which also dated to the 2008 Farm Bill and was a narrow program to improve agricultural conservation practices within the Chesapeake Bay watershed. A total of $188 million of Farm Bill funds were invested over four years to control erosion, improve air and water quality, and reduce sediment and nutrient levels in ground and surface water. Food, Conservation, and Energy Act of 2008, 1069-70; Committee on Conference, Food, Conservation, and Energy Act of 2008, Conference Report to Accompany H.R. 2419, H.R. Rep. No. 110-627, 736–37.

The fourth program was the Great Lakes Basin Program for Soil Erosion and Sediment Control, which dated to the 2002 Farm Bill and invested $60 million over twelve years to provide demonstration grants, technical assistance, and educational campaigns to improve water quality in the Great Lakes basin by reducing soil erosion and improving sediment control. Farm Security and Rural Investment Act of 2002, 267; Food, Conservation, and Energy Act of 2008, 1068–69; American Taxpayer Relief Act of 2012, 2363.

5. Food Security Act of 1985, as Amended through P.L. 115–334, 149–50.

6. Natural Resources Conservation Service, "Regional Conservation Partnership Program," last accessed February 7, 2019, https://www.nrcs.usda.gov/wps/portal/nrcs/main/national/programs/farmbill/rcpp/; Natural Resources Conservation Service, "NRCS Regional Conservation Partnership Program (RCPP) 2014-2018 Project Summaries," last accessed February 7, 2019, https://www.nrcs.usda.gov/wps/PA_NRCSConsumption/download?cid=nrcseprd1386891&ext=pdf. Approximately seventy of these projects were substantially related to wildlife. Steve Riley, Presentation at the Association of Fish and Wildlife Agencies 109th Annual Meeting (September 24, 2019).

7. Food Security Act of 1985, as Amended through P.L. 115–334, 147–57.

8. Natural Resources Conservation Service, "RCPP Partnership Stories," last accessed February 8, 2019, https://www.nrcs.usda.gov/wps/portal/nrcs/detailfull/national/programs/farmbill/rcpp/?cid=stelprdb1253919.

9. Agricultural Act of 2014, 745, 749.

10. For example, National Sustainable Agriculture Coalition, "Regional Conservation Partnership Program: 2018 Awards and Farm Bill Opportunities," February 8, 2018, http://sustainableagriculture.net/blog/rcpp-awards-farm-bill-fy2018/.

11. Agricultural Improvement Act of 2018, 112.

12. Natural Resources Conservation Service, "RCPP: Partner-Led Solutions," December 2016, https://www.nrcs.usda.gov/wps/PA_NRCSConsumption/download?cid=nrcseprd1309909&ext=pdf.

13. Environmental Defense Fund, "How the Farm Bill Changes the Future of Climate and Water Conservation," Deember 12, 2018, http://blogs.edf.org/growingreturns/2018/12/12/farm-bill-climate-water-conservation/.

14. Jimmy Bullock, interview by Christopher E. Segal, March 7, 2019.

15. Ibid. The Forest Service provided two Forest Legacy grants. For more on that program, see chapter 10.

16. Dan Chapman, "Coastal Headwaters Project in Florida Is a Major Step for Longleaf Pine Restoration," April 24, 2019, https://www.fws.gov/southeast/articles/coastal-headwaters-project-in-florida-is-a-major-step-for-longleaf-pine-restoration/.

17. Bullock. This includes both direct and indirect employment.

18. Ibid.

19. Ibid. For further discussion of compensatory mitigation, see chapter 5.

20. Natural Resources Conservation Service, "RCPP: Partner-Led Solutions," 6; Dan Chapman, "A Harmonious Future for Profits, Pine and at-Risk Species Along the Florida-Alabama Line," May 9, 2017, https://www.fws.gov/southeast/articles/a-harmonious-future-for-profits-pine-and-at-risk-species-along-the-florida-alabama-line/; The Conservation Fund, "Coastal Headwaters Forest–Longleaf Conservation and Restoration," last accessed February 8, 2019, https://www.conservationfund.org/projects/coastal-headwaters-forest-longleaf-conservation-and-restoration; The Conservation Fund, "The Conservation Fund and Resource Management Service, LLC Announce Partnership and Initial Funding for Largest, Single Longleaf Pine Landscape Conservation and Restoration Effort on Private Lands," press release, January 14, 2015, https://www.conservationfund.org/news/press-releases/585-the-conservation-fund-and-resource-management-service-llc-announce-partnership-and-initial-funding-for-largest-single-longleaf-pine-landscape-conservation-and-restoration-effort-on-private-lands; Walmart Acres for America, "2014 Grant Fact Sheet Coastal Headwaters Forest, Longleaf Conservation and Restoration (AL, FL)," November 12, 2014, https://www.nfwf.org/acresforamerica/Documents/acres2014longleaf.pdf.

21. The Longleaf Alliance, "The Longleaf Alliance," last accessed October 3, 2019, https://www.longleafalliance.org/; Natural Resources Conservation Servce, "Longleaf Pine Initiative," last accessed October 3, 2019, https://www.nrcs.usda.gov/wps/portal/nrcs/detailfull/national/programs/initiatives/?cid=nrcsdev11_023913; America's Longleaf Restoration Initiative, "America's Longleaf Restoration Initiative," last accessed October 3, 2019, http://www.americaslongleaf.org/; Riley, Abernethy, and McKenzie, "Case Studies of Scalable Wildlife Conservation in Working Landscapes."

22. Theodore Roosevelt Conservation Partnership, "Sportsmen's Priorities for Conservation and Access in the 2018 Farm Bill Voluntary Public Access & Habitat Incentive Program (VPA-HIP)," 2018, http://www.trcp.org/wp-content/uploads/2018/02/Public-Access-fact-sheet.pdf.

23. Natural Resources Conservation Service, "2014–Farm Bill–VPA-HIP 2014 Funding Grantees," last accessed February 8, 2019, https://www.nrcs.usda.gov/wps/portal/nrcs/detail/national/programs/farmbill/?cid=stelprdb1258890; Natural Resources Conservation Service, "2015 VPA-HIP Grantees," last accessed February 8, 2019, https://www.nrcs.usda.gov/wps/portal/nrcs/detail/national/programs/farmbill/?cid=nrcseprd388655.

24. Natural Resources Conservation Service, "Working Lands for Wildlife," last accessed October 4, 2019, https://www.nrcs.usda.gov/wps/portal/nrcs/detailfull/national/programs/?cid=stelprdb1046975; North American Bird Conservation Initiative U.S. Committee, *2014 Farm Bill Field Guide to Fish and Wildlife Conservation*, 17; U.S. Fish and Wildlife Service, Director's Order No. 217: Collaboration with the U.S. Department of Agriculture's Natural Resources Conservation Service (NRCS) to Conserve Listed, Candidate, and Other at-Risk Species (2016). See chapter 13 for a discussion of CCAAs and other ESA tools for establishing regulatory predictability.

25. See chapter 7 for more information on the Sage Grouse Initiative and the Lesser Prairie Chicken Initiative.

26. Natural Resources Conservation Service, "Working Lands for Wildlife Obligations, All Years All Species"; Working Lands for Wildlife, "2018 Fact Sheet."

27. Michael Martinez, "Working Lands for Wildlife: Targeted Landscape-Scale Wildlife Habitat Conservation," *Natural Resources & Environment* 29, no. 3 (2015): 37.

28. Hall and Griffiths.

29. Ibid.

30. Agricultural Improvement Act of 2018, 87–88.

31. Hall and Griffiths.

NON–FARM BILL INCENTIVES FOR PRIVATE LAND CONSERVATION IV

"Like winds and sunsets, wild things were taken for granted until progress began to do away with them. Now we face the question whether a still higher 'standard of living' is worth its cost in things natural, wild and free. For us of the minority, the opportunity to see geese is more important than television."

–ALDO LEOPOLD, *A SAND COUNTY ALMANAC*

10

Other Federal Resources for Conservation

The Farm Bill is the largest and most accessible source of federal funding for private land conservation, but a myriad of other programs are also available. This chapter describes many of the most important, although it is impossible to capture all of them. Because these programs are constantly evolving, the best source of information on them is a trusted conservation partner, such as agency and non-governmental organization (NGO) field staff. The following discussion includes federal assistance programs as well as federally backed partnership and research organizations that support assistance programs.

U.S. Fish and Wildlife Service Small Wetlands Acquisition Program

In addition to supporting the Farm Bill easement programs discussed in chapter 8, the Fish and Wildlife Service (FWS) also operates its own easement program to protect waterfowl habitat. Many of the conservation programs discussed in this chapter have a focus on waterfowl, because waterfowl are highly sought-after game species that present unique conservation challenges due to their migratory nature. They have also been a focus of federal conservation programs for a long time, ranging from the establishment of the first National Wildlife Refuge on Pelican Island, Florida, by President Theodore Roosevelt in 1903, the Migratory Bird Treaty Act of 1918, and the 1934 Duck Stamp Act, to the establishment of the Wetlands Reserve Program (WRP) in 1990 and its modern descendant, Wetland Reserve Easements.[1] From 1999 to 2017, an average of 1.4 million hunters harvested an average of 14.3 million ducks and 4.4 million geese in the United States and

Figure 10.1. The first Duck Stamp was issued on August 14, 1934, and featured a design by Jay N. "Ding" Darling.
Wikipedia.

Figure 10.2. Prairie potholes are small wetlands that are incredibly productive breeding grounds for ducks and other species of waterfowl. The wetlands shown here are part of the Kulm Wetland Management District in North Dakota.
Courtesy of the U.S. Fish and Wildlife Service.

Canada each year.[2] Under 1958 amendments to the 1934 Duck Stamp Act, FWS may use Duck Stamp funds to acquire small wetlands and potholes or easements on small wetlands and potholes to protect waterfowl. These small areas are designated as Waterfowl Production Areas (WPAs) and managed by Wetland Management Districts.[3]

Today, there are approximately four million acres of WPAs, including 3.25 million acres of easements on private land,[4] protecting grasslands and wetlands in roughly equal proportion.[5] Over thirty-four thousand easements have been acquired, and over eighteen hundred more landowners are waiting in line for easements, offering another 250,000 acres of land.[6] A total of 95 percent of WPAs are in the Prairie Pothole region of Minnesota, Montana, North Dakota, and South Dakota, whereas other WPAs have been established in Idaho, Iowa, Maine, Michigan, Nebraska, and Wisconsin.[7] Some WPA easements are donated, but the majority, 2.85 million acres, were purchased by FWS at a cost of $565 million.[8] This funding comes from Federal Duck Stamps and other sources, including the North American Wetlands Conservation Act, the Land and Water Conservation Fund, NGOs, and private donors.[9]

WPA easements protect wetlands and adjacent upland grasslands used by waterfowl. Landowners may not drain, burn, level, or fill wetlands covered by easements, but lands in agricultural production may continue to be cultivated as far as is feasible. For grassland easements, landowners must keep the land in grass. Grazing is permitted at any time; mowing, haying, and grass seed harvesting on these easements is permitted but must be delayed until after July 15 of each year. All WPA easements are perpetual.[10] Interested landowners should contact their local FWS field office or Wetland Management District for more information.[11]

U.S. Forest Service Programs

The U.S. Forest Service is best known for managing 154 national forests and twenty grasslands in forty-three states and Puerto Rico, a total of 193 million acres, and for fighting forest fires. But it also provides research, funding, and tools to support sustainable management of the five hundred million acres of private, state, and tribal forests in the nation.[12] Of these, privately owned forest land makes up 441 million acres, with eleven million different owners.[13] Forest Service programs that support conservation on private land include the Forest Legacy Program, Community Forest and Open Space Program, Landscape Scale Restoration Program, and Forest Stewardship Program. These programs work with state forestry agencies and private partners to provide services including financial support to landowners. Landowners interested in them should contact their state forestry agency.[14]

The Forest Legacy Program: Keeping Working Forests Working

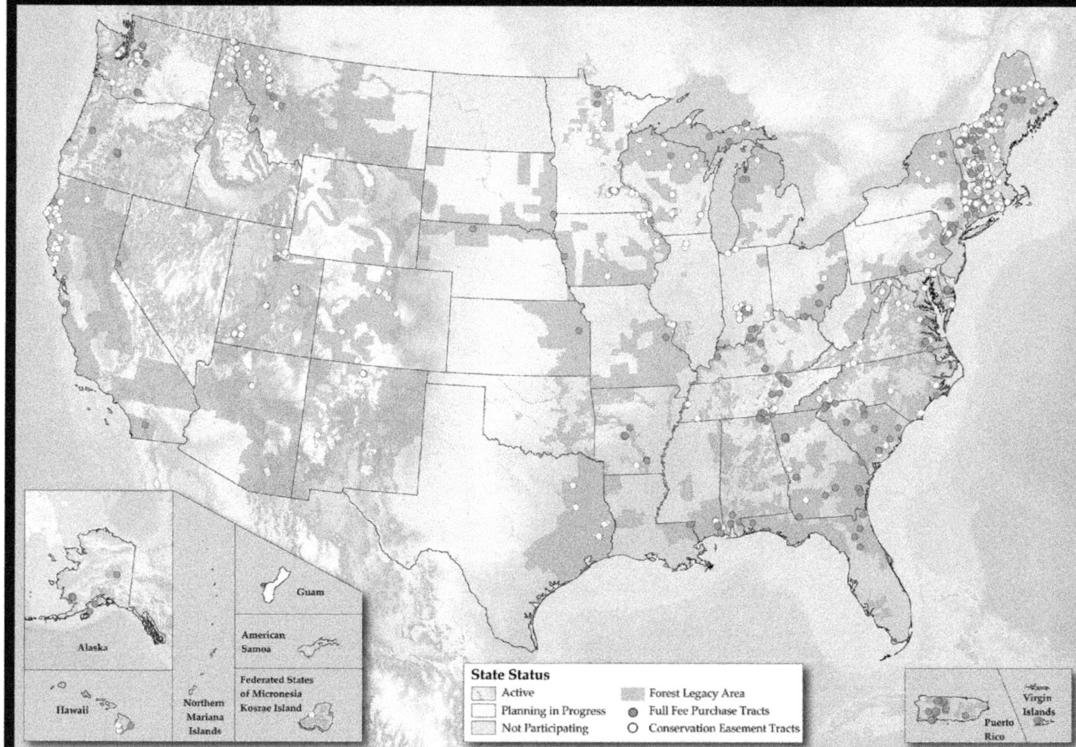

Summary of Completed Tracts as of 9/30/18

State	Assessment of Need Approval Date	Number of Tracts	Total Acres	Forest Legacy Contribution	Non-Federal Cost Share	Total Project Costs*
Alabama	March 22, 2002	8	13,819	$ 11,614,688	$ 9,349,779	$ 20,964,467
Alaska	November 1, 2002	7	4,928	$ 2,841,451	$ 2,845,144	$ 5,686,595
American Samoa	January 23, 2004	0	-	-	-	-
Arizona	August 5, 2005	6	1,242	$ 1,932,679	$ 778,450	$ 2,711,129
Arkansas	February 15, 2006	5	19,853	$ 9,970,500	$ 14,521,887	$ 24,492,387
California	January 22, 1996	26	111,979	$ 23,107,718	$ 75,870,685	$ 98,978,403
Colorado	March 22, 2002	8	21,000	$ 22,611,000	$ 16,465,001	$ 39,076,001
Connecticut	October 26, 1994	31	8,234	$ 8,610,820	$ 12,011,028	$ 20,621,848
Delaware	December 10, 1998	12	3,336	$ 14,925,000	$ 10,571,903	$ 25,496,903
Federated States of Micronesia	August 20, 2010	1	78	$ 390,000	$ 167,928	$ 557,928
Florida	April 11, 2005	5	11,333	$ 12,749,286	$ 30,911,980	$ 43,661,266
Georgia	March 22, 2002	25	72,051	$ 41,469,502	$ 110,723,033	$ 152,192,535
Hawaii	November 29, 1994	8	47,055	$ 14,753,000	$ 47,966,985	$ 62,719,985
Idaho	November 1, 2002	32	96,624	$ 37,856,855	$ 21,574,359	$ 59,431,214
Illinois	November 29, 1994	10	558	$ 2,516,838	$ 912,946	$ 3,429,784
Indiana	December 10, 1998	25	8,485	$ 9,910,450	$ 3,255,476	$ 13,165,926
Iowa	March 7, 2002	11	1,986	$ 1,684,681	$ 1,222,263	$ 2,906,944
Kansas	August 20, 2010	1	166	$ 499,000	$ 757,264	$ 1,256,264
Kentucky	December 3, 2003	9	10,220	$ 13,185,636	$ 12,755,472	$ 25,941,108
Louisiana	November 18, 2008	1	182	$ 341,250	$ 113,750	$ 455,000
Maine	March 18, 1994	37	741,430	$ 76,061,534	$ 70,541,109	$ 146,602,643
Maryland	January 22, 1996	9	2,014	$ 4,613,000	$ 3,231,164	$ 7,844,164
Massachusetts	August 5, 1993	116	17,484	$ 30,229,171	$ 19,965,454	$ 50,194,625
Michigan	February 20, 2004	13	155,966	$ 15,932,048	$ 22,547,039	$ 38,479,087
Minnesota	February 29, 2000	34	145,665	$ 15,950,950	$ 21,492,272	$ 37,443,222
Mississippi	December 3, 2007	9	4,782	$ 5,160,000	$ 1,973,780	$ 7,133,780
Missouri	August 5, 2005	2	583	$ 2,000,000	$ 2,760,180	$ 4,760,180
Montana	February 29, 2000	39	238,467	$ 77,405,533	$ 73,803,793	$ 151,209,326
Nebraska	March 30, 2004	1	330	$ 382,904	$ 260,000	$ 642,904
Nevada	April 11, 2005	1	111	$ 438,000	$ 147,000	$ 585,000
New Hampshire	February 11, 1994	56	273,252	$ 50,379,123	$ 24,969,846	$ 75,348,969
New Jersey	October 26, 1994	15	7,003	$ 17,239,477	$ 33,371,301	$ 50,610,778
New Mexico	March 22, 2002	5	16,787	$ 7,228,044	$ 5,238,300	$ 12,466,344
New York	October 26, 1994	13	135,820	$ 12,060,400	$ 22,204,424	$ 34,264,824
North Carolina	February 29, 2000	16	11,463	$ 19,206,269	$ 22,746,965	$ 41,953,234
Ohio	August 5, 2005	5	8,468	$ 6,628,000	$ 3,078,131	$ 9,706,131
Oklahoma	August 20, 2010	0	-	-	-	-
Oregon	March 22, 2002	6	21,863	$ 4,759,000	$ 4,518,926	$ 9,277,926
Pennsylvania	March 7, 2002	5	5,544	$ 8,015,833	$ 7,140,000	$ 15,155,833
Puerto Rico	October 30, 1997	19	2,857	$ 2,387,400	$ 2,393,958	$ 4,781,358
Rhode Island	December 30, 1993	22	3,583	$ 8,107,900	$ 14,387,995	$ 22,495,895
South Carolina	February 7, 2000	15	77,994	$ 39,377,217	$ 94,508,133	$ 133,885,350
South Dakota	November 2, 2009	3	249	$ 1,688,000	$ 1,012,000	$ 2,700,000
Tennessee	February 7, 2000	27	53,315	$ 32,931,498	$ 23,301,425	$ 56,232,923
Texas	December 29, 2004	6	30,771	$ 9,895,875	$ 3,683,625	$ 13,579,500
Utah	February 27, 1997	38	74,737	$ 30,806,613	$ 36,121,438	$ 66,928,051
Vermont	February 11, 1994	75	88,679	$ 28,936,235	$ 15,167,734	$ 44,103,969
Virgin Islands	June 20, 2003	4	213	$ 2,118,000	$ 706,000	$ 2,824,000
Virginia	January 16, 2001	23	18,526	$ 15,548,750	$ 9,105,299	$ 24,654,049
Washington	October 26, 1994	52	68,947	$ 45,951,179	$ 52,440,967	$ 98,392,146
West Virginia	December 2, 2003	7	4,608	$ 3,990,593	$ 2,817,531	$ 6,808,124
Wisconsin	January 16, 2001	30	117,909	$ 28,373,405	$ 26,311,590	$ 54,684,995
Wyoming	November 2, 2009	1	237	$ 3,000,000	$ 7,650,000	$ 10,650,000
TOTAL		935	2,762,786	$ 837,772,305	$ 1,002,372,712	$ 1,840,145,017

Figure 10.3. The U.S. Forest Service's Forest Legacy program has completed 935 projects on over 2.7 million acres of private land.
Courtesy of the U.S. Forest Service.

The Forest Legacy Program, established in 1990, uses Land and Water Conservation Fund money to protect forest landscapes, including streams and wetlands, from conversion to other uses. The Forest Service partners with state agencies, NGOs such as The Nature Conservancy and The Conservation Fund, and private landowners to purchase titles to or easements on threatened land.[15] Although there is no required federal/non-federal matching ratio, state and private contributions have allowed recent acquisitions to use anywhere from 50 percent to 70 percent non-federal dollars in totally voluntary contributions.[16] The program is currently funded at $62 million per year and has funded the protection of over 2.7 million acres since 1990.[17] One example of a beneficiary of the Forest Legacy Program is the Coastal Headwaters Longleaf Forest Conservation and Restoration Project, discussed in chapter 9, which has received Forest Legacy grants in both Alabama and Georgia.[18]

Two smaller, newer programs are the Community Forest and Open Space Conservation Program (CFOSC) and the Landscape Scale Restoration Program (LSR). CFOSC, operational since 2012, acquires land for community forests, publicly accessible open spaces often located near water features and other public amenities. With initial funding of $2 million per year, increased to $4 million in 2018, CFOSC has funded thirty-seven projects and created 16,410 acres of community forests that conserve wildlife habitat and provide public benefits.[19] LSR began as an agency initiative in 2008 and provides competitive grants for restoration of forest landscapes and "addresses large scale issues such as improving fish and wildlife habitats, including threatened and endangered species."[20] Grants are made to state agencies and other organizations, which then work with multiple private landowners to improve wildlife habitat. LSR gained its own appropriation in 2014 and was codified by the forestry title of the 2018 Farm Bill.[21] Over the last decade, the program has supported enhancing millions of acres of wildlife habitat.[22] Altogether, Forest Legacy, CFOSC, and LSR have contributed to the protection and sustainable management of more than twenty-seven million acres.[23]

The Forest Service provides technical assistance to private forest landowners through its Forest Stewardship Program (FSP). FSP is implemented through state forestry agencies and other partners. In 2018 alone, state forestry agencies provided technical assistance to 147,801 private landowners and educated nearly 319,000 landowners through field days and other events.[24] FSP is marketed and implemented by partners including state forestry agencies, conservation districts, university cooperative extension, land trusts, the American Tree Farm System, The Nature Conservancy, and other non-profits.[25] FSP can directly provide forest landowners with technical assistance, and it can also connect landowners to other programs, including the Natural Resources

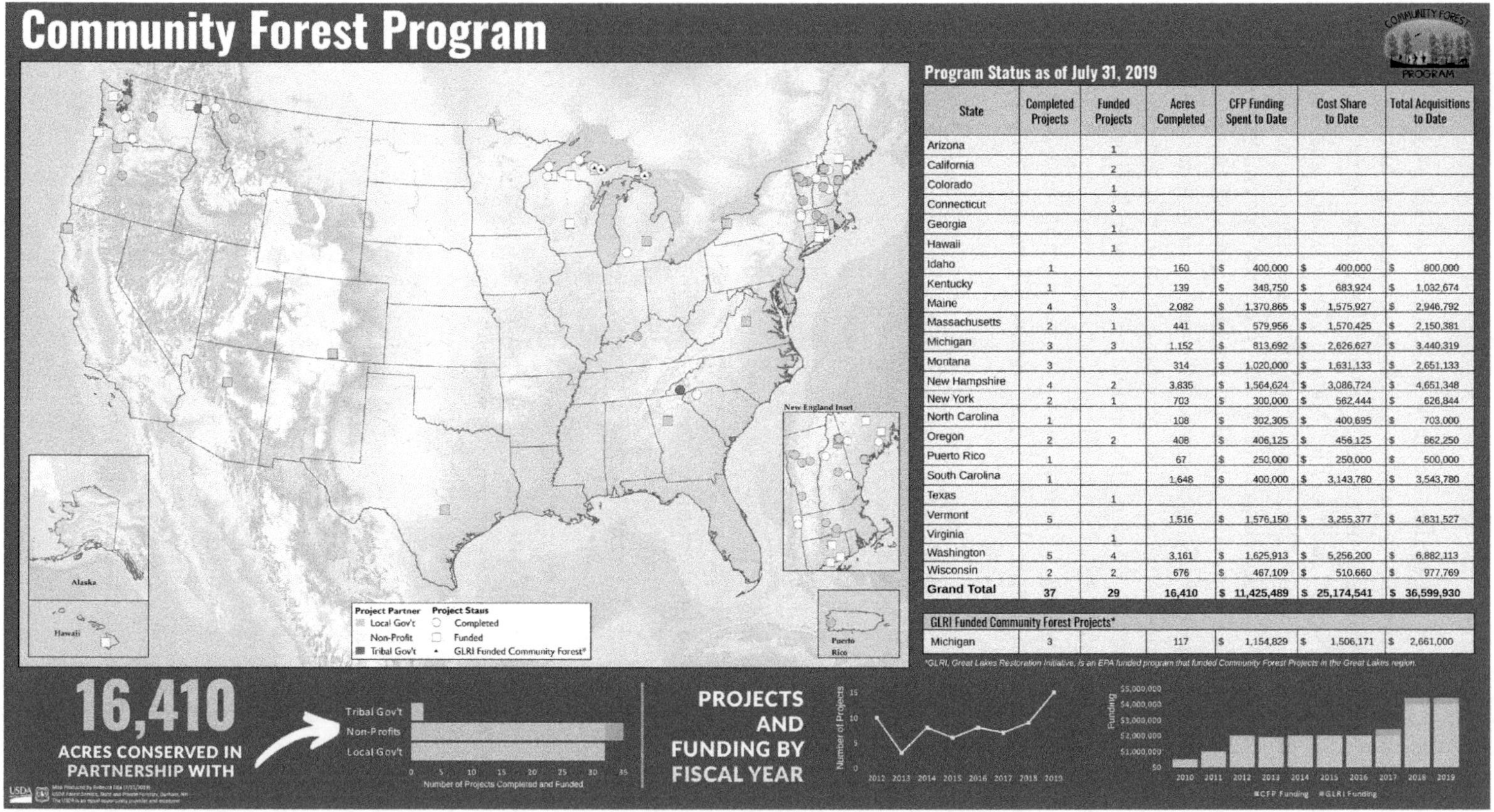

Figure 10.4. The U.S. Forest Service's Community Forest and Open Space Conservation program has competed 27 projects and created over 16,000 acres of community forests.

Courtesy of the U.S. Forest Service.

Figure 10.5. State forestry agencies, NGOs, trade associations, and the U.S. Forest Service all assist private forest landowners in developing and implementing forest stewardship practices.
Photograph by Leslie Robertson/Courtesy of the National Association of State Foresters.

Conservation Service (NRCS) and state forest conservation programs.[26] As of 2019, over 25.8 million acres of private forest lands are managed under plans developed through FSP, and more than forty-one hundred species of at-risk fish and wildlife have benefited from FSP.[27]

Department of Defense Programs

Wildlife conservation may not be first thing that comes to mind when thinking about the U.S. Department of Defense (DoD), but the DoD is one of the largest conservation organizations in the world, managing over twenty-five million acres of land on 425 major military installations. These installations are home to 220 different species listed under the Endangered Species Act of 1973 (ESA), and more than two hundred of these installations host one or more listed or candidate species under the ESA. The DoD has practiced conservation on its installations since 1949, but in the late 1980s and early 1990s its work began to spill over onto adjacent private lands.[28] At that time, off-base development began to impact critical military training operations. Threats included urban development, air pollution, and even light pollution encroaching upon DoD installations.[29] The juxtaposition of off-installation development with on-installation conservation projects further created situations where DoD installations become more and more important to at-risk species.[30] This created an ESA problem for DoD at the start of the 1990s, and no species loomed larger than the red-cockaded woodpecker.

A 1990 FWS finding that operations at Fort Bragg, North Carolina, jeopardized the continued existence of the red-cocked woodpecker virtually shut down training exercises at the base for two years.[31] Then, in early 1992 three civilian DoD employees at Fort Benning, Georgia, were issued criminal indictments for destruction of red-cockaded woodpecker habitat and threatened with prison sentences of up to thirty-six years and fines of up to $650,000.[32] These events prompted a 1992 conference at Fort Bragg, which led in turn to a 1995 plan to conserve the woodpecker on private land around the base.[33] The DoD began to purchase conservation easements on private land around the base. The success of this effort ultimately led to creation of the Readiness and Environmental Protection Integration (REPI) program within the Office of the Secretary of Defense, established by the 2002 National Defense Authorization Act. REPI purchases land or easements around DoD installations from willing landowners to create buffers that limit development incompatible with the mission of the military installation or to preserve wildlife habitat and so prevent future interference with the mission of the installation. REPI focuses primarily on ESA-listed species, because they are the most likely to cause conflicts with military operations.[34]

As of 2019, there are REPI programs on 106 DoD installations in thirty-three states. Figure 10.6 provides a map of active REPI programs. From 2002 to 2018, REPI executed 1,922 transactions with landowners, protecting 586,665 acres of habitat at a cost of over $1.6 billion. Roughly $500 million of this was invested by the DoD, with FWS contributing $350 million and

Figure 10.6. Map of U.S. Department of Defense installations with active Readiness and Environmental Protection Integration (REPI) programs. Individual REPI programs may focus on preserving wildlife and habitat, reducing incompatible off-installation development, or both.

Courtesy of the Office of the Secretary of Defense–REPI Program Office and Texas A&M Natural Resources Institute.

partner organizations $788 million.[35] REPI is implemented through partnership agreements with state fish and wildlife agencies and private conservation organizations, and the federal government shares the cost of acquiring land or easements with these groups. Land and easements are transferred to federal ownership to the minimum extent necessary, with a preference for the land or easements to remain in the hands of the partner organization.[36] Partner organizations, not the DoD, have responsibility for identifying land for the program and working with landowners to establish easements.[37] Therefore, landowners near a DoD installation with an active REPI program who are interested in participating should contact their local land trust or other conservation organization.

REPI has been an unqualified success for the DoD and for wildlife, and as a result the military began to look for conservation opportunities at a larger,

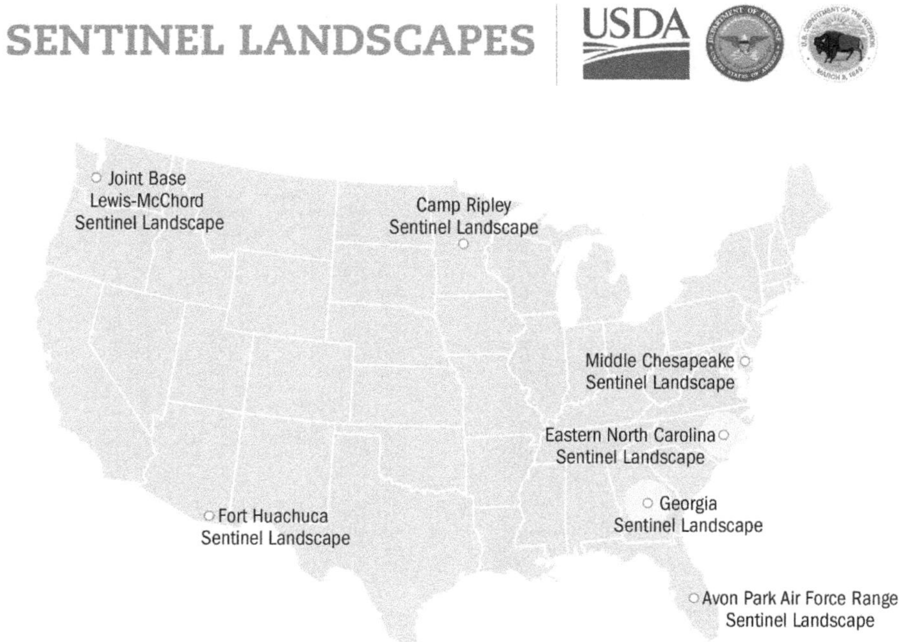

Figure 10.7. Map of Sentinel Landscapes designated by the U.S. Department of Defense, Department of Agriculture, and Department of the Interior.
Courtesy of the Office of the Secretary of Defense–REPI Program Office and Texas A&M Natural Resources Institute.

landscape scale. This led to establishment of the Sentinel Landscapes program in 2013.[38] Sentinel Landscapes is a partnership between the DoD, the U.S. Department of Agriculture (USDA), and the Department of the Interior, with each contributing resources, expertise, and access to existing conservation programs.[39] Designation as a Sentinel Landscape makes additional resources available for conservation within that landscape and can give that landscape priority under other federal conservation programs.[40] As of 2019, seven Sentinel Landscapes have been designated, including locations around the country.[41] Figure 10.7 shows the locations of existing Sentinel Landscapes.

Sentinel Landscapes are working landscapes, and the program provides economic benefits as well as conservation and military readiness. Sentinel Landscapes must feature:

1. An anchor military installation with a military mission that benefits from compatible land uses outside of the installation's boundaries;
2. A defined landscape associated with the anchor installation where federal, state, local, and private programs and efforts can

be coordinated to support voluntary conservation and landowner involvement; and

3. Articulated goals and outcomes that promote and sustain compatible land uses for military operations while providing tangible benefits to conservation and working lands within the defined landscape.[42]

Sentinel Landscapes provide expanded options for conservation around DoD installations, coordination between multiple programs, and participation by landowners, which adds up to greater benefits for at-risk species. Designation of a Sentinel Landscapes prompts coordination between other programs and makes resources available for conservation, but it carries no dedicated funding of its own.[43] The DoD does not directly engage landowners within Sentinel Landscapes. Landowners within a Sentinel Landscape interested in conservation should instead contact their local USDA Service Center, FWS field office, or other conservation organizations.

National Oceanic and Atmospheric Administration Coastal Grants Programs

In addition to managing marine and anadromous species under the ESA, the National Oceanic and Atmospheric Administration (NOAA) regulates commercial fishing, manages the National Weather Service, coordinates disaster preparedness, and carries out research on marine resources, atmospheric conditions and phenomena, coasts, wetlands, and more. NOAA's responsibility includes implementing the Coastal Zone Management Act and conserving coastal ecosystems, which provide substantial economic and climate resiliency benefits. Two grant programs that help accomplish this are the Coastal and Estuarine Land Conservation Program (CELCP) and the Coastal Resilience Grants Program (CRGP). Both make grants to state and local governments, and CRGP is also available to non-profit organizations and higher education institutions. Landowners in coastal areas interested in these programs should contact state, county, and municipal government agencies and NGOs.

CELCP, established in 2002, purchases easements to protect coastal ecosystems, wetlands, corals, shorelines, dunes, barrier islands, water quality, historic and cultural sites, scenic vistas, and public access, and provide human benefits through storm risk mitigation. All CELCP grants must be matched by state and local funds at a one-to-one ratio, and all land purchased under CELCP must remain in public ownership. Participating states must have an established CELCP plan approved by NOAA.[44] One area where CELCP has devoted substantial resources is the Great Lakes region, including Indiana, Michigan,

Figure 10.8. An example of a coastal wetland in Louisiana.
Roschetzky/iStock by Getty Images.

Minnesota, New York, Ohio, and Wisconsin. Since 2002, twenty-six projects have been located along the Great Lakes, representing an investment of almost $42 million in federal funds. These grants have protected coastlines, conserved vulnerable wetlands, and protected storm buffers from the threat of development.[45] Nationwide, from 2002 to 2017 CELCP invested $280 million in twenty-nine states, with more than one hundred thousand acres of coastal lands protected.[46]

CRGP has a narrower mission of improving coastal resilience to natural disasters and climate change, regardless of land ownership.[47] CRGP funds projects that improve coastal resilience to natural disasters, erosion, and flooding. Often, these projects also benefit wildlife by improving habitat or restoring fish passage through former dam sites. Between 2015 and 2018, CRGP funded forty-eight projects with $35.8 million in federal funds and $22.3 million in matching funds. Many Coastal Resilience Grants focus on funding to develop strategic conservation plans, including in California, Georgia, Massachusetts, Oregon, Washington, and Wisconsin. Other grants focus on implementation of conservation plans, stream and wetland restoration, wildlife habitat, and dam removals.[48]

Figure 10.9. The Elm Street Dam on the Jones River in Massachusetts, built in 1920, was removed in 2019 due to the efforts of the Jones River Watershed Association and other partners. Funding for the dam removal was provided in part by a NOAA Coastal Resilience Grant.
Courtesy of Jimmy Powell/Jones River Watershed Association.

U.S. Fish and Wildlife Service Coastal Program

FWS funds conservation work in coastal regions through its Coastal Program, established in 1985. The program works to conserve wildlife habitat on the Atlantic and Pacific Oceans, Gulf of Mexico, Great Lakes, and in the Caribbean by providing technical and financial assistance for habitat conservation planning, restoration, and protection.[49] Financial assistance comes mostly in the form of grants of up to $1 million to state and territorial governments, which are funded by the Sport Fish Restoration and Boating Trust Fund.[50] In 2019, for example, over $20 million was awarded to eleven states, funding projects that ranged from acquiring barrier islands in North Carolina to restoring coastal wetlands in California.[51] The program's technical support work assists grant recipients, the Estuary Restoration Act Council, the National Marine Protected Areas Center, and other partnerships.[52]

A large part of the Coastal Program's work is promoting conservation partnerships, including intergovernmental, public-private, and multi-stakeholder efforts to conserve coastal ecosystems at the landscape-level scale. Focal areas include climate change resiliency, habitat connectivity and ecological integrity, and conservation of federal trust and other priority species.[53] The program is guided by a national Strategic Plan and seven Regional Work Plans, all of which focus on conserving habitat by developing and strengthening partnerships.[54] Over the years, it has worked with over sixty-four hundred partners to protect over 2.1 million acres of habitat, restore over 550,000 acres of habitat and twenty-six hundred miles of streams, and contributed to the down- or delisting of twenty species under the ESA. Federal funding for the Coastal Program is matched by other stakeholders at a ratio of eight to one, resulting in a cumulative investment of over $1.6 billion and supporting twenty-seven hundred businesses and forty-one hundred jobs each year.[55] Information about current Coastal Program projects is available from program staff.[56]

Environmental Protection Agency Grants

The Environmental Protection Agency (EPA) operates a number of grant programs. Information about these programs, including a searchable database of grants, is available at the EPA's grants website, https://www.epa.gov/grants.[57] Many EPA programs, especially water quality programs, provide significant wildlife benefits that are not always recognized by the wildlife conservation community, including benefits to at-risk species.[58] For example, the EPA's 319 Program, referring to Section 319 of the Clean Water Act, provides grants to states, territories, and tribes to improve water quality on private land.[59] The program is intended to address non-point pollution sources such as agricultural runoff and erosion, and it often targets projects in upland and riparian areas that provide important wildlife habitat.[60] From 1990 to 2017, the EPA awarded over $4.3 billion in 319 program grants.[61] Other water quality programs with significant wildlife benefits include the Clean Water Act State Revolving Fund and the Safe Drinking Water Act State Revolving Fund. Since 1987, these funds have generated investments totaling $155 billion, mostly through low-cost financing, to protect waterways and drinking water supplies and to improve water quality. Although the wildlife-specific impacts of this staggering investment cannot be quantified, many state government agency projects supported by these funds have created or restored green infrastructure that provides clean water for fish and wildlife as well as for humans.[62] Other EPA programs have wildlife benefits as well. Wetland Program Development Grants build capacity for state, tribes, and local governments to operate their own wetlands conservation programs, whereas Multipurpose Grants to States and Tribes advance environmental objectives from wetlands and water quality to air quality or climate change adaptation. The EPA is also an important partner for DoD's REPI program.

Figure 10.10. Green infrastructure projects include the preservation and restoration of natural riverbanks and wetlands that act as buffers between water resources and terrestrial sources of soil erosion and pollution. This wetland is part of The Conservation Fund's Greenseams program, operated in partnership with the Milwaukee Metropolitan Sewerage District to protect water resources within the Milwaukee metropolitan area.
Photograph by Ivan LaBianca.

Emergency Conservation Funding Through the U.S. Department of Agriculture

Natural disasters impact wildlife as well as humans, and disaster relief funding can be a valuable resource for conservation. Emergency conservation funding may be available through established USDA programs and also one-time recovery packages passed by Congress following natural disasters. Under authority of the Agricultural Credit Act of 1978, USDA operates three disaster relief programs for agricultural producers. Information about all of them is available at USDA Service Centers. In addition, the Environmental Quality Incentives Program can provide assistance in recovering from natural disasters, even though it is not designed to be an emergency response program.[63]

The Emergency Conservation Program, established in 1978 and operated by the Farm Service Agency (FSA), provides payments to "agricultural producers who carry out emergency measures to control wind erosion on farmlands or to rehabilitate farmlands damaged by wind erosion, floods, hurricanes, tornadoes, or other natural disasters."[64] FSA ordinarily provides 75 percent of the cost of these measures, but select classes of landowners, such as socially disadvantaged or beginning farmers and ranchers, may receive 90 percent funding, whereas projects for the repair or replacement of fencing are limited to 25 percent federal funding. In all cases, payments are limited to a total of 50 percent of the value of the land.[65] The 2008 Farm Bill created a second FSA program, the Emergency Forest Restoration Program. This is a variation on the Emergency Conservation Program that makes "payments to an owner of nonindustrial private forest land who carries out emergency measures to restore the land after the land is damaged by a natural disaster."[66]

The third program, the Emergency Watershed Program, was also established in 1978 but is administered by NRCS. This program provides for "emergency measures for runoff retardation and soil-erosion prevention . . . to safeguard lives and property from floods, drought, and the products of erosion on any watershed whenever fire, flood, or any other natural occurrence is causing or has caused a sudden impairment of that watershed."[67] In addition to a recovery program where NRCS will cover 75 percent of producers' recovery costs, the program also authorizes NRCS to purchase Floodplain Easements, which are permanent conservation easements.[68] NRCS then bears the cost of restoring the land under the easement. Easements may be placed on open lands or on land developed for residential use, in which case NRCS will destroy existing structures. To qualify, land must have been damaged by flooding at least once in the previous year or twice in the previous ten years. Land can also qualify if it would be subject to inundation by a dam breach or if it contributes to restoration of floodwater storage and flow, offers a way to control erosion, or improves the practical management of another Floodplain Easement. Landowners retain the right to quiet enjoyment of the land, the right to control public access, and the right to undeveloped recreational uses of the land such as hunting and fishing, and on a case-by-case basis may continue other compatible uses such as timber harvest, haying, or grazing. In general, Floodplain Easements are not a way for landowners to preserve a working family farm or ranch, but rather a way to minimize their losses when flooding deprives their land of value.[69]

Conservation can also be funded on a case-by-case basis following major natural disasters. For example, following Hurricane Katrina and the other storms of the 2005 hurricane season, Congress passed a disaster relief package that included a supplemental appropriation of $199,800,000 for the Agricultural Credit Act of 1978 programs discussed previously.[70] In addition, it created the Emergency Forestry Conservation Reserve Program. The program provided $404.1 million for a one-time enrollment of private non-industrial land

Figure 10.11. Forest in Hancock County, Mississippi, destroyed by Hurricane Katrina in 2005. Photograph by James L. Young.

into ten-year conservation contracts. Qualifying land was limited to the 261 counties that suffered from declared disasters caused by 2005 hurricanes and must have suffered at least a 35 percent loss of merchantable timber. Contracts provided 50 percent cost sharing for cleanup and restoration of tree stands, with the requirement that the landowner prepare, with assistance from FSA, a conservation plan addressing soil conservation, water quality improvements, and wildlife habitat restoration as well as tree stand restoration.[71]

Following Hurricane Sandy in 2012, Congress passed a relief package that included a supplemental appropriation of $218 million for the Agricultural Credit Act of 1978 programs discussed previously.[72] It also appropriated $829.2 million for the Department of the Interior to repair and rebuild coastal assets and promote coastal resiliency. One hundred million dollars of this money was given to the National Fish and Wildlife Foundation (NFWF), which created the Hurricane Sandy Coastal Resiliency Competitive Grant Program. As with other NFWF programs (discussed later in this chapter), grants were awarded to state and local governments, tribes, non-profits, and universities. This funded fifty-four projects and was combined with $3.1 million in private funding and $55 million in matching funds from grantees for a total conservation investment of over $158 million. Projects addressed "coastal storm surge, wave velocity, and sea-level rise," "strengthen[ed] the ecological integrity and functionality of [affected] ecosystems," and "enhance[ed] understanding of the impacts of storm events and identified cost-effective resilience tools to help mitigate the effects of future storms." The program ultimately restored "more than 6,600 acres of marsh and wetland, more than 225 acres of beach and dune, and over 200 miles of river and stream connectivity," while building "green infrastructure to reduce over 182 million gallons of stormwater runoff." Projects benefited twelve states and the District of Columbia, although most were located in New Jersey, New York, and Rhode Island.[73]

Partners for Fish and Wildlife

Partners for Fish and Wildlife (Partners) is the leading FWS program dedicated to providing voluntary habitat restoration and enhancement projects for ESA-listed and other federal trust species on private land. Private land is defined as any land other than state- or federal-owned property. Federal trust species are "migratory birds, threatened species, endangered species, interjurisdictional fish, marine mammals, and other species of concern."[74] The philosophy for Partners began in the mid-1970s in Minnesota as a pilot program focused on wetlands and nesting cover for declining waterfowl.[75] In 1987 the program was formally recognized; it expanded into a nationwide partnership with FSA to provide support for wetland conservation on Conservation Reserve Program–enrolled lands.[76] The initiative soon received its first $100,000 appropriation and began to expand into grass restoration, upland habitat conservation, stream restoration, fish habitat, and work dedicated to endangered species conservation.[77] In 2006, Congress gave the program its own legislative mandate, the Partners for Fish and Wildlife Act.[78] The Partners for Fish and Wildlife Act was reauthorized in March 2019 as part of the passage of the John D. Dingell, Jr. Conservation, Management, and Recreation Act.[79] Today, Partners' mission is to provide habitat improvement, habitat restoration, habitat enhancement, and habitat establishment for federal trust species on private land; the program has grown to approximately $52 million per year, which funds a combination of technical and cost-sharing assistance.[80] Partners' goals are to "[p]romote and implement habitat improvement projects that benefit Federal Trust Species; [p]rovide conservation leadership and promote partnerships; [e]ncourage public understanding and participation; and [w]ork with U.S. Department of Agriculture . . . to implement conservation programs."[81] Providing "technical assistance in the development, implementation, and evaluation" of Farm Bill programs remains a major part of this work.[82]

Since 1987, Partners has engaged over fifty thousand private landowners and over sixty-six hundred partner groups, leading to the voluntary restoration of over 6.4 million acres of land, including over 4.5 million acres of upland habitat and over 1.5 million wetland acres. Partners' work has also resulted in enhancement of over 13,500 miles of stream habitat.[83] Partners has provided extensive support for efforts to conserve grassland species such as the lesser prairie chicken and greater sage-grouse, coordinating its work with NRCS' Lesser Prairie Chicken Initiative and Sage Grouse Initiative. In 2017, for example, Partners supported the conservation of over thirty thousand acres of sage-grouse habitat.[84] Examples of other recent projects include work on prairie management in Kansas, bull trout recovery in Montana, grassland restoration in New Mexico, and pollinator habitat across the nation.[85] Landowners interested in working with Partners should contact the Partners for Fish and Wildlife Coordinator in their local FWS state office. Contact information is available at https://www.fws.gov/partners/contactUs.html.[86]

Bureau of Land Management Programs

The Bureau of Land Management (BLM) manages 245.6 million acres of public land, overwhelmingly in the West, using a multiple-use approach under the Federal Land Policy and Management Act of 1976.[87] Common land uses include oil and gas operations, other energy development, mining, ranching, logging, and recreation. BLM has historically focused its wildlife management activities on its own land, providing relatively few resources for private landowners. However, under the 1996 Wyden Amendment, BLM has the authority to fund work on private land if that work benefits resources on public lands within the same watershed.[88] Such resources include water quality and air quality, wildlife habitat and connectivity, and public access for hunting and recreation. Between 2008 and 2017, BLM funds for private land work were centrally allocated through the agency's Healthy Lands Initiative. Since 2017, authority over these projects has been delegated to state and district offices, meaning that the number and types of projects funded can vary widely from place to place and year to year. Private landowners interested in applying for funds from BLM should contact their local office.[89] A directory is available at https://www.blm.gov/office/national-office.[90]

BLM also works with private landowners that are also BLM permit holders and lessees in order to address wildlife conservation issues across the fence line, on BLM and adjacent private land.[91] As BLM permittees, these landowners should already be connected to staff at their local office. Another resource is BLM's national Threatened and Endangered Species program,

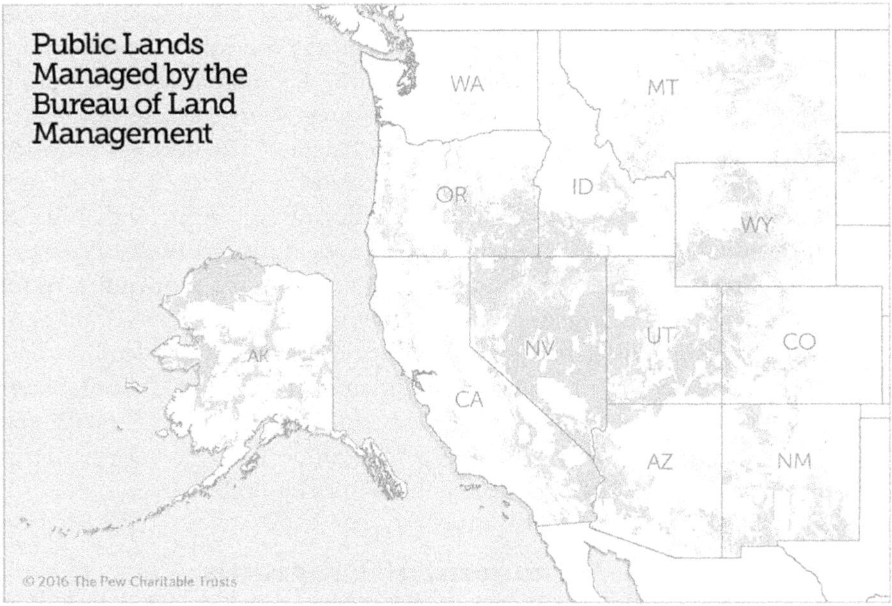

Figure 10.12. The Bureau of Land Management's land is overwhelmingly in the Western United States due to the land acquisition, settlement, and disposition patterns discussed in Chapter 1. Courtesy of The Pew Charitable Trusts.

online at https://www.blm.gov/programs/fish-and-wildlife/threatened-and-endangered.[92] BLM has been working in recent years to increase its wildlife conservation outreach, for example by participating in the Partners in the Sage coalition for greater sage-grouse.[93] As the agency's involvement in wildlife conservation partnerships continues to grow, more BLM resources will hopefully become available to private landowners in the future.[94]

U.S. Fish and Wildlife Service Funding for State Fish and Wildlife Agencies

State fish and wildlife agencies can be extremely valuable resources for private landowners interested in conservation. These agencies are funded by a combination of license and tag fees paid by hunters and anglers, federal funding, state appropriations and bonds, and grants. Federal funding comes largely from FWS, which operates several programs dedicated to funding state fish and wildlife agencies, including the Wildlife and Sport Fish Restoration Funds, the State Wildlife Grants (SWGs) program, and the Cooperative Endangered Species Fund. Although these programs do not directly fund private landowners, they fund many state programs that ultimately benefit private landowners, including funding conservation on private property and providing technical assistance. Landowners interested in these programs should contact their state fish and wildlife agency. Information about state programs for private landowners can be found in chapter 11, and a directory of state agencies is included in appendix A.

The Wildlife and Sport Fish Restoration Funds

The Pittman-Robertson Wildlife Restoration Act of 1937 and the Dingell-Johnson Sport Fish Restoration Act of 1950 are two bedrocks of American wildlife conservation. These laws, originally lobbied for and still strongly supported by hunters and anglers, levy federal excise taxes on equipment used in hunting and fishing. A series of amendments over the years, most notably the 1984 Wallop-Breaux amendment to the Dingell-Johnson Act,[95] expanded the scope of these laws, with each amendment a reaffirmation of the commitment of hunters and anglers to conservation. Today, taxes are levied on firearms, ammunition, bows, quivers, arrow tips, fishing tackle, motorboat and small engine fuel, outboard boat motors, boats less than twenty-five feet (excluding sail boats, kayaks, and hydroplanes), imported yachts and pleasure craft, and other boating equipment. Any American who purchases one of these products is paying for wildlife conservation.[96] Under Pittman-Robertson, which benefits wildlife, $12.2 billion was disbursed to state agencies between 1939 and 2019, including $673 million in 2019.[97] Under Dingell-Johnson, which benefits fish, $9.75 billion was disbursed to state agencies between 1952 and 2019, including $365 million in 2019.[98]

These funds, collectively known as "PR/DJ" or the "Wildlife and Sport Fish Restoration Funds" (WSFR), are disbursed to state and territorial fish and wildlife agencies according to an equation that sets a minimum and maximum percentage any single state may receive. In addition, every seventy-five dollars of federal funding must be matched by twenty-five dollars of state funding.[99] WSFR funds are used on both public and private lands. A total of 85 percent of sportsmen hunt on private lands and 64 percent hunt exclusively on private lands, so private land conservation is important to state fish and wildlife agencies for game management.[100] Pittman-Robertson funds are used to "restore, enhance, and manage wildlife resources" and fund hunter education programs, whereas Dingell-Johnson funds are used to "restore, enhance, and manage sport fishery resources." Although most WSFR money is ultimately expended on game species important to hunters and anglers, some is used to directly benefit non-game and at-risk species, for example the endangered Delmarva Peninsula fox squirrel, which was delisted due to recovery in 2015. In addition, non-game and at-risk species benefit from any wildlife

Figure 10.13. The Delmarva Peninsula fox squirrel (*Sciurus niger cinereus*) was listed as endangered in 1967, as part of the U.S. Fish and Wildlife Service's very first list of endangered species. After decades of intense conservation efforts, it was delisted in 2015.
Moose Henderson/iStock Getty Images.

conservation within their ecosystems.[101] However, the extent to which at-risk species have benefited from WSFR cannot be quantified.

WSFR also includes sub-programs that target specific resources or initiatives. For example, under the Multistate Conservation Grant Program, \$6 million of grant funding per year is allocated to wildlife and sport fish restoration projects that affect states on a regional or national, rather than individual, level.[102] Another program, the National Coastal Wetlands Conservation Grant program, provides grants of up to \$1 million each "to protect, restore and enhance coastal wetland ecosystems and associated uplands." In 2019, twenty-two projects were funded in eleven states, representing a WSFR investment of over \$20 million and matching funds from state and local governments, private landowners, conservation groups, and other partners of \$26.7 million.[103] As with the main WSFR program, these sub-programs provide funding to state fish and wildlife agencies and the funds ultimately reach local landowners through their participation in the programs operated by state agencies. Information on state fish and wildlife agency programs for private landowners can be found in chapter 11, and a directory of state agencies is included in appendix A.

State Wildlife Grants

A much newer federal grant program is squarely focused upon non-game species, including at-risk species. In 2001, Congress created SWGs "for the development and implementation of programs for the benefit of wildlife and their habitat, *including species that are not hunted or fished*."[104] SWGs are funded by appropriations and like WSFR are distributed to states and territories according to a set formula. SWGs must be matched with state funds at a ratio of thirty-five dollars in state funds for every sixty-five dollars in federal funds.[105] Between 2002 and 2019, the SWG program disbursed grants totaling $1.02 billion, including $52 million in 2019.[106]

In order to qualify for SWGs, states must adopt State Wildlife Action Plans (SWAPs). A SWAP is a "comprehensive wildlife conservation plan . . . that considers the broad range of the State, territory, or other jurisdiction's wildlife and associated habitats, with appropriate priority placed on those species with the greatest conservation need."[107] Collectively, SWAPs act as a national assessment of conservation needs.[108] SWAPs must be revised at least every ten years, and every state both developed an initial plan by 2005 and revised it in 2015.[109] Development and revision of SWAPs may itself be funded by SWGs, and these planning grants are subject to a federal/state matching ratio of seventy-five to twenty-five.[110]

SWAPs are developed based on criteria jointly established by FWS and the Association of Fish and Wildlife Agencies, which represents the state agencies of all fifty states and the several territories. Each SWAP identifies species of greatest conservation need within that state, and collectively the states have identified 16,420 such species.[111] SWGs have become the "primary source for States and Tribes to fund proactive actions to address the needs of declining species."[112] The benefits of SWGs have been immense. Seven species that were federally listed as threatened or endangered when SWGs were created have now been delisted after directly benefiting from SWG funds. A further nineteen species (at minimum) have been removed from the list of candidate species because of conservation efforts utilizing SWG funds. One example of this is the Arctic grayling, a Salmonid that was removed from the candidate list in 2014 due to conservation efforts led by the state of Montana.[113]

Although the Arctic grayling is secure through a substantial range in Canada, by 1991 the only significant remaining population in the United States was in the Big Hole River watershed in Montana.[114] In that year, the Montana De-

Figure 10.14. The arctic grayling (*Thymallus arcticus*), a candidate for listing under the Endangered Species Act of 1973, is recovering due to state-led voluntary conservation efforts. Photograph by Patrick Clayton/Engbretson Underwater Photography.

partment of Fish, Wildlife and Parks began working on Arctic grayling conservation in an effort to restore the species without it being listed under the ESA, and in 1994 FWS made it a candidate for listing.[115] When Montana prepared its first SWAP in 2005, the Arctic grayling was identified as a species of greatest conservation need. Subsequent state conservation efforts included habitat protection and restoration, invasive species control, fish passage construction, breeding and stocking, and reintroducing the species to other waterways—all on private lands. As a result of these and similar conservation efforts on tribal and public lands, the species' population has begun to recover.[116] Information on state fish and wildlife agency programs like this one that work with private landowners can be found in chapter 11, and a directory of state agencies is included in appendix A.

Cooperative Endangered Species Fund

Back in 1973, the ESA recognized that one way to conserve threatened, endangered, and at-risk species was to join federal funding with local expertise. Section 6 of the ESA authorizes financial assistance for state fish and wildlife agencies "to assist in development of programs for the conservation of endangered and threatened species or to assist in monitoring the status of candidate species . . . and recovered species." These Section 6 grants come from the federal appropriation known as the Cooperative Endangered Species Conservation Fund.

As with WSFR and SWG, Section 6 grants are made to state agencies but may ultimately finance projects on private land. Funding for Section 6 grants is assigned to each FWS region, and the regions then solicit project proposals. Section 6 requires twenty-five dollars in state or private funds for every seventy-five dollars in federal funds awarded, except that for multi-state projects, just ten dollars in state or private funds may match ninety dollars in federal funds. Most Section 6 grants are used for "species and habitat conservation actions on non-federal lands, including habitat acquisition, conservation planning, habitat restoration, status surveys, captive propagation and reintroduction, research, and education."[117]

Within the Section 6 grant program there are sub-programs that are subject to their own appropriations, and some of these focus on private land conservation. Sub-programs include Habitat Conservation Planning Assistance Grants for state and local governments to develop regional, multiple species Habitat Conservation Plans (HCPs), Habitat Conservation Plan Land Acquisition Grants for acquiring habitat that complements (without replacing) mitigation carried out under HCPs, and Recovery Land Acquisition Grants providing funding for habitat acquisition.[118] These sub-programs make up a large percentage of Section 6 grants. For example, in 2016 FWS made 221 traditional conservation grants totaling $16 million, seven HCP planning assistance grants totaling $5 million, ten HCP land acquisition grants totaling $12 million, and twenty recovery land acquisition grants totaling $6 million.[119] Benefiting species included various trout, sturgeon, bats, rattlesnake, songbirds, steelhead, salmon, and turtles, as well as American burying beetle, greater sage-grouse, Columbian sharp-tailed grouse, desert tortoise, peninsular bighorn sheep, Canada lynx, grizzly bear, and countless others.[120] Information on state fish and wildlife agency programs like this one that work with private landowners can be found in chapter 11, and a directory of state agencies is included in appendix A.

Figure 10.15. The American burying beetle (*Nicrophorus americanus*), listed as endangered since 1989, was proposed for downlisting to threatened in 2019.
Nature and Science/Alamy Stock Photo.

The National Fish and Wildlife Foundation

Up to this point, this chapter has focused on federal agencies that provide funding to landowners and states for wildlife conservation. This section and the two following deal with quasi-independent organizations that rely on federal appropriations for their programs but operate alongside, rather than within, the structure of government agencies.

NFWF is a 501(c)(3) non-profit organization chartered by Congress in 1984. It is funded by a combination of federal appropriations and donations from other government agencies, corporations, foundations, and individuals.[121] NFWF works with fifteen different federal agencies and more than forty-five corporations and foundations to identify and fund conservation projects in all fifty states and the U.S. territories.[122] Projects range from protecting wildlife habitat and improving water quality to restoring marine ecosystem after oil spills, preserving working land, and much more.[123] Since 1984, NFWF has invested $5.34 billion in conservation, just $1.29 billion of which came from the federal treasury, matched by $1.68 billion in non-federal funds and $2.37 billion in grantee funds.[124] Over 17,250 projects have been funded.[125]

NFWF operates through focused programs that work on a specific issue or area year after year. Examples include its Chesapeake Bay Stewardship Fund, Central Appalachia Habitat Stewardship Program, Longleaf Stewardship Fund, and Gulf Environmental Benefit Fund (GEBF). NFWF frequently partners with businesses ranging from Walmart to oil and gas companies, as well as NRCS. Walmart's work with NFWF is primarily through its Acres for America program, which is the largest public-private corporate partnership program in the country, having now contributed and preserved 1.5 million acres since 2005, and generated $774 million in funds matching their original $35 million contribution. A full list of NFWF's conservation programs, including contact information for each program, is available online at https://www.nfwf.org/whatwedo/programs/Pages/home.aspx.[126]

Individual awards are made to public and private projects of all sizes. In 2017 more than 730 projects were funded, with recipients including The Conservation Fund, Ducks Unlimited, the Wyoming Stock Growers Land Trust, the River Alliance of Wisconsin, Audubon Florida, American Rivers, and state agencies such as the Montana Department of Fish, Wildlife and Parks.[127] NFWF grants are made to government agencies at all levels, conservation districts, non-profit organizations, educational institutions, and commodity organizations such as stockgrowers associations.[128]

One example of a public-private partnership funded by NFWF is the recently created Pecos Watershed Conservation Initiative (PWCI). Launched in 2017, PWCI is NFWF's first multi-partner program, involving six oil and gas companies.[129] NFWF and the group of companies each contributed $3.5

million over three years, for a total of $7 million in funding. PWCI focuses on aquatic and grassland species and habitats within the Pecos River watershed in Texas and New Mexico, with a secondary focus on aquatic habitats upstream on the Rio Grande where some of the same species are present. Eligible projects include efforts to improve stream and riparian habitat, restore and manage grassland habitat, and provide outreach and technical assistance. For certain species, funding is also available for projects to fill information gaps. Aquatic focal species are the Pecos pupfish, Pecos gambusia, Rio Grande (western) cooter, southwestern willow flycatcher, and Texas hornshell. Grassland focal species are the chestnut-collared longspur, Sprague's pipit, pronghorn, and the desert massasauga.[130]

PWCI made its first grants in 2018, and the impact was immediate. PWCI grants, which are already half federal and half private money, must be matched at a minimum ratio of one to one. In 2018, $1.4 million in PWCI grants generated $2.3 million in matching contributions, for a total impact of $3.7 million. Grants ranged from $63,000 to $250,000, and were made to state fish and wildlife agencies, conservation districts, conservation foundations, and NGOs. Work funded included stream restoration, watershed management improvement, invasive fish barriers, water infrastructure construction, brush management, improved grassland management, and more on private lands in

Figure 10.16. The Pecos River in Texas.
Photograph by Daniel Rodriguez/Mestengo Productions.

New Mexico and Texas and BLM land in New Mexico. PWCI funding for 2019 grants increased from $1.4 million to $2.16 million, with $2.87 million in matching contributions for a total impact of $5.03 million.[131] Information on PWCI, including its history, funding, accomplishments, application process, and evaluation criteria are all available on NFWF's website.[132]

Separate from its other programs, NFWF also operates the GEBF. Financed with $2.544 billion paid by BP and Transocean as part of legal settlements following the 2010 Deepwater Horizon explosion and oil spill, the GEBF is dedicated to conserving coastal resources in the Gulf states, including Florida, Alabama, Mississippi, Louisiana, and Texas. GEBF projects have restored coastal habitats, beaches and dunes, wetlands, and barrier islands, and benefited species ranging from marine mammals and sea turtles to fish and migratory birds. Between its establishment in 2013 and 2018, the GEBF supported 114 distinct projects with $1.3 billion.[133] A continuously updated list of GEBF projects is available online at https://www.nfwf.org/gulf/Pages/gulf-projects.aspx. Each project page includes the government agency, NGO, or other entity that received the GEBF award, which landowners in the area should contact for information on becoming engaged with the project.[134]

Migratory Bird Joint Ventures

Technical support for many conservation programs comes from Migratory Bird Joint Ventures (JVs) established under the North American Waterfowl Management Plan (NAWMP), which work to develop science to support conservation programs, equip partners with that science, and build partnerships between government agencies, conservation organizations, and landowners.[135] JVs are self-directed public-private partnerships, funded by FWS under the 1989 North American Wetlands Conservation Act (NAWCA), and they either work to conserve wildlife habitat within a region or work nationwide on a specific species of waterfowl. As of 2019 there are eighteen habitat-focused and three species-focused JVs in the United States, and four additional habitat-focused JVs in Canada.[136]

The history of JVs is a story of gradual expansion, driven by the recognition that coordination is needed for multiple conservation investments to have an impact greater than the sum of their parts.[137] The NAWMP was first signed by the United States and Canada in 1986, in response to a continent-wide crash in waterfowl populations and in recognition of the fact that waterfowl populations would not be restored without conservation efforts focused on private lands.[138] JVs were intended to coordinate between public and private entities to plan, fund, and implement specific conservation projects.[139] The NAWMP has evolved considerably over the years, with passage of NAWCA in 1989, Mexico joining the NAWMP in 1994, and official amendments to the NAWMP in 1994, 1998, 2004, 2012, and 2018.[140]

As the mission of JVs has expanded, so has their geographic presence. JVs are established on as as-needed basis. The oldest, the Arctic Goose JV, was established in 1986, with the newest, the Oaks and Prairies JV, in 2008.[141] JVs now cover all of the United States except for a small strip of southern California coastline, virtually all of subarctic Canada, and most of northern Mexico.[142] The 1998 update to the NAWMP encouraged greater cooperation between waterfowl programs other migratory bird initiatives, and in response JVs expanded their scope to encompass all birds.[143] JVs now work on waterfowl, upland birds, shore birds, and migratory birds of all kinds, drawing upon resources that include Partners in Flight's Landbird Conservation Plan and the North American Bird Conservation Initiative's State of the Birds reports.[144] Some JVs are beginning to expand their focus even further and evaluate wetland health in terms of ecosystem services provided for humans as well as wildlife.[145]

Today, "JVs leverage Federal contributions with non-federal funds and in-kind contributions, and foster collaboration at the international to local scale, thereby improving the effectiveness of conservation."[146] One function of JVs

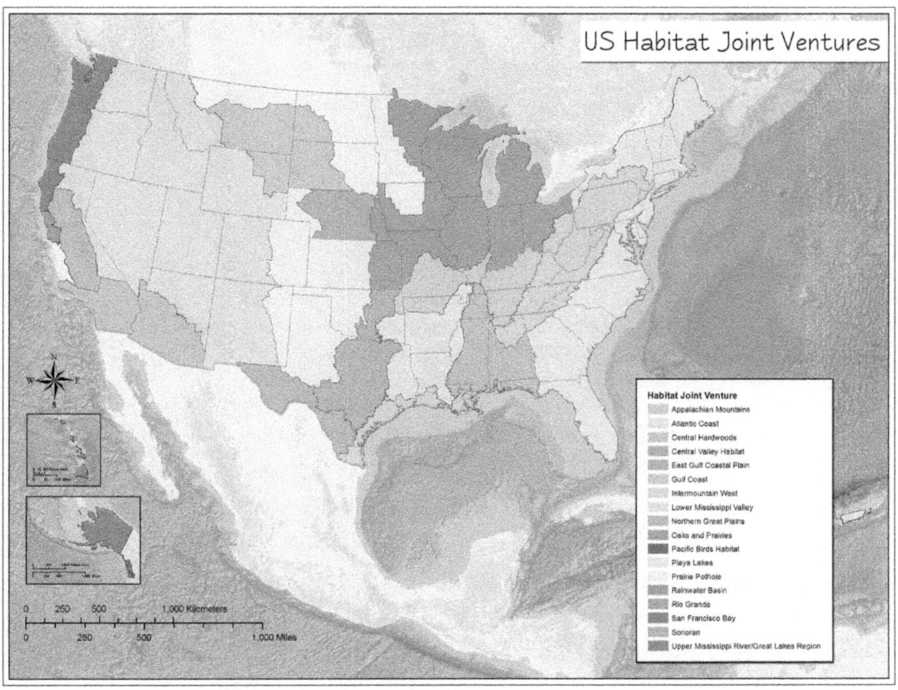

Figure 10.17. Map of U.S. Migratory Bird Joint Ventures.
Courtesy of the U.S. Fish and Wildlife Service.

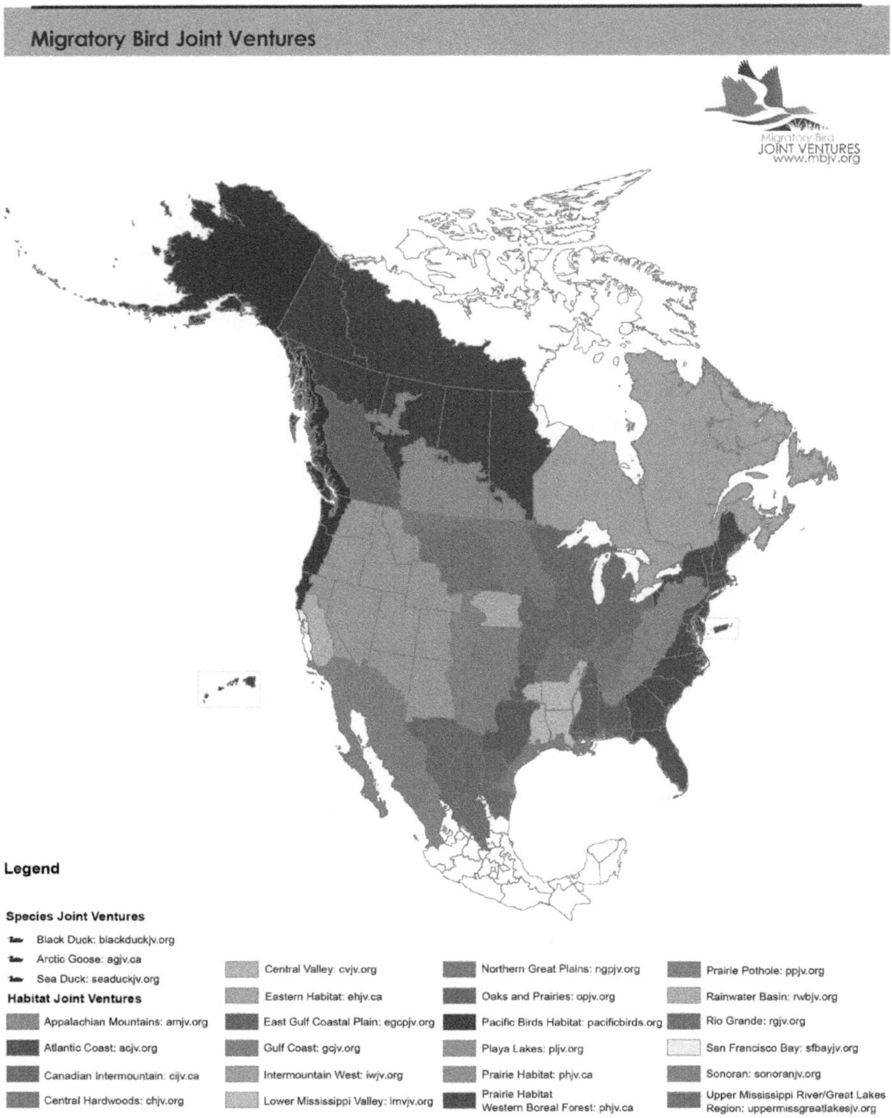

Figure 10.18. Map of North American Migratory Bird Joint Ventures.
Courtesy of the U.S. Fish and Wildlife Service.

is the application of science to rank different lands and potential projects to allow efficient targeting of Farm Bill and other conservation programs.[147] Another is the cultivation of public-private partnerships.[148] One example of such a partnership is in South Carolina, where the Ashepoo, Combahee, and Edisto rivers empty into St. Helena Sound. Since 1988, a coalition of partners led by the Atlantic Coast JV has worked to protect coastal and forested wetlands. By combining corporate donations, state government funding, NAWCA grant funding, and conservation easement donations, the partnership has protected almost three hundred thousand acres from development, including 218,000 acres of conservation easements.[149] JVs receive approximately $13 million in funding from FWS each year,[150] and for every federal dollar received JVs raise an average of thirty-one dollars from other partners.[151] To date, JVs have worked with over fifty-seven hundred partners and supported conservation of nearly twenty-seven million acres of land.[152]

A directory of JVs is available online at https://mbjv.org/who-we-are/directory/. Each individual JV is different and has its own implementation plan that sets priorities and defines the scope of its work. Some JVs, such as the Rainwater Basin JV in Nebraska, work directly with landowners one-on-one, helping to create conservation plans and secure grant funding for conservation work.[153] Others may not work directly with private landowners, instead focusing on fostering cooperation between different federal agencies and NGOs. Different JVs also focus on different species. For example, one of the priority species for the Intermountain West JV is the greater sage-grouse.[154]

The Intermountain West Joint Venture

Established in 1994, the Intermountain West Joint Venture (IWJV) is geographically the largest JV, covering a 486 million acre region in eleven states, including 142 million acres of private land.[155] Between 1994 and 2017, IWJV worked with over two hundred partners to invest $574 million in conservation of 3.5 million acres of land, leveraging thirty-nine dollars for every federal dollar invested.[156] IWJV's governing board includes four federal agencies, seven states, six NGOs, and four landowners, and the organization works with all of these partners on a regular basis.[157] Its staff are funded by a combination of FWS, Pheasants Forever, the Western Association of Fish and Wildlife Agencies, and the Wildlife Management Institute.[158]

IWJV works to develop both partnerships and science to guide those partnerships. Building partnerships involves both one-on-one work with organizations and landowners and also identifying practices for agencies and other organizations to become more partnership oriented.[159] IWJV's science program seeks to make conservation programs within its area more effective, for example by translating population goals into the quantity, quality, and location of habitat needed to achieve those goals.[160] Other work includes identifying and filling data gaps, assisting partners in delivering Farm Bill programs, and identifying sources of matching funds for partners.[161]

One example of how Farm Bill programs have benefited from IWJV's involvement is the use of the Wetlands Reserve Program (WRP) and later Agricultural Conservation Easement Program–Wetland Reserve Easements to conserve waterfowl habitat in northeastern California, southern Oregon, and northwestern Nevada. In this region, wetlands are often the most productive parts of ranches, making permanent WRP and Agricultural Conservation Easement Program–Wetland Reserve Easements that restrict grazing rights unattractive to landowners. In 2013, IWJV helped NRCS and other partners to implement a pilot project that allowed enrollees to retain grazing rights on WRP easements and also leveraged other resources, including cost-shared biologist positions and NAWCA funding, to make the program more cost-effective.[162] This project is now being expanded into IWJV's Working Wetlands & Water in the West (Water 4) Initiative.[163]

An area where IWJV has been a leader is in greater sage-grouse conservation, promoting conifer removal, wetland restoration, and rangeland fire management to improve habitat within the sagebrush sea. In 2016, IWJV launched the Partnering to Conserve Sagebrush Rangelands initiative in partnership with NRCS, BLM, and the U.S. Forest Service. This new effort seeks to bring lessons from successful partnerships onto public lands and help federal land management agencies become more proactive in working with private landowners to achieve conservation.[164] Although IWJV does not usually work one-on-one with private landowners, its governing board includes four private landowners, and its work fostering effective partnerships helps other organizations work with private landowners more effectively.[165]

Supporting Conservation with the Best Science: U.S. Geological Survey Cooperative Research Units and U.S. Forest Service Research & Development

The work that so many organizations do with private landowners is only effective because it is grounded in strong science. Many partners, including government agencies, universities, NGOs, and others, work to develop and refine this science. Two government agencies in particular have a long history of scientific collaborations to benefit at-risk species, including the greater sage-grouse.

The U.S. Geological Survey (USGS) operates forty Cooperative Research Units (CRUs) in thirty-eight states. The brainchild of the cartoonist, political satirist, and conservationist J. N. "Ding" Darling, the first CRUs were established in 1935, strategically placed to represent the variety of ecosystems in the nation.[166] Each CRU is located at a state land-grant university, and CRU researchers, each of whom is both a federal government employee and a university faculty member, carry out graduate education and research, while also

Figure 10.19. Jay N. "Ding" Darling, 1876–1962. Ding Darling was a Pulitzer Prize–winning cartoonist and conservationist whose accomplishments included creating the Federal Duck Stamp in 1934, serving as director of the U.S. Biological Survey (now the U.S. Fish and Wildlife Service) from 1934-1935, and founding the National Wildlife Federation in 1936.
Courtesy of the Jay N. Darling Papers, The University of Iowa Libraries.

providing technical assistance to real-world conservation projects.[167] CRUs work with their partner-stakeholders including USGS, FWS, the Wildlife Management Institute, state fish and wildlife agencies, their host universities, and others to address regional scientific needs.[168] Examples of CRU projects benefiting private landowners as of 2019 include research into habitat, genetics, and disease threats to the endangered Wyoming toad in Wyoming; the hydrology needed to support the threatened Arkansas River shiner in Oklahoma; and the forest habitat needs of species including the delisted (due to recovery) northern flying squirrel in West Virginia, the delisted (due to recovery) Kirtland's warbler in Wisconsin, and the threatened northern long-eared bat throughout the eastern United States. A recently concluded project looked at the impact of grazing practices on greater sage-grouse in Idaho.[169]

Another important research program is run by the U.S. Forest Service's Research & Development (R&D) program, which operates seven research stations and eighty-one experimental forests and ranges, employing five hundred scientists.[170] The R&D program collaborates with other agencies, NGOs, academic institutions, and industry, with products published in journals and online, including on Treesearch, the program's free, online publication portal.[171] Much of the program's recent work has focused on the greater sage-grouse. In the early 2010s, the program experimented with different grazing practices in order to guide Forest Service and BLM plans to conserve sage-grouse. After FWS found the species did not warrant listing under the ESA, research continued. As of 2019, Treesearch contains over forty-three hundred publications on greater sage-grouse, and more continue to be published.

Even though USGS CRUs and the U.S. Forest Service R&D program do not work directly with private landowners, their contributions are critical to conservation programs on private land.

Notes

1. Brasher et al., "The History and Importance of Private Lands for North American Waterfowl Conservation."

2. Ibid.

3. An Act to Amend the Migratory Bird Hunting Stamp Act of March 16, 1934, as Amended, Pub. L. No. 85-585, 72 Stat. 486 (1958); U.S. Fish and Wildlife Service National Wildlife Refuge System, "The Small Wetlands Program: A Half Century of Conserving Prairie Habitat," November 7, 2012, https://www.fws.gov/refuges/whm/fwn.html. A map of Wetland Management Districts is available online at https://www.fws.gov/refuges/smallwetlands/WPAs/WMDs2008.pdf. U.S. Fish and Wildlife Service, "Wetland Management Districts of the National Wildlife Refuge System," 2007, https://www.fws.gov/refuges/smallwetlands/WPAs/WMDs2008.pdf.

4. U.S. Fish and Wildlife Service, "Statistical Data Tables for Fish & Wildlife Service Lands (as of 9/30/2018)," https://www.fws.gov/refuges/land/PDF/2018_Annual_Report_of_Lands_Data_Tables.pdf.

5. Brasher et al., "The History and Importance of Private Lands for North American Water fowl Conservation."

6. Ibid.

7. U.S. Fish and Wildlife Service National Wildlife Refuge System, "Waterfowl Production Areas," October 15, 2015, https://www.fws.gov/refuges/about/wpas.html.

8. U.S. Fish and Wildlife Service, "Statistical Data Tables for Fish & Wildlife Service Lands (as of 9/30/2018)," 1.

9. U.S. Fish and Wildlife Service National Wildlife Refuge System, "The Small Wetlands Program: A Half Century of Conserving Prairie Habitat"; Prairie Pothole Joint Venture, "U.S. Fish and Wildlife Service's Small Wetland Acquisition Program," last accessed July 10, 2019, http://ppjv.org/newsletters/summer-2016/us-fish-and-wildlife-services-small-wetland-acquisition-program. In 1961, Congress passed the Wetlands Loan Act, authorizing the advance of funds against future Duck Stamp revenue in order to accelerate habitat acquisition. An Act to Promote the Conservation of Migratory Waterfowl by the Acquisition of Wetlands and Other Essential Waterfowl Habitat, and for Other Purposes, Pub. L. No. 87-383, 75 Stat. 813 (1961); Brasher et al., "The History and Importance of Private Lands for North American Waterfowl Conservation." In 2015, Congress provided an increase in funding for WPAs by requiring that all Federal Duck Stamp revenue over fifteen dollars per stamp must be used to purchase easements for migratory bird conservation. Federal Duck Stamp Act of 2014, Pub. L. 113-264, 128 Stat. 2939 (2014).

10. U.S. Department of the Interior Office of Inspector General, U.S. Fish and Wildlife Service Conservation Easement Enforcement and Monitoring (2012), 2; U.S. Fish and Wildlife Service Division of Realty, "Grassland Easements," March 31, 2010, https://www.fws.gov/mountain-prairie/realty/grassesmt.htm.

11. For example, Kelly Reed, interview by Christopher E. Segal, July 12, 2019. FWS National Wildlife Refuges and Conservation Areas sometimes have borders that include private land. FWS places a high priority on acquiring these lands or conservation easements on them from willing landowners.

12. U.S. Forest Service, "By the Numbers," November 2013, https://www.fs.fed.us/about-agency/newsroom/by-the-numbers; U.S. Forest Service, "Our History," last accessed July 3, 2019, https://www.fs.fed.us/learn/our-history.

13. Schoonhoven.

14. The Forest Legacy Program, Community Forest and Open Space Conservation Program, and Landscape Scale Restoration Program collectively make up the agency's Cooperative Forestry programs, which fund state forestry agencies that help landowners sustain healthy forests and wildlife habitat. Ibid.

15. U.S. Forest Service, *FY 2018 Budget Justification* (2017), 60.

16. Ibid., 60–61.

17. Ibid., 60; Schoonhoven.

18. Bullock.

19. U.S. Forest Service, *FY 2018 Budget Justification*, 62–63; U.S. Forest Service, *FY 2020 Budget Justification* (2019), 36, 48; Schoonhoven; U.S. Forest Service, "How the Community Forest Program Works," last accessed August 14, 2019, https://www.fs.fed.us/managing-land/private-land/community-forest/program.

20. Schoonhoven.

21. U.S. Forest Service, Fiscal Year 2020 Landscape Scale Restoration Competitive Process National Guidance (May 29, 2019), 1; Agricultural Improvement Act of 2018, 356; Schoonhoven.

22. Schoonhoven. Exact figures for this program are not available from the U.S. Forest Service.

23. American Forests, *Partnering for Forests: A Look at the U.S. Forest Service's Cooperative Forestry Program and Its Partners* (2017), 3.

24. Schoonhoven.

25. Ibid.

26. Ibid.

27. Ibid.; American Forests, *Partnering for Forests: A Look at the U.S. Forest Service's Cooperative Forestry Program and Its Partners*, 7.

28. The Sikes Act of 1949 authorized investments in conservation at Eglin Air Force Base in Florida. The success of this program led to an Air Force collaboration with FWS in 1956 and a formal nationwide conservation program in 1958. The Sikes Act of 1960, still in effect today after many amendments, authorized all DoD installations to enter into cooperative wildlife management agreements with state fish and wildlife agencies and FWS. 1982 amendments expanded the Sikes Act to cover all ESA-listed species, and the 1997 Sikes Act Improvement Act required the DoD to develop an Integrated Natural Resources Management Plan (INRMP) on every installation. In 2003, Congress exempted lands covered by INRMPs from designation as critical habitat under the ESA, provided that the individual INRMP provides conservation benefits to the species in question. See generally Jean A. Mansavage, *Natural Defense: U.S. Air Force Origins of the Department of Defense Natural Resources Conservation Program*, Air Force Historical Studies (Washington, DC: Air Force History and Museums Program, 2014). National Defense Authorization Act for Fiscal Year 2004, § 318, Pub. L. No. 108-136, 117 Stat. 1392, 1433 (2003).

29. Beard.

30. U.S. Department of Defense Readiness and Environmental Protection Integration Program, *13th Annual Report to Congress* (2019), 2.

31. Mark D. Ahner, *Can the United States Army Adjust to the Endangared Species Act of 1973?* (Carlisle Barracks, PA: U.S. Army War College, 1992), 15–18.

32. Ibid., 1. All charges were ultimately dropped.

33. This subsequently evolved into FWS's Safe Harbor Agreements, discussed in chapter 13.

34. Bob Stump National Defense Authorization Act for Fiscal Year 2003, Pub. L. No. 107-314, 116 Stat. 2458 (2002), 2705–07; Beard.

35. The majority of REPI acres and dollars have benefited Army installations, but Air Force, Marine Corps, and Navy installations also operate REPI programs. Since 2006, REPI has submitted an annual report to Congress, and the most recent report includes information on individual projects and data on the types of projects executed by each of the armed services within the DoD. U.S. Department of Defense Readiness and Environmental Protection Integration Program, *13th Annual Report to Congress*; Beard.

36. Bob Stump National Defense Authorization Act for Fiscal Year 2003, 2706.

37. Beard.

38. Ibid.

39. The DoD provides conservation planning, conflict reduction, and funding through REPI. USDA provides technical and financial assistance through Farm Bill programs such as the Regional Conservation Partnership Program, the Conservation Stewardship Program, the Environmental Quality Incentives Program, and the Agricultural Conservation Easement Program, as well as the Forest Legacy Program. Interior provides funding through the North American Wetlands Conservation Act. Other resources include Wildlife and Sport Fish Restoration funds, and programs at the EPA, Department of Transportation, and NOAA. Sentinel Landscapes are overseen by a central Federal Coordinating Committee, which designates Sentinel Landscapes and works to facilitate cooperation between the various stakeholders. USDA, DoD, and U.S. Department of the Interior, *Sentinel Landscapes 2016 Report* (2016), 1–4.

40. Beard.

41. Sentinel Landscapes, "About," last accessed July 16, 2019, https://sentinellandscapes.org/about/; Beard.

42. Sentinel Landscapes, "About."

43. Ibid.

44. National Oceanic and Atmospheric Administration Office for Coastal Management, "Coastal and Estuarine Land Conservation Program," last accessed February 26, 2019, https://coast.noaa.gov/czm/landconservation/media/celpfactsheet.pdf.

45. National Oceanic and Atmospheric Administration Office for Coastal Management, "Preserving the Pristine," last accessed July 18, 2019, http://noaa.maps.arcgis.com/apps/Cascade/index.html?appid=30535f92095d4a25b0ba26ec04f5af57. Since 2010, some of this funding has come from EPA's Great Lakes Restoration Initiative. National Oceanic and Atmospheric Administration Office for Coastal Management, "About Coastal and Estuarine Land Conservation Program," last accessed July 18, 2019, https://coast.noaa.gov/czm/landconservation/about/.

46. National Oceanic and Atmospheric Administration Office for Coastal Management, "Coastal and Estuarine Land Conservation Program"; National Oceanic and Atmospheric Administration Office for Coastal Management, "About Coastal and Estuarine Land Conservation Program."

47. CRGP began in 2015 as two separate coastal resiliency programs, one under the National Ocean Service and one under the National Marine Fisheries Service. In 2017, the two programs were combined under the National Ocean Service's Office for Coastal Management, and in 2018 NOAA enlisted the expertise of NFWF to help implement the program.

48. National Oceanic and Atmospheric Administration Office for Coastal Management, "NOAA Coastal Resilience Grants Program," last accessed February 26, 2019, https://coast.noaa.gov/resilience-grant/; National Oceanic and Atmospheric Administration Office for Coastal Management, "Coastal Resilience Grants for Coastal Communities," last accessed February 26, 2019, https://www.coast.noaa.gov/data/resilience/factsheet-resilience-grants.pdf; National Oceanic and Atmospheric Administration Office for Coastal Management, "Coastal Resilience Grant Program Project Summaries," last accessed February 26, 2019, https://www.coast.noaa.gov/resilience-grant/projects/.

49. U.S. Fish and Wildlife Service Coastal Program, "What Is the Coastal Program?" August 5, 2019, https://www.fws.gov/coastal/about.html.

50. This is one of the two funds that make up the WSFR, discussed later in this chapter.

51. U.S. Fish and Wildlife Service, "National Coastal Wetlands Conservation Grants," August 20, 2019, https://www.fws.gov/coastal/CoastalGrants/index.html; U.S. Fish and Wildlife Service, "2019 National Coastal Wetland Conservation Grants Project Summaries," 2019, https://www.fws.gov/coastal/CoastalGrants/pdfs/2019-Coastal-Wetlands-Grants-Project-Summaries-Final.pdf.

52. U.S. Fish and Wildlife Service Estuary Restoration Act Council, "Estuary Restoration Act (ERA) of 2000," February 28, 2019, https://www.fws.gov/coastal/era.html; National Marine Protected Areas Center, "Fact Sheet," July 2013, https://nmsmarineprotectedareas.blob.core.windows.net/marineprotectedareas-prod/media/archive/pdf/helpful-resources/mpa_center_fact_sheet071213.pdf.

53. U.S. Fish and Wildlife Service Coastal Program, "What Is the Coastal Program?".

54. U.S. Fish and Wildlife Service Coastal Program, "Strategic Plan," February 28, 2019, https://www.fws.gov/coastal/strategic_plan.html.

55. U.S. Fish and Wildlife Service Coastal Program, "Coastal Conservation: Key to a Thriving Economy and a Healthy Environment," last accessed September 20, 2019, https://www.fws.gov/coastal/pdfs/Coastal_Program_Brochure-Online_Version.pdf.

56. U.S. Fish and Wildlife Service Coastal Program, "Regional Coastal Contacts," July 24, 2019, https://www.fws.gov/coastal/contacts.html.

57. EPA, "EPA Grants," last accessed September 17, 2019, https://www.epa.gov/grants.

58. Echols, Front, and Cummins, "Broadening Conservation Funding."

59. EPA, "319 Grant Program for States and Territories," last accessed July 18, 2019, https://www.epa.gov/nps/319-grant-program-states-and-territories.

60. Steve Jester, email to Christopher E. Segal, July 17, 2019.

61. EPA, "319 Grant Program for States and Territories."

62. Echols, Front, and Cummins, "Broadening Conservation Funding."

63. Natural Resources Conservation Service, "Natural Disaster Recovery Assistance through EQIP," last accessed February 20, 2019, https://www.nrcs.usda.gov/Internet/FSE_MEDIA/nrcseprd1429025.pdf.

64. Agricultural Credit Act of 1978, Pub. L. No. 95-334, 92 Stat. 420 (1978), 433–34.

65. Agricultural Credit Act of 1978–Title IV (Emergency Conservation Program), Public Law 95–334, 92 Stat. 420, 16 U.S.C. 2201–2205, as Amended through P.L. 115–334, Enacted December 20, 2018.

66. Food, Conservation, and Energy Act of 2008, 1290. "Nonindustrial private forestland" is defined as "rural land . . . [that] has existing tree cover (or had tree cover immediately before the natural disaster and is suitable for growing trees); and . . . is owned by any nonindustrial private individual, group, association, corporation, or other private legal entity, that has definitive decision-making authority over the land." Ibid.

67. Agricultural Credit Act of 1978, 434; Agricultural Credit Act of 1978–Title IV (Emergency Conservation Program), 3. The 1996 Farm Bill amended the Emergency Watershed Program to make it clear that the authorized emergency watershed protection measures included the purchase of floodplain easements. Federal Agriculture Improvement and Reform Act of 1996, 1016.

68. In rare cases where the fair market value of the land is less than the value of the easement or a landowner offers to sell an easement for less than its value, NRCS will pay the lowest of the figures.

69. Agricultural Credit Act of 1978–Title IV (Emergency Conservation Program); Natural Resources Conservation Service, "EWP Floodplain Easement (EWP-FPE)," last accessed February 20, 2019, https://www.nrcs.usda.gov/wps/portal/nrcs/detail/national/programs/financial/ewp/?cid=nrcs143_008225.

70. Department of Defense, Emergency Supplemental Appropriations to Address Hurricanes in the Gulf of Mexico, and Pandemic Influenza Act, 2006, Pub. L. No. 109-148, 119 Stat. 2680, 2747.

71. Farm Service Agency, "Fact Sheet: Emergency Forestry Conservation Reserve Program 2005 Hurricane Assistance," July 2006, https://www.fsa.usda.gov/Internet/FSA_File/efcrp06.pdf.

72. Disaster Relief Appropriations, Pub. L. No. 113-2, 127 Stat. 4 (2013), 16.

73. National Fish and Wildlife Foundation, "Hurricane Sandy Coastal Resiliency Competitive Grant Program," last accessed February 1, 2019, https://www.nfwf.org/hurricanesandy/Pages/home.aspx.

74. Partners for Fish and Wildlife Act, Public Law No. 109-294, 120 Stat. 1351 (2006), 1352.

75. U.S. Fish and Wildlife Service Partners for Fish and Wildlife Program, "Background and History," last accessed July 8, 2019, https://www.fws.gov/mountain-prairie/refuges/partners.php#.

76. U.S. Fish and Wildlife Service Partners for Fish and Wildlife Program, "Partners for Fish and Wildlife Program: Program History," last accessed December 18, 2018, https://www.fws.gov/partners/programHistory.html; U.S. Fish and Wildlife Service Partners for Fish and Wildlife Program, "Background and History."

77. U.S. Fish and Wildlife Service Partners for Fish and Wildlife Program, "Partners for Fish and Wildlife Program: Program History"; U.S. Department of the Interior, *Budget Justifications and Performance Information Fiscal Year 2019: Fish and Wildlife Service* (2018), HC-2.

78. Partners for Fish and Wildlife Act.

79. John D. Dingell, Jr. Conservation, Management, and Recreation Act, Public L. No. 116-9 (2019).

80. Partners for Fish and Wildlife Act; U.S. Department of the Interior, *Budget Justifications and Performance Information, Fiscal Year 2018: U.S. Fish and Wildlife Service* (2017), HC-1.

81. U.S. Fish and Wildlife Service Partners for Fish and Wildlife Program, "Partners for Fish and Wildlife Program," last accessed August 10, 2018, https://www.fws.gov/partners/aboutus.html.

82. U.S. Department of the Interior, *Budget Justifications and Performance Information, Fiscal Year 2018: U.S. Fish and Wildlife Service*, HC-1-2.

83. Ibid., HC-4-5; U.S. Department of the Interior, *Budget Justifications and Performance Information Fiscal Year 2020: Fish and Wildlife Service* (2019), HC-3.

84. U.S. Fish and Wildlife Service Partners for Fish and Wildlife Program, *Mountain-Prairie Region Annual Narrative FY 2017* (2018), 4.

85. U.S. Department of the Interior, *Budget Justifications and Performance Information, Fiscal Year 2018: U.S. Fish and Wildlife Service*, HC-5-7; U.S. Fish and Wildlife Service Partners for Fish and Wildlife Program, *Mountain-Prairie Region Annual Narrative FY 2017*, 4.

86. U.S. Fish and Wildlife Service Partners for Fish and Wildlife Program, "Regional and State Contact Information."

87. Bureau of Land Management, *Public Land Statistics 2017*, 7.

88. An Act Making Omnibus Consolidated Appropriations for the Fiscal Year Ending September 30, 1997, and for Other Purposes, Pub. L. No. 104-308 § 124, 110 Stat. 3009 (1996), 3204.

89. Karen Prentice, interview by Christopher E. Segal, September 30, 2019.

90. Bureau of Land Management, "National Office," last accessed September 19, 2019, https://www.blm.gov/office/national-office.

91. Smith and Duvall.

92. Bureau of Land Management, "Threatened and Endangered Species," last accessed September 19, 2019, https://www.blm.gov/programs/fish-and-wildlife/threatened-and-endangered.

93. "Partners in the Sage," last accessed September 19, 2019, https://www.partnersinthesage.com/.

94. Smith and Duvall.

95. Deficit Reduction Act of 1984, Pub. L. No. 98-369, 98 Stat. 494, 1017–19.

96. U.S. Fish and Wildlife Service, "Items Taxed to Support Wildlife and Sport Fish Restoration in America," January, 2018, https://wsfrprograms.fws.gov/subpages/aboutus/ItemsTaxedJan2018.pdf.

97. U.S. Fish and Wildlife Service, "1939 through 2019 Wildlife Restoration Apportionments," 2019, https://wsfrprograms.fws.gov/Subpages/GrantPrograms/WR/WRApportionmentsHE-1939-2019.xlsx.

98. U.S. Fish and Wildlife Service, "1952 through 2019 Sport Fish Restoration Apportionments," 2019, https://wsfrprograms.fws.gov/Subpages/GrantPrograms/SFR/SFRApportionments-1952-2019.xlsx.

99. To qualify, states also must either maintain five-year fish and wildlife management plans updated at least every three years or submit to FWS a statement of every planned wildlife restoration project in the state. 16 U.S.C. § 669e; 16 U.S.C. § 777e; 50 CFR Part 80.

100. Morgan et al., "A State Assessment of Private Lands Wildlife Conservation in the United States."

101. U.S. Department of the Interior, *Budget Justifications and Performance Information, Fiscal Year 2018: U.S. Fish and Wildlife Service*, WR-3, 6.

102. FWS works with the Association of Fish and Wildlife Agencies to implement this program, which targets projects the Association of Fish and Wildlife Agencies identifies as priorities. U.S. Fish and Wildlife Service Wildlife & Sport Fish Restoration Program, "Multistate Conservation Grant Program–Overview," February 2, 2018, https://wsfrprograms.fws.gov/Subpages/GrantPrograms/MultiState/MS.htm; Association of Fish and Wildlife Agencies, "Multistate Conservation Grant Program Overview," last accessed July 10, 2019, https://www.fishwildlife.org/afwa-informs/multi-state-conservation-grants-program.

103. U.S. Fish and Wildlife Service National Coastal Wetlands Conservation Grant Program, "National Coastal Wetlands Conservation Grants," May 8, 2019, https://www.fws.gov/coastal/CoastalGrants/index.html; U.S. Fish and Wildlife Service, "Interior Secretary Awards States More Than $20 Million in Grants Supporting Coastal Wetlands Conservation, Recreation and Communities," press release, May 8, 2019, https://www.fws.gov/news/ShowNews.cfm?ref=interior-secretary-awards-states-more-than-$20-million-in-grants-&_ID=36408.

104. Department of the Interior and Related Agencies Appropriations Act, Pub. L. No. 107-63, 115 Stat. 414, 422 (2002). (Emphasis added.)

105. This was initially fifty-fifty; it was changed to sixty-five–twenty-five in 2012.

106. U.S. Fish and Wildlife Service, "State Wildlife Grant Program Apportionment History," 2019, https://wsfrprograms.fws.gov/Subpages/GrantPrograms/SWG/SWGAppt2002-2019.pdf.

107. Department of the Interior and Related Agencies Appropriations Act, 422.

108. Morgan et al., "A State Assessment of Private Lands Wildlife Conservation in the United States."

109. Association of Fish and Wildlife Agencies, "State Wildlife Action Plans," last accessed December 13, 2018, https://www.fishwildlife.org/afwa-informs/state-wildlife-action-plans.

110. Department of the Interior and Related Agencies Appropriations Act, 422.

111. U.S. Geological Survey, "New Database Available: USGS Releases 'Species of Greatest Conservation Need' Lists," press release, October 25, 2017, https://www.usgs.gov/news/new-database-available-usgs-releases-species-greatest-conservation-need-lists.

112. U.S. Department of the Interior, *Budget Justifications and Performance Information, Fiscal Year 2018: U.S. Fish and Wildlife Service*, STWG-4.

113. Ibid., STWG-5, 9. This was subsequently overturned in 2018 by the 9th Circuit Court of Appeals, which ruled among other things that FWS did not fully explain how it had documented Arctic grayling population increases. Center for Biological Diversity v. Zinke, 900 F.3d 1053 (9th Cir. 2018). A new ESA listing decision is expected in 2020, and Montana and its partners (both public and private) remain committed to Arctic grayling conservation.

114. Leanne Roulson, "The Culture of Conservation and the Arctic Grayling: A State Wildlife Management Success Story," July 14, 2018, http://greatergallatin.org/the-culture-of-conservation-and-the-arctic-grayling-a-state-wildlife-management-success-story/.

115. Finding on a Petition to List the Fluvial Population of the Arctic Grayling as Endangered, 59 Fed. Reg. 37738 (July 25, 1994).

116. Revised 12-Month Finding on a Petition to List the Upper Missouri River Distinct Population Segment of Arctic Grayling as an Endangered or Threatened Species, 79 Fed. Reg. 49384 (August 20, 2014); U.S. Department of the Interior, *Budget Justifications and Performance Information, Fiscal Year 2018: U.S. Fish and Wildlife Service*, STWG-9.

117. U.S. Department of the Interior, *Budget Justifications and Performance Information, Fiscal Year 2018: U.S. Fish and Wildlife Service*, CESCF-3.

118. Ibid. For more on HCPs, see chapter 13.

119. Ibid., CESCF-4, CESCF-9.

120. Ibid., CESCF-4-CESCF-8.

121. David Gagner, interview by Christopher E. Segal, May 28, 2019.

122. National Fish and Wildlife Foundation, "About the National Fish and Wildlife Foundation," last accessed February 15, 2019, https://www.nfwf.org/whoweare/Pages/home.aspx.

123. Ibid.; National Fish and Wildlife Foundation, "What We Do," last accessed February 15, 2019, https://www.nfwf.org/whatwedo/Pages/home.aspx.

124. National Fish and Wildlife Foundation, "2017 Annual Report," last accessed October 5, 2019, https://www.nfwf.org/whoweare/mediacenter/Documents/2017_annual_report.pdf; National Fish and Wildlife Foundation, "2018 Annual Report," last accessed October 5, 2019, https://www.nfwf.org/whoweare/mediacenter/Documents/2018-annual-report.pdf; Rob Blumenthal, email to Christopher E. Segal, August 20, 2019.

125. National Fish and Wildlife Foundation, "2017 Annual Report," 7; National Fish and Wildlife Foundation, "2018 Annual Report," 7.

126. National Fish and Wildlife Foundation, "Conservation Programs," last accessed February 15, 2019, https://www.nfwf.org/whatwedo/programs/Pages/home.aspx.

127. National Fish and Wildlife Foundation, "2017 Annual Report."

128. Gagner; National Fish and Wildlife Foundation, "Applicant Information," last accessed February 15, 2019, https://www.nfwf.org/whatwedo/grants/applicants/Pages/home.aspx.

129. Anadarko Petroleum, Chevron, Noble Energy, Occidental Petroleum, Shell Oil Company, and XTO Energy.

130. National Fish and Wildlife Foundation, "Pecos Watershed Conservation Initiative 2019 Request for Proposals," 2018, https://www.nfwf.org/swrivers/pecos/Documents/2019rfp.pdf.

131. Ibid.; National Fish and Wildlife Foundation, "National Fish and Wildlife Foundation Partners with Permian Basin Oil and Natural Gas Companies to Conserve Pecos Watershed in Texas and New Mexico," press release, November 8, 2017, https://www.nfwf.org/swrivers/pecos/Documents/nfwf-partners-with-permian-basin-oil-natural-gas-companies-to-conserve-pecos-watershed-in-texas-new-mexico-2017-1109.pdf; National Fish and Wildlife Foundation, "NFWF's Partnership with Permian Basin Oil and Natural Gas Companies and Federal Agencies Leads to $3.7 Million in Conservation Impact in First Year," press release, March 22, 2018, https://www.nfwf.org/swrivers/pecos/Documents/partnership-with-permian-basin-oil-natural-gas-companies-federal-agencies-leads-to-3-7-million-in-conservation-2018-0322.pdf; National Fish and Wildlife Foundation, "Pecos Watershed Conservation Initiative 2018 Grant Slate," last accessed August 14, 2019, https://www.nfwf.org/swrivers/pecos/Documents/2018-grant-slate.pdf; National Fish and Wildlife Foundation, "Pecos Watershed Conservation Initiative 2019 Grant Slate," last accessed August 14, 2019, https://www.nfwf.org/swrivers/pecos/Documents/2019grantslate.pdf.

132. National Fish and Wildlife Foundation, "Pecos Watershed Conservation Initiative," last accessed August 14, 2019, https://www.nfwf.org/swrivers/pecos/Pages/home.aspx.

133. National Fish and Wildlife Foundation, "Gulf Environmental Benefit Fund," last accessed July 2, 2019, https://www.nfwf.org/gulf/Pages/home.aspx.

134. National Fish and Wildlife Foundation, "Gulf Environmental Benefit Fund Projects," last accessed August 14, 2019, https://www.nfwf.org/gulf/Pages/gulf-projects.aspx.

135. "Migratory Bird joint ventures are collaborative, regional partnerships of government agencies, non-profit organizations, corporations, tribes, and individuals that conserve habitat for the benefit of priority bird species, other wildlife, and people. Joint ventures bring these diverse partners together under the guidance of national and international bird conservation plans to design and implement landscape-scale conservation efforts in support of the North American Waterfowl Management Plan and the other bird management plans." U.S. Fish and Wildlife Service, "Migratory Bird Joint Ventures," last accessed August 13, 2018, https://www.fws.gov/birds/management/bird-conservation-partnership-and-initiatives/migratory-bird-joint-ventures.php.

136. Migratory Bird Joint Ventures, "Joint Venture Directory," last accessed March 25, 2019, https://mbjv.org/who-we-are/directory/.

137. Smith and Duvall.

138. Brasher et al., "The History and Importance of Private Lands for North American Waterfowl Conservation"; Smith and Duvall.

139. The NAWMP identified specific projects in the Prairie Pothole region, the Canadian prairie, the lower Mississippi valley, the Great Lakes and St. Lawrence river, the Atlantic coast, and the Central Valley of California as suitable for JVs. U.S. Fish and Wildlife Service and Canadian Wildlife Service, *North American Waterfowl Management Plan* (May 1986).

140. Migratory Bird Joint Ventures, "Joint Venture Timeline," last accessed March 26, 2019, http://mbjv.org/who-we-are/jv-timeline/. NAWCA provides funding for JVs and federal matching funds for wetlands conservation programs. U.S. Fish and Wildlife Service, "North American Wetlands Conservation Act: Protecting, Restoring and Enhanging Wetland Habitats for Birds," last accessed October 4, 2019, https://www.fws.gov/birds/grants/north-american-wetland-conservation-act.php; Donald J. Barry, interview by Christopher E. Segal and Lowell E. Baier, February 27, 2019.

The successive revisions to the NAWMP have extended the timeframe of the document; recognized changed environmental conditions; updated population goals; set new habitat objectives; promoted landscape-scale planning; broadened collaboration with larger, more diverse sets of partners; recognized new threats to waterfowl such as climate change and increased human demand for ecological services and recreational opportunities; and adopted adaptive management and integrated decision-making. North American Waterfowl Management Plan, *1994 Update to the North American Waterfowl Management Plan: Expanding the Commitment* (1994); North American Waterfowl Management Plan, *Expanding the Vision: 1998 Update North American Waterfowl Management Plan* (1998); North American Waterfowl Management Plan, *Strengthening the Biological Foundation: 2004 Strategic Guidance* (2004); North American Waterfowl Management Plan, *North American Waterfowl Management Plan 2012: People Conserving Waterfowl and Wetlands* (2012); North American Waterfowl Management Plan, *2018 North American Waterfowl Management Plan (NAWMP) Update: Connecting People, Waterfowl, and Wetlands* (2018).

141. Migratory Bird Joint Ventures, "Joint Venture Timeline."

142. Migratory Bird Joint Ventures, "Migratory Bird Joint Ventures Map," last accessed March 26, 2019, https://mbjv.org/joint-venture-map/.

143. Jeffrey P. Cohn, "Joint Ventures: A Different Approach to Conservation," *BioScience* 55, no. 10 (October 2005); Migratory Bird Joint Ventures, "Joint Venture Timeline." "Plan partners should seek to participate in the development of other landscape-level, migratory bird conservation plans to ensure that waterfowl needs are considered. Likewise, representatives of other migratory bird initiatives may be invited to participate in planning efforts to identify habitat needs of those species that should be considered by Plan partners. Eventually, a joint venture/regional partnership may be affiliated with more than one of these initiatives." North American Waterfowl Management Plan, *Expanding the Vision: 1998 Update North American Waterfowl Management Plan*, 11–12.

144. Smith and Duvall; Partners in Flight, "Partners in Flight Landbird Conservation Plan (2016)," last accessed August 12, 2019, https://www.partnersinflight.org/resources/the-plan/; North American Bird Conservation Initiative U.S. Committee, "State of the Birds," last accessed August 12, 2019, http://nabci-us.org/how-we-work/state-of-the-birds/.

145. Smith and Duvall.

146. U.S. Department of the Interior, *Budget Justifications and Performance Information, Fiscal Year 2018: U.S. Fish and Wildlife Service*, MB-12.

147. Cummins.

148. Smith and Duvall.

149. Brasher et al., "The History and Importance of Private Lands for North American Waterfowl Conservation."

150. U.S. Department of the Interior, *Budget Justifications and Performance Information, Fiscal Year 2018: U.S. Fish and Wildlife Service*, MB-11.

151. Todd Fearer, *Migratory Bird Joint Ventures* (Presentation at the Bird Conservation Alliance 116th Congress Legislative Briefing, Washington, DC, April 2, 2019).

152. Migratory Bird Joint Ventures, "Migratory Bird Joint Ventures Fact Sheet: Conserving Birds and Their Habitats Throughout North America," 2017, https://www.fws.gov/migratory birds/pdf/management/national_jv_factsheet.pdf.

153. Andy Bishop, interview by Christopher E. Segal, July 23, 2019.

154. Intermountain West Joint Venture, "Priority Species List," last accessed October 5, 2019, https://iwjv.org/priority-species-list.

155. Intermountain West Joint Venture, "Where We Work," last accessed March 27, 2019, https://iwjv.org/where-we-work; Intermountain West Joint Venture, "Mission & Scope," last accessed March 27, 2019, https://iwjv.org/mission-scope.

156. Intermountain West Joint Venture, "Intermountain West Joint Venture Factsheet," 2017, https://iwjv.org/sites/default/files/iwjv_fact_sheet.pdf.

157. Smith and Duvall.

158. Ibid.

159. Ibid.

160. Intermountain West Joint Venture, "Mission & Scope"; Intermountain West Joint Venture, *2013 Implementation Plan—Strengthening Science and Partnerships.* (Missoula, MT: IWJV, 2013), ii.2.

161. Intermountain West Joint Venture, *2013 Implementation Plan—Strengthening Science and Partnerships*, ii.3; Intermountain West Joint Venture, "Western Association of Fish & Wildlife Agencies and the IWJV," last accessed March 28, 2019, https://iwjv.org/node/1028.

162. Brasher et al., "The History and Importance of Private Lands for North American Waterfowl Conservation."

163. Intermountain West Joint Venture, "Water 4 Initiative," last accessed October 3, 2019, https://iwjv.org/water-4-initiative.

164. Smith and Duvall; Intermountain West Joint Venture, "Partnering to Conserve Sagebrush Rangelands," last accessed September 20, 2019, https://iwjv.org/sites/default/files/sage brush_parntership_four_pages.pdf.

165. Smith and Duvall.

166. In 1960 Congress formally organized the CRUs into a program of the U.S. Fish and Wildlife Service, and responsibility for CRUs was transferred to USGS in the mid-1990s. USGS Cooperative Research Units, "History of the Unit Program," last accessed July 1, 2019, https://www.coopunits.org/Headquarters/History/.

167. U.S. Geological Survey Cooperative Research Units, "About the Cooperative Research Units Program," last accessed July 1, 2019, https://www.coopunits.org/Headquarters/About_Us/.

168. John Organ, interview by Christopher E. Segal, April 5, 2019.

169. Ibid.

170. U.S. Forest Service Research & Development, "About R&D," August 13, 2018, https://www.fs.fed.us/research/about/.

171. U.S. Forest Service, "Treesearch," last accessed August 15, 2019, https://www.fs.usda.gov/treesearch/.

Non-Federal Resources for Conservation 11

THE PREVIOUS CHAPTERS PRESENTED Farm Bill and other federal resources available to assist landowners in conservation. A theme that runs through all of them is the importance of partnerships and the availability of assistance from state government agencies and non-governmental organizations (NGOs) in funding, promoting, and implementing conservation. This chapter turns to these non-federal organizations and highlights them and the programs they operate. Individual approaches vary and can include developing conservation plans, identifying funding, and providing other assistance. Some organizations function as facilitators that move funding from large grants down to individual landowners, and some conduct research to identify conservation needs and strategies to meet them. Some organizations may do all of these things and more. This chapter provides a non-exhaustive summary of some of the most significant national programs and organizations that do this kind of work. Many organizations are omitted due to lack of space, but can be identified through resources listed in chapter 3 or through other local partners.

State Governments

State governments and especially state fish and wildlife agencies play central roles in wildlife conservation in the United States. In our federalist system, primary authority over fish and wildlife resides with state governments. Even for federal trust species such as migratory birds, bald and golden eagles, and listed threatened and endangered species, state governments retain co-management authority alongside the federal government.[1] Much of the most important work to benefit at-risk species is being done by state agencies, with most of their work on private land. This is clear from the federal programs discussed in chapter 10: virtually every program is implemented with state agencies as partners providing technical support or operating the program with federal funding. Accordingly, one of the first resources that landowners should consider when investing in conservation is their state agencies. Resources in chapter 3, such as the LandCAN website, can also connect landowners to programs in their state. The following discussion provides examples and case studies documenting private land conservation programs run by state agencies.

State Fish and Wildlife Agencies

In most states, the state fish and wildlife agency is the primary operator of private land wildlife conservation programs. Historically, state governments have been leaders in developing wildlife conservation programs on private land, working in partnership with each other and others through organizations such as the Association of Fish and Wildlife Agencies, the Midwest Association of Fish and Wildlife Agencies, the Western Association of Fish and Wildlife Agencies, the Southeastern Association of Fish and Wildlife Agencies, and the Northeast Association of Fish and Wildlife Agencies. The Midwest has been the regional leader in developing private lands conservation programs, but nearly every state has committed some of its resources to private lands conservation.[2] State fish and wildlife agency private land conservation programs can take many forms. A total of 96 percent of states have programs that provide staff to assist landowners with conservation plans, opportunities for funding, and technical assistance to implement conservation.[3] A total of 70 percent of states provide financial incentives for landowners to improve wildlife habitat on their land through cost-share, seed or herbicide allocations, specialized equipment loans, or other programs.[4] Many also provide financial incentives to open land to public access for hunting and fishing. States typically have programs to reduce conflicts between humans and wildlife by providing compensation for damage caused by wildlife. The following five examples of state fish and wildlife agency programs are intended to represent the variety of programs across the nation designed to connect with, and benefit, private landowners, as of 2019. Landowners should contact their state agency for current information for their state.

ARKANSAS The Arkansas Game and Fish Commission employs biologists who specialize in working with landowners to develop conservation plans, location-specific practices, and connections with other programs. Contact information for these biologists is posted online at https://www.agfc.com/en/wildlife-management/private-lands-program/private-lands-biologist/, together with a more complete list of programs offered.[5] The Arkansas Game and Fish Commission also operates private land conservation programs for specific species, including waterfowl, deer, and bobwhite quail.[6] Several programs offer financial incentives. For example, the Waterfowl Rice Incentive Conservation Enhancement offers payments to rice farmers willing to keep their fields flooded longer in order to benefit waterfowl.[7] The state's flagship program, Acres for Wildlife, provides for five-year enrollments and provides landowners incentive payments of up to $5,000, additional in-kind assistance, and technical assistance for developing conservation plans. Most enrolled land may still be farmed and grazed, although restrictions may be necessary to facilitate restoration of native grasses and forbs.[8]

COLORADO Colorado Parks and Wildlife's Colorado Wildlife Habitat Program purchases land, conservation easements, and public access easements in order to promote wildlife conservation. Eligible resources include public access for hunting, fishing, and wildlife viewing; big game winter range and migration corridors; habitat for species of greatest conservation need identified in the Colorado State Wildlife Action Plan; riparian areas and wetlands; and parcels that protect entire landscapes or provide landscape-scale connectivity.[9] Currently funded at $11 million per year from state revenues, the Colorado Wildlife Habitat Program has invested $164 million in the protection of 257,000 acres since 2007.[10] Colorado Parks and Wildlife also operates the Pheasant Habitat Improvement Program, which partners with Pheasants Forever to install pheasant habitat on property of willing landowners,[11] and the Habitat Partnership Program, which develops and implements projects to reduce conflicts between landowners and deer, elk, pronghorn, and moose.[12] In addition, the agency's website provides technical resources to educate landowners about wildlife-friendly practices.[13]

KANSAS The Kansas Department of Wildlife, Parks and Tourism has worked with private landowners to improve wildlife habitat since 1973. Today the program includes thirty-two field staff, and an easy-to-use online map of the state (available at https://ksoutdoors.com/Services/Private-Landowner-Assistance/Wildlife-Biologists) allows landowners to identify the contact for their county.[14] Kansas operates an incentive program called Habitat First, which pays landowners between ten dollars and sixty-five dollars an acre for conservation practices that range from prescribed burning to maintaining thirty-foot-wide strips of untreated weeds. Other activities such as installing quail shelters or wetlands are incentivized on a per-action basis.[15] Other programs offer incentives for public access for hunting and fishing.[16] Finally, the agency provides technical information to assist landowners with wildlife management.[17]

SOUTH DAKOTA The South Dakota Department of Game, Fish and Parks and Department of Agriculture operate the website Habitat Pays, http://habitat.sd.gov/, providing access to all of their landowner assistance programs.[18] The Habitat Advisors program, operated in partnership with the Natural Resources Conservation Service (NRCS), the American Bird Conservancy (ABC), the Bird Conservancy of the Rockies, the National Wild Turkey Federation (NWTF), and Pheasants Forever, provides twenty-one field biologists that provide technical assistance and educate landowners about state-run and other conservation programs. Contact information for local biologists can be found online at http://habitat.sd.gov/advisors/docs/habitat-advisor-map.pdf.[19] Habitat Pays also provides a Habitat Resources portal with information about private lands conservation programs operated by federal and state agencies, conservation districts, and NGOs.[20] Some Department of Game, Fish and Parks programs provide financial incentives. These include programs promoting nesting cover, pollinator food plots, upland grass restoration, grazing management, and wetland creation, restoration, and enhancement.[21] The department also operates public hunting access programs on private land, which had fourteen hundred landowners and over 1.25 million acres of land enrolled as of 2019.[22]

TEXAS Texas has more private land than any other state, and the Texas Parks & Wildlife Department (TPW) operates a large private lands program. TPW's Private Lands and Habitat Program employs approximately 170 biologists to implement technical assistance and cost-sharing programs for landowners. As of 2019, they had completed over seven thousand active wildlife management plans on approximately 31.2 million acres of land.[23] Landowners can find a local biologist through https://tpwd.texas.gov/landwater/land/technical_guidance/biologists/.[24] TPW's Landowner Incentive Program works to establish, restore, and enhance habitat for at-risk species, using funding from Partners for Conservation, the National Fish and Wildlife Foundation, and various state sources.[25] The Texas Farm and Ranch Lands Conservation Program provides matching funds for Farm Bill Agricultural Conservation Easements and has provided $3.7 million since 2016.[26] Finally, TPW's website provides technical assistance as well.[27]

These are but five examples of how state fish and wildlife agencies are staffed to connect with private landowners interested in wildlife conservation on their lands. Contact information for every state fish and wildlife agency is provided in appendix A.

State Forestry Agencies

As discussed in chapter 10, state forestry agencies are important partners for the U.S. Forest Service and the chief point of contact for all of the Forest Service's private landowner assistance programs. In most states, this means that the state forestry agency is the first point of contact for landowners interested in the Forest Legacy Program or the Forest Stewardship Program.[28] Most forestry agencies also provide technical assistance to forest landowners and assistance with U.S. Forest Service and NRCS programs. Some also provide grants and other financial incentives. As with state fish and wildlife agencies, the following descriptions of four state forestry agencies are examples only, and landowners should contact their state forestry agency directly for current information.

COLORADO The Colorado State Forest Service offers grants for forest restoration and wildfire risk management through fuels reductions and firebreak construction.[29] It also implements the state's Forest Agriculture Classification Program, which certifies landowners producing commercial wood products as agricultural producers eligible for state tax benefits.[30] Finally, the agency works with the U.S. Forest Service and the National Association of State Foresters on the Forest Legacy Program and the Forest Stewardship Program, providing forest stewardship management plans and technical assistance.[31]

GEORGIA The Georgia Forestry Commission's staff provide landowners with technical assistance including forest stewardship management plans, site-specific consultations and written best practices guidelines, and connections to other forestry and conservation resources.[32] Contact information for foresters in every county in the state is available online at http://www.gfc.state.ga.us/about-us/contact-us/county-units/index.cfm.[33] The commission partners with the Forest Legacy program and purchases conservation easements that promote best forest management practices on working landscapes.[34] Its website also provides useful information for landowners.[35]

MISSISSIPPI The Mississippi Forestry Commission's staff provides forest management plans for up to 250 acres, third-party compliance checks for cost-share programs, and technical advice for landowners at no charge. Its website provides contact information for professional and private foresters at https://www.mfc.ms.gov/private-landowner-services.[36] Commission staff also help prepare forest stewardship management plans and provide information on the state's Reforestation Tax Credit,[37] and the commission operates two cost-sharing programs of its own.[38] Finally, commission staff provide services for a fee that include prescribed burning, firebreak establishment and maintenance, road maintenance, and southern pine beetle suppression.[39]

WISCONSIN The Wisconsin Department of Natural Resources is responsible for both forest and fish and wildlife programs. Its forestry website provides contact information for state foresters, state tax law experts, and private foresters and planners who utilize Department of Natural Resources standards.[40] The state's 1985 Managed Forest Law is an incentive program that reduces the property taxes of landowners practicing sustainable forestry.[41] As in other states, Department of Natural Resources staff help landowners with cost-sharing programs, and the state has a cost-sharing program of its own, the Wisconsin Forest Landowner Grant Program.[42] Wisconsin's state foresters prepare written forest management recommendations, connect landowners to other conservation programs, and provide technical assistance.[43]

These four examples are indicative of what each state forestry agency can provide private landowners. Contact information for every state forestry agency is provided in appendix A.

Other State and Local Programs

Although state fish and wildlife agencies and forestry agencies are the largest sources of non-federal government support for private land conservation, they are not the only ones. Every state offers other private lands programs, and the best way to access them is to talk to local conservation partners including state agencies, NGOs, U.S. Department of Agriculture (USDA) Service Centers, and conservation districts. The LandCAN database discussed in chapter 3 includes resources for many states, organized by state, region, and resource type.

For example, every state has some sort of incentive or tax benefit for agricultural producers, and in many cases qualifying practices include some types of wildlife conservation, such as developing pollinator habitat.[44] For example, Texas has recognized beekeeping as an agricultural use enterprise since 2012.[45] Tax benefits such as these can have significant financial value, which has attracted the attention of the private sector, and there are private firms that help landowners qualify for special property valuations based on agricultural enterprise.[46] Landowners can connect to these businesses and other private firms working in conservation through the LandCAN database, state agencies, conservation NGOs, and local business listings. Because biologists at USDA Service Centers, state fish and wildlife, and state forestry agencies may not be familiar with tax benefits, landowners should make a point of asking about them and should consider consulting a tax professional.

County and local government programs make up an even more fragmented landscape. There is really no way to get a sense of programs available in your county without contacting local officials and talking to neighbors, NGOs, and conservation advocates. For example, the people of Missoula, Montana, have voted several times since 1980 to issue bonds to fund open space conservation. These bonds have funded projects that conserve working lands, provide trails and access to rivers, protect water quality and wildlife habitat, and provide open space and scenic landscapes.[47] Many of these projects would qualify for federal, state, or private matching funds under conservation programs discussed in this book. This is an example of how conservation work can be extremely local and dynamic.

Non-Governmental Organization Support for Private Land Conservation

Federal, state, and local government programs are just the tip of the iceberg of private land conservation. A vast network of NGOs also provides funding, technical expertise, and in-kind contributions to private land conservation. As with the earlier discussion of state programs, the following profiles capture only a small number of the highest impact organizations, and those incompletely. It includes multi-species and single-species organizations, but focuses on organizations with on-the-ground programs, meaning many more policy-oriented environmental organizations are omitted. As with state agencies, these are a selection of examples rather than a comprehensive list of resources for landowners.

The Nature Conservancy

The nation's largest conservation organization is The Nature Conservancy (TNC), with $1.18 billion in revenue and $6.59 billion in assets in 2018.[48] Incorporated in 1951, TNC traces its roots earlier, to the formation of the Ecological Society of America in 1915. TNC has forty-eight offices that work in all fifty states, twenty-five foreign offices, and projects in seventy-two countries.[49] Pragmatic and science-driven in its approach to conservation, TNC is relatively decentralized, with state offices driving the organization's agenda in each state. This means that the organization is able to tailor programs to local needs, and the best place for landowners to connect with TNC is by contacting the office in their state.[50]

TNC focuses its national program in four areas: protecting land and water by purchasing land and conservation easements, addressing climate change, promoting food and water sustainability through sound agricultural practices, and building healthy cities. TNC owns over 2.5 million acres of land in the United States, and conservation easements on another three million acres. Easement purchases are guided by science mapping resilient landscapes and migration corridors, which it prioritizes. TNC works with NRCS, the U.S. Fish and Wildlife Service, and many other organizations as well as landowners, and employs nearly five hundred scientists.[51]

TNC's work has been crucial for many listed and at-risk species. For example, TNC was an early partner in red-cockaded woodpecker conservation in North Carolina and promoted private landowner conservation of the species throughout its range, including through Safe Harbor Agreements.[52] Other work on Endangered Species Act–listed species has included Safe Harbor Agreements for the Karner blue butterfly in New York and Indiana, the silverspot butterfly in Oregon, and the Gila topminnow and desert pupfish in Arizona, and Habitat Conservation Plans in California in Contra Costa County, Riverside County, and San Diego.[53] TNC works with Farm Bill programs as well. For example, in 2014 in the Central Valley of California, TNC partnered with the Cornell Laboratory of Ornithology to create BirdReturns, providing payments to rice farmers willing to keep their fields flooded after harvest in order to benefit neotropical migratory birds. NRCS subsequently joined as a partner and made Farm Bill funds available for the program through the Regional Conservation Partnership Program.[54] Other conservation partnerships also formed around this concept, including the Rice Stewardship Partnership created by Ducks Unlimited (DU) and USA Rice and the California Winter Rice Habitat Incentive Program created by the California Department of Fish and Wildlife.[55]

TNC's main business, however, is land transactions. These include buying land, buying conservation easements, and executing buy-protect-sell transactions. TNC also provides technical support and assistance for landowners, and different programs and mechanisms can combine to multiply benefits. One example of an initiative that has expanded over time is the Matador Ranch in Phillips County, Montana. The story began in 2000, when TNC purchased the 114-year-old, sixty thousand-acre ranch.[56] In 2003, after a severe drought, TNC turned the ranch into the Matador Grassbank, which provides local ranchers with supplemental feed at reduced prices provided that they implement

Figure 11.1. Young rice growing in the Sacramento River Delta. Flooded rice fields make ideal habitat for waterfowl, including temperate and neotropical migratory birds.
Dissolve.com stock photos.

Figure 11.2. Cattle on the move near The Nature Conservancy's Matador Ranch in Phillips County, Montana.
Photograph by Ami Vitale.

conservation practices on their own properties, such as controlling noxious weeds and not converting rangeland to agriculture.[57] In 2019, TNC purchased 4,340 more acres in the area, which it intends to place under easements and sell to adjacent landowners.[58] To date, TNC's work on the Matador Ranch has conserved 76,800 acres of grassland that benefit threatened wildlife while preserving a traditional way of life for local ranch families.[59]

The Conservation Fund

Another large, national conservation NGO is The Conservation Fund (TCF), founded in in 1985.[60] In 2017, TCF reported revenues of $235 million and assets of $466.5 million.[61] TCF's focus is on conservation finance, providing funding for others to acquire land or conservation easements, rather than in owning and managing land itself. Most land that TCF buys is eventually sold or transferred to other owners under conservation easements. TCF aims to be ready to invest in conservation on short notice, relying on staff in twenty-eight field offices in twenty-three states as well as at its headquarters in Virginia. Field staff are in constant communication with other conservation organizations, and most TCF investments are initiated by a call from a local land trust. Since its founding, TCF has contributed to the conservation of over eight million acres of land and water.[62]

TCF's Conservation Loans program offers "versatile, low-interest loans to help conservation efforts of all sizes achieve success." Sometimes ecologically significant land becomes available for sale and conservation groups are unable to raise money fast enough to preempt its development. TCF can step in with a loan allowing the land or an easement to be purchased immediately. TCF also provides technical support alongside these loans to make sure that the conservation purposes of the transaction are realized.[63] Since 1993, TCF has made 350 conservation loans to 150 partners, totaling $200 million. These loans have helped to protect 140,000 acres of land in the United States and Canada, with a total land value of nearly $380 million. Today, the Conservation Loan program operates through a $50 million revolving fund.[64]

TCF also promotes green business practices. Its Natural Capital Investment Fund provides financial and technical assistance to small and mid-sized enterprises in nine states in the Rust Belt and the Southeast. Since 2001, the Natural Capital Investment Fund has helped 253 companies create or retain forty-five hundred jobs while also providing over fifteen hundred companies with 19,742 hours of technical assistance. Beneficiaries have ranged from family farms to handmade furniture companies.[65] These are loans to be repaid, designed to inject capital into ecologically significant regions where economic conditions may make it difficult for landowners and businesses to secure commercial funding.[66] Because these grants are made directly to landowners and small businesses, applications are handled through a web portal at https://www.ncifund.org/apply.[67]

TCF also funds conservation easements. TCF easements have benefited many species, including the greater sage-grouse. Working with a coalition of fourteen government agencies and private partners, TCF has funded easements on hundreds of thousands of acres of sage-grouse habitat in Colorado, Idaho, and Wyoming. This includes 58,500 acres on fifteen properties in Wyoming and, through the Pioneers Alliance, eighty thousand acres in central Idaho.[68] TCF also played a critical role in coordinating fundraising for the Coastal Headwaters Longleaf Forest Conservation and Restoration Project discussed in chapter 9.[69]

National Wildlife Federation

The nation's largest membership-based conservation organization, the National Wildlife Federation (NWF) traces its history back to a gathering of more than two thousand conservationists at the Mayflower Hotel in Washington, DC, in 1936, convened by the cartoonist and former head of the U.S. Biological Survey (now the U.S. Fish and Wildlife Service) J. N. "Ding" Darling.[70] Today, the organization claims six million members spread among fifty-one state and territory affiliates.[71] In fiscal year 2017/2018, NWF had a budget of over $85 million and net assets of $71 million.[72] NWF is instantly

Figure 11.3. J. N. "Ding" Darling's 1936 illustration of the diverse constituencies who created the National Wildlife Federation to promote conservation programs in Congress.
Courtesty of the "Ding" Darling Wildlife Society.

recognizable to school children across the country due to its line of Ranger Rick publications, first launched in 1959.[73] NWF's work is driven by its fifty-one affiliates, each of which is an independent 501(c)(3) corporation with its own staff and programs.[74] Local affiliates have diverse names and missions. For example, the Tennessee Wildlife Federation operates a habitat conservation program driven by sportsmen that includes public access for hunting and fishing as a priority, whereas PennFuture primarily focuses on clean air and water policy.[75] Information about affiliates is available at https://www.nwf.org/About-Us/Regional-Centers-and-Affiliates.[76]

As of 2019, many of NWF's national programs involve partnerships that provide financial assistance, technical assistance, and training to landowners. One such program promotes cover crops through a network of "Cover Crop Champions."[77] Another promotes longleaf pine restoration in the southeast, including one-on-one work with forest landowners.[78] NWF also promotes public awareness of pollinator declines and encourages the development of pollinator habitat through initiatives such as its backyard wildlife certification program, the Mayors' Monarch Pledge, and support for Farm Bill programs that conserve native grasslands.[79] NWF has taken the lead in promoting the development of milkweed habitat for the monarch butterfly, whose population has declined 97 percent since the 1980s.[80] A small NWF program, in partnership with the Wild Sheep Foundation, is a good case study of voluntary conservation. The two groups work with woolgrowers to prevent disease transmission from their domestic sheep to vulnerable wild populations. Landowners are encouraged to relocate their operations or convert from sheep to cattle, and if neither of these are possible the two groups will offer to buy

Figure 11.4. A Rocky Mountain bighorn sheep (*Ovis canadensis canadensis*). Photograph by Robbie George.

out the woolgrower for a fair price. This partnership works with landowners rather than against them, while protecting iconic wild species.[81]

Ducks Unlimited

With a history that stretches back to 1937, Ducks Unlimited (DU) is among the oldest and largest conservation organizations. As of 2019, the organization claimed almost eight hundred thousand members in the United States, Mexico, and Canada,[82] and over twenty-six hundred chapters.[83] Its 2018 tax filings indicate annual revenue of $211 million and net assets of $239 million.[84] DU works in every U.S. state and has carried out a total of over twenty thousand conservation projects across North America that have directly conserved 14,478,372 acres of land. An additional 177,514,637 acres in Canada have been conserved under government programs supported by DU.[85]

DU does not focus on any Endangered Species Act–listed species, but listed and at-risk species invariably benefit from its programs. DU focuses on wetlands and riparian areas, which provide valuable habitat for many species. Some of DU's strongest programs are in the Great Plains and Prairie Pothole regions, where it conserves prairie wetlands and grasslands, and conserves working lands through the Farm Bill and other programs.[86] DU does not make grants directly to landowners but provides matching funds for federal conservation programs, purchases easements, and provides technical support.

Figure 11.5. A wood duck (*Aix sponsa*).
Oral Zirek/Shutterstock.

Rarely, DU purchases land, restores it, and sells or donates it to another landowner. Like Joint Ventures, DU also has a strong science program that translates new research about a species into geographic information that can help target high-priority areas for conservation.[87] DU's programs are operated through the organization's four regional offices, which are the best point of contact for landowners. Information on regional offices can be found online at https://www.ducks.org/about-ducks-unlimited/regional-offices.[88] DU also has offices in Canada and Mexico, each with their own national staff and conservation programs.[89] For example, DU Canada has worked with the Prairie Habitat Joint Venture to provide technical assistance and economic incentives, including seed cost reductions, to landowners willing to convert former cropland into forage for hay harvest.[90]

National Audubon Society

With history dating back to 1905, the National Audubon Society (Audubon) is one of the most widely recognized conservation organizations in the nation. Originally formed as a national coalition of independent state-based societies, Audubon has grown to encompass twenty-three state programs, forty-one nature centers, nearly five hundred local chapters, and 1.7 million members.[91] Contact information for offices, chapters, centers, and sanctuaries can be found online at https://www.audubon.org/about/audubon-near-you.[92] In 2017/2018 tax filings, the organization reported $133.5 million of revenue and $480 million in net assets.[93] Audubon is dedicated to conserving birds of all types, with a historical focus on waterfowl and migratory birds.

Best known as a national advocacy and policy organization, and for its nature centers, preserves, magazine, and photography awards, Audubon also has a robust and expanding private lands conservation portfolio. Audubon is involved in countless conservation partnerships like the ones described in this book, including for example with the Intermountain West Joint Venture and other partners on greater sage-grouse conservation. It also participates in many of the Farm Bill programs described in this book in partnership with other groups like DU and Pheasants Forever, often as a financial backer for easement purchases, private cost-shares, and other conservation initiatives.[94] In the Prairie Pothole states of North and South Dakota, Audubon recently launched its own conservation enrollment program, the Prairie Management Toolbox Program. In this program, landowners agree to maintain working grasslands as prairie for no less than ten years, and in exchange Audubon provides 60 percent of the cost for improvements such as fencing, grazing plan development, water resources development, invasive species removal, and seeding and restoring grassland, all of which improve wildlife habitat and also enhance the value of working ranchland. Landowners are responsible for 40 percent of costs but may meet that obligation through in-kind contributions of labor. In the first eighteen months of this relatively small pilot program, 183 landowners have enrolled over ten thousand acres.[95]

One of Audubon's most innovative programs is the Conservation Ranching program, which uses free market incentives to promote grassland range conservation. Launched in 2015, the program uses the Audubon name to certify beef products as "bird friendly." To qualify, cattle must be pasture-raised, grass-fed, and grass-finished, and ranch operations must meet Audubon standards for forage, feeding, habitat management, environmental stewardship, and animal husbandry. Audubon staff provide participating ranchers with assistance in preparing ranch management plans and connect ranchers to participating slaughterhouses that process grass-fed and -finished animals. Audubon also finances bird monitoring, soil carbon level monitoring, and independent third-party verification of compliant grazing practices. Ranchers absorb any increase in operational costs, although in many cases environmentally friendly ranching practices actually reduce operating costs. Participating ranchers are able to market their beef as an Audubon-certified bird-friendly product and command retail price premiums based on the willingness of consumers to pay more for that product. In its first 2.5 years of operation, this program has enrolled two million acres of land in thirteen states, including not only private land but also public rangelands leased from the Bureau of Land Management, U.S. Forest Service, and U.S. Fish and Wildlife Service. To support this program, Audubon employs forty field staff and invests two to three million dollars annually. Because of the Conservation Ranching program's rapid growth, Audubon is exploring the possibility of exporting this product certification model of conservation on working lands to other ecosystems such as eastern hardwood forests.[96]

For both Conservation Ranching and the Prairie Management Toolbox Program, the lead implementing office is Audubon Dakota, https://dakota.audubon.org/.[97] Various conservation programs are run by different Audubon offices around the country, making them responsive to local needs. Audubon's Working Lands portal, https://www.audubon.org/conservation/working-lands, provides information about working lands programs across the country.[98]

Mule Deer Foundation

Founded in 1988, the Mule Deer Foundation (MDF) has chapters in twenty-two states and over forty thousand members. According to 2018 tax filings, the organization has close to $15 million in revenue and $4 million in assets.[99] Mule deer are found across the western United States, and all mule deer populations are either stable or declining. Although the species is not widely

Figure 11.6. A mule deer (*Odocoileus hemionus*).
Gray Photo Online/Shutterstock.

regarded as at-risk, some states have included it in their State Wildlife Action Plans. Like the greater sage-grouse, the mule deer is an indicator species for the imperiled sagebrush ecosystem.[100]

Mule deer live on both public and private land, according to migration patterns dictated by local climate and topography. MDF devotes a majority of its programs to mule deer habitat on public land, but it also works with private landowners providing financial and technical assistance. MDF chapters and volunteers also work directly with landowners on their land. MDF also owns conservation easements, although it does not generally purchase them with its own funds.[101] MDF primarily works through its chapters and partners to make connections to private landowners. Information about chapters is available at https://muledeer.org/state/.[102] Although there is some technical assistance available on the organization's website, MDF's main outreach is through its free magazine. Some MDF projects involve proactively reaching out to landowners for specific purposes, such as educating landowners on the importance of migration corridors.[103]

National Wild Turkey Federation

NWTF, founded in 1973, today claims sixteen hundred local chapters and 230,000 members in all fifty U.S. states.[104] In tax year 2018, it reported

Figure 11.7. A wild turkey (*Meleagris gallopavo*).
Tory Kallman/Shutterstock.

$40 million in revenue and net assets of $23 million.[105] Wild turkeys can be found in every U.S. state except for Alaska,[106] and NWTF's work spans the nation. The organization is currently in the middle of a ten-year initiative called "Save the Habitat. Save the Hunt." This program seeks to conserve four million acres of wildlife habitat, recruit 1.5 million hunters, and secure access to five hundred thousand new acres for hunting.[107]

NWTF focuses on assisting government agencies and private landowners in active management using scientific techniques, and it also holds some conservation easements in critical locations.[108] Tools used include grants to state fish and wildlife agencies, wildlife food plot seed distribution, and technical support to NGOs, landowners, and other partners. NWTF's website includes a wealth of technical information accessible to the public.[109] Most of NWTF's work with landowners is through the Farm Bill. The organization employs thirty biologists funded by a cost-sharing agreement with NRCS that work with private landowners, some located in NRCS offices and some in NWTF offices. NWTF also provides support to Partners for Fish and Wildlife.[110] Finally, NWTF chapters each raise money for cost-sharing payments and in-kind labor needed by state fish and wildlife agencies and conservation districts. This money provides local matches for federal programs and ultimately benefits landowners.[111] Contact information for NWTF chapters can be found at http://www.nwtf.org/about/nation.[112]

Rocky Mountain Elk Foundation

The Rocky Mountain Elk Foundation (RMEF), founded in 1984, claims more than 234,000 members in over five hundred chapters, and over the years has marshalled over twelve thousand volunteers, protected more than 7.4 million acres, opened or secured hunting access to over 1.2 million acres, and carried out close to twelve thousand projects.[113] In 2017, the organization reported income of $53 million and net assets of $85 million.[114]

Unlike many organizations profiled here, RMEF focuses on acquiring conservation easements and does very little technical assistance. RMEF is an "eligible entity" under the Farm Bill's Agricultural Conservation Easement Program and participates in that program in partnership with NRCS. As an easement holder, RMEF operates as a land trust, and as of 2019 owned

Figure 11.8. An elk (*Cervus canadensis*).
Peter Wey/Shutterstock.

215 easements on four hundred thousand acres in sixteen states. In order to execute these transactions, the organization maintains legal, real estate, and tax experts on staff, as well as grant writers to help secure funding. RMEF primarily raises outside money for conservation easements, using its own resources to bring the various parties and funders together. RMEF provides technical assistance for easement management, although RMEF's easements generally do not have formal land management plans. RMEF identifies landowners interested in donating or selling easements by word of mouth, through neighbors and other local organizations.[115]

RMEF's other main focus is on partnerships with federal land management agencies and state fish and wildlife agencies to protect, restore, and enhance elk habitat. It has been a partner in successful programs to reintroduce elk in Kentucky, Missouri, North Carolina, Tennessee, Virginia, West Virginia, Wisconsin, and Ontario, Canada, and has provided technical assistance supporting evaluations for reintroductions in Illinois, Maryland, and New York.[116] RMEF chapters raise money to support a variety of conservation grants, include grants directly to landowners, but most grants go to federal and state agencies for elk conservation on public lands open to hunting.[117] Local chapters can be found through http://www.rmef.org/NewsandMedia/PressRoom/AboutRMEF/Chapters/RMEFChapterMap.aspx.[118]

Trout Unlimited

Trout Unlimited (TU), founded in 1959, is dedicated to the conservation of trout, salmon, and other coldwater species and their habitats. With three hundred thousand members, 387 chapters, and thirty-six state councils, TU is the nation's largest coldwater fisheries conservation organization.[119] In 2017/2018, it reported revenue of $53.6 million and net assets of $22.3 million.[120]

TU's habitat conservation work, although focused on coldwater fisheries, impacts overall ecosystem health and benefits many species, including nongame species of fish, wildlife, and birds. Conservation programs are implemented through TU's "Protect, Reconnect, Restore, Sustain" framework. "Protect" focuses on mountain headwaters and primarily involves political advocacy. "Reconnect" involves removing barriers to fish movement up and down a waterway, whereas "Restore" improves riparian habitat. "Sustain" refers to the organization's volunteer and financial resources.[121] In 2017, TU invested 734,824 volunteer hours into 1,013 projects, protected 1,164 miles of streams, reconnected 737 miles of streams, and restored 361 miles of streams.[122]

Riparian areas tend to be privately owned, and much of TU's work on reconnecting and restoring waterways involves private landowners. Reconnecting rivers and streams may require dam removal, culvert improvements,

Figure 11.9. Cutthroat trout (*Oncorhynchus clarkii*) are a variety of trout popular with anglers. Several subspecies are listed as threatened, but the subspecies seen here, a coastal cutthroat trout (*Oncorhynchus clarkii clarkii*), is not.
Wikipedia.

acquiring water rights, and improving irrigation efficiency. Restoring riparian habitats can involve re-meandering streams and fencing cattle out of riparian corridors. These activities rely on partnerships with private landowners and also benefit landowners by reducing water use, improving water retention, and reducing erosion.[123]

TU's work with private landowners takes many forms. Some projects are simply carried out on private land, with TU volunteers providing labor and equipment. Other projects are carried out in partnership with other NGOs and government agencies, including through the NRCS Regional Conservation Partnership Program.[124] TU also provides landowners with private matching funds for Farm Bill and National Fish and Wildlife Foundation programs. Landowners connect to TU in a variety of ways, including through other local organizations and TU's own chapters. Local chapters can be found through https://www.tu.org/find-your-chapter/.[125]

Pheasants Forever and Quail Forever

Pheasants Forever and Quail Forever (PF), founded as Pheasants Forever in 1982, has 149,000 members in over seven hundred local chapters in forty-five states.[126] Its Quail Forever division was established in 2005, and today the organization works to conserve both pheasant and quail by focusing on upland bird habitat. In 2018 tax filing documents, PF reported revenue of $76 million and net assets of $35 million.[127] Altogether, PF has created or

Figure 11.10. Ring-necked pheasants (*Phasianus colchicus*).
Marcin Perkowski/Shutterstock.

enhanced wildlife habitat on more than 15.8 million acres in the United States and Canada.[128]

PF works with private landowners through both the Farm Bill and its local chapters. As discussed in chapter 6, PF is a leading organization in assisting with the delivery of Farm Bill conservation programs. Its Farm Bill Biologist program began in 2005 with three positions in Ohio and has grown to 188 biologists in thirty states. These biologists work at USDA Service Centers, and since 2003 have provided information to over 231,500 landowners and impacted ten million acres of private land.[129] PF also employs professional Habitat Teams that carry out on-the-ground conservation such as restoring habitat, treating for invasive species, etc.[130] The majority of PF's other work with landowners is done at the chapter level. Chapters provide financial assistance by raising money for matching funds for federal programs, foundation grants, and other large conservation investments. Chapters also provide technical assistance, loans of equipment to members and landowners in their communities, and in some cases conservation planning. Chapters also connect landowners to PF's national conservation program and other partners such as state and federal agencies and other NGOs.[131] The best way for landowners to connect with PF is through its biologists working at USDA Service Centers or by contacting their local chapter. Contact information for chapters is available at https://pheasantsforever.org/Participate/Find-a-Chapter.aspx.[132]

American Bird Conservancy

ABC was established in 1994 with a mission of conserving native birds and their habitats throughout the Americas.[133] In 2017, it reported revenue of $17.5 million and net assets of $10.7 million. ABC is based in Virginia, with over half of its staff working in small offices around the country. The organization is involved in many of the partnerships described in this book, including the delivery of Farm Bill conservation programs and Joint Ventures.[134]

ABC focuses its private lands work on bird populations not addressed by larger conservation programs. This has led to the development of ABC's "BirdScapes" initiatives, which focus on landscapes where migratory birds serve as indicator species for ecological systems in decline. One of the goals of BirdScapes is to prevent the need for listing species under the Endangered Species Act of 1973.[135] For example, Michigan, Minnesota, and Wisconsin are home to ABC's Golden-winged Warbler BirdScape Initiative, which also benefits the American woodcock, Kirtland's warbler, and approximately twenty other at-risk species that rely on early and mixed successional forest habitat. ABC is the lead partner in implementing the voluntary Golden-winged Warbler Conservation Plan in the Great Lakes. The initiative is a $10 million project funded by a combination of a 2015 Regional Conservation Partnership Program grant from NRCS, grants from the National Fish and Wildlife Foundation, and the Minnesota Outdoor Heritage Fund. It works through grassroots implementing

Figure 11.11. A golden-winged warbler (*Vermivora chrysoptera*).
Frode Jacobsen/Shutterstock.

committees that set local priorities in each of the three states and six professional foresters that meet with landowners, conduct trainings for landowners and private consulting foresters, and prepare individual land management plans. ABC then provides technical support for landowners to hire contractors and implement forest thinning and regeneration activities. Moreover, NRCS reimburses the landowners for their expenses. To date, the initiative has established approximately six thousand acres of high-quality habitat, and work is ongoing. Indiana University of Pennsylvania and Cornell University support the initiative with annual site monitoring.[136]

Quality Deer Management Association

The Quality Deer Management Association (QDMA) was founded in 1988 to "promote sustainable, high-quality deer populations, wildlife habitats and ethical hunting experiences through research, education, advocacy, and hunter recruitment."[137] In 2017, the organization reported $7 million in revenue and $2 million in net assets.[138] As of 2019, QDMA has two hundred branches and sixty thousand members.[139] QDMA is primarily a hunter advocacy organization. However, 60 percent of its members

Figure 11.12. A whitetail deer (*Odocoileus virginianus*).
Andrew Kandel/Alamy Stock Photos.

own, maintain, and improve land and habitat for whitetail deer, and these activities benefit myriad wildlife species. QDMA supports this work by providing technical support to its members and others. It hosts seminars and workshops around the country, publishes a magazine, and maintains a website with resources for landowners, information on branches (https://www.qdma.com/get-involved/find-a-branch/), and events (https://www.qdma.com/get-involved/events/).[140]

Ruffed Grouse Society & American Woodcock Society

The Ruffed Grouse Society & American Woodcock Society (RGS/AWS) started in 1961 as the Ruffed Grouse Society. Like Pheasants Forever, the Ruffed Grouse Society expanded its mission in 2014 when it created the American Woodcock Society. Although a chapter might be a member of one organization or the other, the two are a single national entity, which claims twenty thousand members in 130 chapters.[141] Its 2018 tax filings show revenues of $5 million and net assets of $6.6 million.[142]

RGS/AWS provides private landowners with technical assistance and conservation planning. Although RGS/AWS staff do not prepare land management plans directly, they can provide landowners with resources that will help them identify priorities, improve management practices, and connect with other conservation organizations.[143] RGS/AWS' work is driven by local part-

Figure 11.13. A male ruffed grouse (*Bonasa umbellus*) "drums" to attract a mate.
Agnieszka Bacal/Shutterstock.

nerships. Most is in the Great Lakes region, where RGS/AWS is part of multiple Farm Bill Regional Conservation Partnership Program projects.[144] Most RGS/AWS programs focus on public lands, but its local programs vary greatly. Landowners in need of assistance should contact their local chapter. Chapter information can be found at https://ruffedgrousesociety.org/chapter-info/.[145]

Land Trusts

For many of the conservation programs discussed in this book, local land trusts are critical partners. Technically, a land trust is any charitable organization that owns or manages land or conservation easements for conservation purposes.[146] The history of land trusts can be traced all the way back to 1891, when the landscape architect Charles Eliot established The Trustees of Reservations, dedicated to natural and historical places in Massachusetts.[147] The Trustees of Reservations is still operating today and has conserved 76,868 acres of land.[148]

In a 2015 survey, the Land Trust Alliance, a national advocacy group representing land trusts, identified 1,363 active land trusts. As of 2015, these land trusts protected 56,434,181 acres of land through a combination of easements, fee simple ownership, and other mechanisms, including 16,784,077 acres of conservation easements.[149] Land trusts range in size from TNC, with 21,558,908 acres conserved,[150] all the way down to local organizations with just a few hundred acres conserved.[151] Most land trusts are small, local organizations that work largely within a single state or county and are intimately familiar with the landscape in that area. Landowners interested in conservation easements should contact a local land trust, which can be found through the Land Trust Alliance at https://www.findalandtrust.org/.[152] These land trusts purchase easements, raise money to finance easements, accept donations of easements, and can advise landowners on all matters relating to conservation easements. The number of these organizations is growing. In North Carolina, for example, in the last forty years the land trust community has grown from just one to twenty-seven organizations.[153]

One of those North Carolina land trusts is the North Carolina Coastal Land Trust, established in 1992 and now the largest land trust in the state, conserving more than 79,000 acres in thirty-one counties within North Carolina's coastal plain.[154] The trust's assets include nature preserves that it owns, manages, and opens to the public, as well as conservation easements on private land. To maintain these properties, the trust has established an endowment that funds annual site monitoring on all of its properties, ensuring that landowners abide by conservation easements and utilize appropriate land management practices. This site monitoring is a requirement of both conservation easements and the governance of land trusts themselves.[155] Additional programs provide landowners with outreach and education.[156]

Figure 11.14. Charles Eliot, 1859–1897, was an American landscape architect who pioneered the concept of a land trust, establishing The Trustees of Reservations in Massachusetts in 1891. Courtesy of Southwest Harbor Public Library Digital Archive.

North Carolina's Salmon Creek State Natural Area is an example of how the North Carolina Coastal Land Trust works with multiple stakeholders to secure public benefits. The preserve is 995 acres of riverfront, freshwater marsh, forested swamps, and higher land, containing eighteen significant archeological sites, including one suspected to be associated with the lost

Figure 11.15. View of Salmon Creek State Natural Area in Bertie County, North Carolina. The land making up Salmon Creek was purchased by the North Carolina Coastal Land Trust in 2016, using a loan from The Conservation Fund, and subsequently transferred to the state.
Courtesy of North Carolina Coastal Land Trust.

colony of Roanoke. It is also on the training approach to the U.S. Air Force's Dare County Bombing Range, the protection of which falls within the purview of the U.S. Department of Defense's Readiness and Environmental Protection Integration program. The property's owners had planned to develop it, but plans were put on hold by the 2008 economic crash. In 2016, they instead approached the Coastal Land Trust about purchasing the property. The trust worked with TCF to secure a multi-million dollar loan, and later raised the money to repay it. In November 2018, the trust transferred the property to the State of North Carolina, which will permanently protect it for the benefit of future generations.[157]

Notes

1. The development of federal trust species occurred in the twentieth century, with passage of the Migratory Bird Treaty Act of 1918, the Bald Eagle Protection Act of 1940 (expanded to the Bald and Golden Eagle Protection Act in 1962), and the Endangered Species Act of 1973. The species these laws placed under federal jurisdiction include many of the most iconic,

imperiled, and controversial, including the polar bear, gray wolf, grizzly bear, northern spotted owl, red-cockaded woodpecker, and whooping crane.

2. Morgan et al., "A State Assessment of Private Lands Wildlife Conservation in the United States."

3. Ibid.

4. Ibid.

5. Arkansas Game and Fish Commission, "Find a Private Lands Biologist," last accessed July 27, 2019, https://www.agfc.com/en/wildlife-management/private-lands-program/private-lands-biologist/.

6. Arkansas Game and Fish Commission, "Private Lands Program," last accessed July 24, 2019, https://www.agfc.com/en/wildlife-management/private-lands-program/.

7. This program is in a pilot phase as of 2019. Arkansas Game and Fish Commission, "Arkansas Waterfowl Rice Incentive Conservation Enhancement," 2019, https://drive.google.com/file/d/1n9Bimc7FaQ_7Y26FCcP_IupemrHe6iQA/view.

8. Arkansas Game and Fish Commission, "Acres for Wildlife Program," last accessed July 24, 2019, https://www.agfc.com/en/wildlife-management/private-lands-program/acres-for-wildlife/.

9. Colorado Parks and Wildlife, "Colorado Wildlife Habitat Program," last accessed July 24, 2019, https://cpw.state.co.us/aboutus/Pages/LandWaterCWHP.aspx.

10. Ibid.; Colorado Parks and Wildlife, "Colorado Wildlife Habitat Program 2019 Request for Proposals (RFP)," 2019, https://cpw.state.co.us/Documents/LandWater/COWildlifeHabitatProtectionProgram/2019-CWHP-RFP_ProposalInstructions.pdf. The Cooperative Habitat Improvement Program allows third-party ownership of conservation easements and requires land management plans for all easements.

11. Colorado Parks and Wildlife, "Pheasant Habitat Improvement Program," last accessed July 24, 2019, https://cpw.state.co.us/aboutus/Pages/PheasantHabitatImprovementProgram.aspx.

12. Colorado Parks and Wildlife, "Habitat Partnership Program (HPP)," last accessed July 24, 2019, https://cpw.state co.us/aboutus/Pages/HabitatPartnershipProgram.aspx.

13. Colorado Parks and Wildlife, "Private Land Program," last accessed July 24, 2019, https://cpw.state.co.us/aboutus/Pages/PrivateLand.aspx.

14. Kansas Department of Wildlife Parks and Tourism, "Wildlife Biologists," last accessed July 25, 2019, https://ksoutdoors.com/Services/Private-Landowner-Assistance/Wildlife-Biologists; Kansas Department of Wildlife Parks and Tourism, "Wildlife Biologists & Bio-Technicians," April 2019, https://ksoutdoors.com/content/download/26765/182561/version/11/file/Biologists+Counties_April_2019.pdf.

15. Payment is reduced if agency staff provide in-kind services or the same conservation action is enrolled in a federal conservation program. Kansas Department of Wildlife Parks and Tourism, "Habitat First Habitat Practices and Payments," January 2016, https://ksoutdoors.com/content/download/46850/478951/version/1/file/Habitat+Practices+and+Payments.pdf.

16. The Walk-in Hunting Access program pays $9–18 per acre per year and the Fishing Impoundments and Stream Habitats program pays $65–125 per pond acre per year, with larger payments for stream or river access. Kansas Department of Wildlife Parks and Tourism, "Public Hunting Access & Your Land," 2018, https://ksoutdoors.com/content/download/39527/378058/version/2/file/WIHA+Flyer+-+2018+Update.pdf; Kansas Department of Wildlife Parks and Tourism, "Fishing Impoundments and Stream Habitats (F.I.S.H.) Program," last accessed July 25, 2019, https://ksoutdoors.com/Fishing/Special-Fishing-Programs-for-You/F.I.S.H.-Program.

17. For example, Kansas Department of Wildlife Parks and Tourism, "Pond Management Program," last accessed July 25, 2019, https://ksoutdoors.com/Fishing/Special-Fishing-Programs-for-You/Pond-Management-Program.

18. South Dakota Habitat Pays, "SD Habitat Pays," last accessed July 27, 2019, http://habitat.sd.gov/.

19. South Dakota Habitat Pays, "Habitat Advisors," May 28, 2019, http://habitat.sd.gov/advisors/docs/habitat-advisor-map.pdf.

20. Programs are categorized as follows: grassland habitat, wetland habitat, woody habitat, food plots, pollinator habitat, wildlife habitat management, hunting access programs, watershed protection, forestry programs, resources conservation, and other. South Dakota Habitat Pays, "Habitat Resources for Landowners," last accessed July 25, 2019, http://habitat.sd.gov/resources/default.aspx.

21. South Dakota Habitat Pays, "Dense Nesting Cover," last accessed July 25, 2019, http://habitat.sd.gov/resources/nestingcover.aspx; South Dakota Habitat Pays, "Pollinator Habitat Assistance," last accessed July 25, 2019, http://habitat.sd.gov/resources/pollinator-habitat.aspx; South Dakota Habitat Pays, "Working Grassland Habitat Programs," last accessed July 25, 2019, http://habitat.sd.gov/resources/workinggrassland.aspx.

22. South Dakota Habitat Pays, "Hunting Access Programs," last accessed July 25, 2019, http://habitat.sd.gov/resources/access.aspx; South Dakota Department of Game Fish and Parks, "2019 Walk-in Area Program Information," 2019, http://habitat.sd.gov/WIA%20INFO%20PACK%202019.pdf.

23. Justin Dreibelbis, email to Christopher E. Segal, July 26, 2019.

24. Texas Parks & Wildlife, "Find a Wildlife Biologist," last accessed July 30, 2019, https://tpwd.texas.gov/landwater/land/technical_guidance/biologists/. The TPW website also provides a form landowners can use to make a request for technical assistance. Texas Parks & Wildlife, "Landowner Request for Technical Guidance," 2013, https://tpwd.texas.gov/publications/pwdforms/media/pwd_0153_w7000_landowner_request_technical_guidance.pdf.

25. Texas Parks & Wildlife, "Landowner Incentive Program (LIP)," last accessed July 30, 2019, https://tpwd.texas.gov/landwater/land/private/lip/; Dreibelbis; Meredith Longoria, email to Christopher E. Segal, July 29, 2019.

26. Texas Parks & Wildlife, "Texas Farm and Ranch Lands Conservation Program," last accessed July 30, 2019, https://tpwd.texas.gov/landwater/land/private/farm-and-ranch/; Christopher Abernathy, email to Christopher E. Segal, July 26, 2019.

27. This includes information on land management practices, Farm Bill programs, state tax benefits, conservation easements, prescribed burning, and more. Texas Parks & Wildlife, "Private Lands and Habitat Program," last accessed July 30, 2019, https://tpwd.texas.gov/landwater/land/private/.

28. There are exceptions to this, and in some states the state fish and wildlife agency instead implements the Forest Legacy Program. For example, Henning.

29. Information on this program is available on the agency's website. Colorado State Forest Service, "Grants & Funding Assistance," last accessed July 27, 2019, https://csfs.colostate.edu/funding-assistance.

30. In order to participate in this program, forest landowners must have a professionally prepared forest management plan. Colorado State Forest Service, "Forest Ag Program," last accessed July 26, 2019, https://csfs.colostate.edu/forest-ag-program/.

31. Other partners include other state agencies, federal agencies, and NGOs such as Pheasants Forever, TNC, and RMEF. Colorado State Forest Service, "Forest Stewardship Program," last accessed July 26, 2019, https://csfs.colostate.edu/forest-stewardship-program/;

Colorado State Forest Service, "Colorado's Forest Stewardship Program," 2016, https://csfs.colostate.edu/media/sites/22/2016/05/stewardship-brochure-web.pdf.

32. Georgia Forestry Commission, "Landowner Services," last accessed July 26, 2019, http://www.gfc.state.ga.us/forest-management/private-forest-management/landowner-programs/.

33. Georgia Forestry Commission, "Contact Us," last accessed July 27, 2019, http://www.gfc.state.ga.us/about-us/contact-us/county-units/index.cfm.

34. Georgia Forestry Commission, "Forest Legacy & Conservation Easements," last accessed July 26, 2019, http://www.gfc.state.ga.us/forest-management/private-forest-management/landowner-programs/forest-legacy/index.cfm.

35. This includes Farm Bill programs, other cost-share and incentive programs, prescribed burning, firebreak construction, tax and estate planning, water quality issues, forestry best management practices, and the management of invasive weeds and pests. Georgia Forestry Commission, "Landowner Services"; Georgia Forestry Commission, "Cost Share & Incentive Programs," last accessed July 26, 2019, http://www.gfc.state.ga.us/forest-management/private-forest-management/landowner-programs/other-landowner-programs/index.cfm; Georgia Forestry Commission, "Forest Health in Georgia," last accessed July 26, 2019, http://www.gfc.state.ga.us/forest-management/forest-health/index.cfm.

36. Mississippi Forestry Commision, "Private Landowner Services," last accessed July 26, 2019, https://www.mfc.ms.gov/private-landowner-services.

37. Mississippi Forestry Commision, "Forest Stewardship Program," last accessed July 26, 2019, https://www.mfc.ms.gov/Forest-Stewardship; Mississippi Forestry Commision, "Reforestation Tax Credit," last accessed July 26, 2019, https://www.mfc.ms.gov/reforestation-tax-credit.

38. The Forest Resource Development Program provides a 50 percent to 75 percent cost-share of up to $7,000 per landowner per year to support long-term timber production. Enrolling landowners must protect the area from fire and grazing for a minimum of ten years. Mississippi Forestry Commision, "Forest Resource Development Program," last accessed July 26, 2019, https://www.mfc.ms.gov/FRDP. The Southern Pine Beetle Prevention Program provides up to $100 per acre to support forest thinning, subject to various conditions on the specific property. Mississippi Forestry Commision, "Southern Pine Beetle Prevention Program," last accessed July 26, 2019, https://www.mfc.ms.gov/SPB-Prevention.

39. Mississippi Forestry Commision, "Private Landowner Services."

40. Wisconsin Department of Natural Resources, "Professional Forestry Assistance," last accessed July 26, 2019, https://dnr.wi.gov/topic/ForestLandowners/assist.html.

41. Eligible properties must be twenty or more contiguous acres of non-industrial private forest, 80 percent covered by forest growing commercial timber products, unimproved, and with a forest management plan of twenty-five or fifty years. The Managed Forest Law replaced the earlier 1927 Forest Crop Law, but some Forest Crop Law contracts are still ongoing and the Department of Natural Resources manages them as well. Wisconsin Department of Natural Resources, "Managed Forest Law," last accessed July 26, 2019, https://dnr.wi.gov/topic/forestlandowners/mfl/; *Wisconsin's Managed Forest Law: A Program Summary, PUB-FR-295* (November 2017); Wisconsin Department of Natural Resources, "Forest Crop Law," last accessed July 26, 2019, https://dnr.wi.gov/topic/ForestLandowners/fcl/.

42. This program pays up to 50 percent of landowner costs to implement priority forest stewardship practices, provided that the landowner owns between ten and five hundred contiguous acres of non-industrial private forest. Wisconsin Department of Natural Resources, "Wisconsin Forest Landowner Grant Program (WFLGP)," last accessed July 26, 2019, https://dnr.wi.gov/Aid/ForestLandowner.html.

43. Topics include timber production, forest health, wildlife management opportunities, protecting threatened and endangered species, erosion control, recreation opportunities, and forest aesthetics. Wisconsin Department of Natural Resources, "DNR Service Foresters," last accessed July 26, 2019, https://dnr.wi.gov/topic/ForestLandowners/dnrForesters.html.

44. Pollinator Stewardship Council Inc., "Funding Programs for Pollinator Habitat," last accessed July 26, 2019, http://pollinatorstewardship.org/index.php/funding-programs-for-pollinator-habitat/.

45. Laura Miller, "Beekeeping Qualifies as an Agricultural Use Enterprise," April 13, 2012, Texas A&M AgriLife Extension, https://agrilife.org/tcch/2012/04/13/beekeeping-qualifies-as-an-agricultural-use-enterprise/.

46. Jackass Honey Farms, "How to Get Qualified for an Ag Valuation in Texas," 2018, https://www.jackasshoneyfarms.com/bee-tax-exemption.

47. Projects considered for funding under the most recent 2018 bond issuance included stream and river restoration and cleanup, trail construction, and acquisitions of trails and open spaces, including through conservation easements. City of Missoula, "2019 Open Space Bond," last accessed July 26, 2019, https://www.ci.missoula.mt.us/2429/2018-Open-Space-Bond; Missoula County, Montana, "2006 Open Space Bond Program for Missoula County," last accessed July 26, 2019, https://www.missoulacounty.us/government/community-development/community-planning-services/open-lands/open-space-bond-program#ad-image-0; "Report to Missoula City Council: Projects for the 2018 Proposed County Wide General Obligation Bond and Proposed City Wide Conservation and Responsible Stewardship Mill Levy," 2018, https://www.ci.missoula.mt.us/DocumentCenter/View/46387/2018-06-25-Projects-for-the-2018-Proposed-County-Wide-GO-Bond.

48. U.S. Internal Revenue Service, *990 Form Filed by the Nature Conservancy* (2018), 1.

49. The Nature Conservancy, "Contact Us," last accessed February 21, 2019, https://www.nature.org/en-us/connect/contact-us/.

50. Brent Keith, interview by Christopher E. Segal, June 28, 2019.

51. Ibid.

52. Jackson; Annand. For more information on Safe Harbor Agreements, see chapter 13.

53. Keith.

54. Ibid.; The Nature Conservancy, "BirdReturns Program Overview," last accessed July 26, 2019, http://birdreturns.org/home-page/about/about/; The Nature Conservancy, "BirdReturns Bird Response," last accessed July 26, 2019, http://birdreturns.org/bird-response/; Brasher et al., "The History and Importance of Private Lands for North American Waterfowl Conservation."

55. Brasher et al., "The History and Importance of Private Lands for North American Waterfowl Conservation."

56. The actual transaction was complicated. TNC purchased the 31,000 acres of the ranch that were privately owned. Of these, TNC retained 14,400 acres while placing a conservation easement on the other 16,600 acres and reselling them to a local rancher. TNC also took over management of 29,000 acres of public land that had historically been leased to the ranch. Associated Press, "Nature Conservancy to Preserve Historic Ranch," *Daily Journal of Commerce*, January 20, 2000.

57. Grazing fees drop further for additional wildlife-friendly practices, including protecting prairie dog habitat, protecting sage-grouse leks, and modifying fences to make them wildlife-friendly. The Nature Conservancy, "The Matador Ranch," last accessed February 25, 2019, https://www.nature.org/en-us/get-involved/how-to-help/places-we-protect/matador

-ranch/; Brian Martin, "Guest Opinion: TNC's Vision for Montana Prairie Conservation," *Billings Gazette*, February 23, 2019.

58. The Nature Conservancy, "Another Vision for Conservation on the Prairie: The Nature Conservancy Buys Land for Partnership with Neighboring Ranchers," press release, February 5, 2019, https://www.nature.org/en-us/explore/newsroom/montana-rancher-land-partnership/; Martin, "Guest Opinion: TNC's Vision for Montana Prairie Conservation."

59. The Nature Conservancy, "Another Vision for Conservation on the Prairie: The Nature Conservancy Buys Land for Partnership with Neighboring Ranchers"; Martin, "Guest Opinion: TNC's Vision for Montana Prairie Conservation."

60. TCF was founded by Patrick F. Noonan, who had previously been the president of TNC from 1973 to 1980.

61. U.S. Internal Revenue Service, *990 Form Filed by the Conservation Fund* (2017), 1.

62. Reed; The Conservation Fund, "Contact Us," last accessed March 20, 2019, https://www.conservationfund.org/about-us/contact-us; The Conservation Fund, "Financials," last accessed March 20, 2019, https://www.conservationfund.org/about-us/financials.

63. Reed.

64. The Conservation Fund, "Conservation Loans: Flexible Financing and Tailored Support to Advance Conservation," last accessed September 20, 2019, https://www.conservationfund.org/images/resources/Conservation_Loans_Program.pdf. A similar program, the Working Forest Fund, acquires forest land directly. The forest is then restored, placed under conservation easements, and sold to resume its role as a working forest. This fund has protected more than six hundred thousand acres of forests in seventeen states. The Conservation Fund, "Working Forest Fund: Forestry," last accessed March 20, 2019, https://www.conservationfund.org/our-work/working-forest-fund.

65. The Conservation Fund, "Natural Capital Investment Fund: Business Finance," last accessed March 20, 2019, https://www.conservationfund.org/our-work/natural-capital-investment-fund; The Conservation Fund, "Lavender Helps a Family Farm Bloom," last accessed March 20, 2019, https://www.conservationfund.org/projects/lavender-helps-a-family-farm-bloom; The Conservation Fund, "Hand-Built in Asheville," last accessed March 20, 2019, https://www.conservationfund.org/projects/hand-built-in-asheville.

66. Reed.

67. Natural Capital Investment Fund, "Borrow from NCIFund," last accessed July 26, 2019, https://www.ncifund.org/apply. Another program, the Strategic Conservation Planning program, provides technical assistance and counseling to larger entities such as cities and counties, and has worked in close to forty states. Examples include a rails-to-trail project to Durham, North Carolina, and a project to improve water quality and riparian habitat along the Milwaukee River in Milwaukee, Wisconsin. The Conservation Fund, "Cities: Green Infrastructure," last accessed March 20, 2019, https://www.conservationfund.org/our-work/cities-program; Steve Schewel and David Proper, "Cities: Building an Urban Greenspace and Trail in Durham, North Carolina," March 11, 2019, https://www.conservationfund.org/blog/cities/2005-building-an-urban-greenspace-and-trail-in-durham-north-carolina; The Conservation Fund, "Milwaukee River Watershed Conservation Partnership," last accessed March 20, 2019, https://www.conservationfund.org/projects/milwaukee-river-watershed-conservation-partnership.

68. The Conservation Fund, "Supporting the Sage Grouse and Family Ranches in the West," last accessed March 21, 2010, https://www.conservationfund.org/projects/support-for-wyoming-family-ranches-and-the-sage-grouse; The Conservation Fund, "Conservation Working for Ranchers and Sage Grouse in the West," press release, February 17, 2016, https://www.conservationfund.org/news/press-releases/1220-conservation-working-for-ranchers

-and-sage-grouse-in-the-west; The Pioneers Alliance, "The Pioneers Alliance," last accessed March 21, 2019, http://www.pioneersalliance.org/.

69. Chapman, "Coastal Headwaters Project in Florida Is a Major Step for Longleaf Pine Restoration"; Reed.

70. National Wildlife Federation, "History & Heritage," last accessed July 27, 2019, https://www.nwf.org/About-Us/History.

71. U.S. Internal Revenue Service, *990 Form Filed by National Wildlife Federation* (2017), 70.

72. Ibid., 1.

73. National Wildlife Federation, "History & Heritage."

74. Julie Sibbing, interview by Christopher E. Segal, July 8, 2019.

75. Tennessee Wildlife Federation, "Habitat Conservation," last accessed July 27, 2019, https://tnwf.org/conservation/#tab-id-1; PennFuture, "Our Priorities," last accessed July 27, 2019, https://www.pennfuture.org/our-priorities.

76. National Wildlife Federation, "Regional Centers & Affiliates," last accessed July 27, 2019, https://www.nwf.org/About-Us/Regional-Centers-and-Affiliates.

77. Jessica Espenshade, "Become a Cover Crop Champion!", *National Wildlife Federation Blog*, September 21, 2017, https://blog.nwf.org/2017/09/become-a-cover-crop-champion/. Cover crops improve soil biodiversity, water runoff capture, and nutrient retention, and decrease the need for fertilizers.

78. National Wildlife Federation, "Southeast Forestry Program," last accessed July 27, 2019, https://www.nwf.org/Our-Work/Our-Lands/Southeast-Forestry-Program. NWF is also a member of the America's Longleaf Pine Restoration Initiative, a program working in nine states to develop conservation plans, restore and maintain longleaf pine stands, and place easements on longleaf ecosystems. America's Longleaf Restoration Initiative, "America's Longleaf Restoration Initiative," November 2015, https://www.nwf.org/~/media/PDFs/Misc/ALRI-General-Messaging_FINAL.ashx.

79. Mary Phillips, "Pollinator Protection: Pledges, Policy and Proclamations," *National Wildlife Federation Blog*, June 10, 2019, https://blog.nwf.org/2019/06/pollinator-protection-pledges-policy-and-proclamations/.

80. For example, Eve Watling, "Monarch Butterflies Are Going Extinct: How Can We Save Them?" *Newsweek*, January 26, 2019.

81. Kevin Hurley, interview by Lowell E. Baier and Christopher E. Segal, June 25, 2019; Wild Sheep Foundation, "Wild Sheep Foundation Honors Missoula National Wildlife Federation Leaders and BC Hunter-Conservationist," press release, January 25, 2017, https://www.wildsheepfoundation.org/blog/wild-sheep-foundation-honors-missoula-national-wildlife-federation-leaders-and-bc-hunter-conservationist.

82. Ducks Unlimited, "Fact Sheet," January 1, 2019, https://www.ducks.org/media/_global/_documents/stateFactSheets/NationalFactSheet.pdf.

83. Ducks Unlimited, "Ducks Unlimited Recognizes Top Fundraising Chapters across the Nation," press release, March 22, 2016, https://www.ducks.org/press-room/news-releases/ducks-unlimited-recognizes-top-fundraising-chapters-across-the-nation.

84. U.S. Internal Revenue Service, *990 Form Filed by Ducks Unlimited* (2017), 1.

85. Ducks Unlimited, "Where Ducks Unlimited Works," last accessed August 6, 2019, https://www.ducks.org/conservation/where-ducks-unlimited-works; Ducks Unlimited, "Fact Sheet."

86. Johann Walker, interview by Christopher E. Segal, August 2, 2019.

87. Ibid.; Ducks Unlimited, "Conserving Wetlands & Waterfowl," last accessed August 6, 2019, https://www.ducks.org/conservation/how-du-conserves-wetlands-and-waterfowl.

88. Walker; Ducks Unlimited, "DU National & Regional Offices," last accessed August 6, 2019, https://www.ducks.org/about-ducks-unlimited/regional-offices.

89. Ducks Unlimited Canada, "Ducks Unlimited Canada," last accessed September 18, 2019, https://www.ducks.ca/; Ducks Unlimited De México, "Ducks Unlimited De México," last accessed September 18, 2019, http://www.dumac.org/dumac/habitat/ing/index.htm.

90. Brasher et al., "The History and Importance of Private Lands for North American Waterfowl Conservation."

91. National Audubon Society, "About Us," last accessed October 2, 2019, https://www.audubon.org/about; Marshall Johnson, interview by Christopher E. Segal, October 1, 2019.

92. National Audubon Society, "Audubon Near You," last accessed October 2, 2019, https://www.audubon.org/about/audubon-near-you.

93. U.S. Internal Revenue Service, *990 Form Filed by National Audubon Society* (2017), 1.

94. Johnson.

95. National Audubon Society, "Prairie Management Toolbox Program," last accessed October 2, 2019, https://dakota.audubon.org/conservation/prairie-management-toolbox-program; Johnson.

96. National Audubon Society, "Conservation Ranching," last accessed October 2, 2019, https://www.audubon.org/conservation/ranching; Johnson.

97. Audubon Dakota, "Audubon Dakota," last accessed October 3, 2019, https://dakota.audubon.org/.

98. National Audubon Society, "Working Lands," last accessed October 3, 2019, https://www.audubon.org/conservation/working-lands.

99. U.S. Internal Revenue Service, *990 Form Filed by Mule Deer Foundation* (2018), 1.

100. Moretti.

101. Ibid.; Mule Deer Foundation, "Conservation Projects," last accessed July 30, 2019, https://muledeer.org/projects/.

102. Mule Deer Foundation, "MDF in Your State," last accessed October 5, 2019, https://muledeer.org/state/.

103. Moretti. Migration corridors can be as short as a few miles long, giving individual landowners great influence over them.

104. Pedersen and Hatfield.

105. U.S. Internal Revenue Service, *990 Form Filed by National Wild Turkey Federation* (2017), 1.

106. Pedersen and Hatfield.

107. National Wild Turkey Federation, "Save the Habitat. Save the Hunt," last accessed July 27, 2019, http://www.nwtf.org/about/save-the-habitat-hunt.

108. Pedersen and Hatfield.

109. National Wild Turkey Federation, "Manage Your Property," last accessed July 27, 2019, http://www.nwtf.org/conservation.

110. Pedersen and Hatfield.

111. Ibid.

112. National Wild Turkey Federation, "In Your Area," last accessed July 27, 2019, http://www.nwtf.org/about/nation.

113. Rocky Mountain Elk Foundation, "RMEF Mission Statement & Fast Facts," last accessed July 27, 2019, http://www.rmef.org/NewsandMedia/PressRoom/AboutRMEF/MissionFastFacts.aspx.

114. U.S. Internal Revenue Service, *990 Form Filed by Rocky Mountain Elk Foundation* (2017), 1.

115. Henning.

116. Rocky Mountain Elk Foundation, "Eastern Elk Initiative," last accessed July 27, 2019, http://www.rmef.org/Conservation/Initiatives/EasternelkInitiative.aspx.

117. Examples of projects supported on both private and public land include fire and noxious weed control and water facility installation.

118. Rocky Mountain Elk Foundation, "Chapter Web Sites," last accessed July 27, 2019, http://www.rmef.org/NewsandMedia/PressRoom/AboutRMEF/Chapters/RMEFChapterMap.aspx.

119. Robert Masonis, interview by Christopher E. Segal, September 13, 2019; Trout Unlimited, "About Trout Unlimited," last accessed September 16, 2019, https://www.tu.org/about/; Trout Unlimited, "Find Your Chapter," last accessed September 16, 2019, https://www.tu.org/find-your-chapter/.

120. U.S. Internal Revenue Service, *990 Form Filed by Trout Unlimited* (2017), 1.

121. Masonis; Trout Unlimited, "Fishing, Conservation and Science, Community: Trout Unlimited 2017 Annual Report," last accessed October 4, 2019, https://www.tu.org/wp-content/uploads/2019/02/Trout_Unlimited_Annual_Report_2017-1.pdf.

122. Trout Unlimited, "Fishing, Conservation and Science, Community: Trout Unlimited 2017 Annual Report," 1–3.

123. Masonis.

124. See chapter 9.

125. Trout Unlimited, "Find Your Chapter"; Masonis.

126. Pheasants Forever, "Unique Model," last accessed July 27, 2019, https://pheasantsforever.org/getdoc/5dee6cc7-d6ca-475c-aa05-6195cdcd1942/Unique-Model.aspx.

127. U.S. Internal Revenue Service, *990 Form Filed by Pheasants Forever* (2017), 1.

128. Pheasants Forever, "Why Habitat?" last accessed July 27, 2019, https://pheasantsforever.org/Habitat/Why-Habitat.aspx.

129. Pheasants Forever, "Farm Bill and Conservation Technical Assistance Summary"; Pheasants Forever, "Farm Bill Biologists"; Inglis.

130. Inglis; Pheasants Forever, "Habitat Forever," last accessed October 1, 2019, https://pheasantsforever.org/getdoc/c5e1e18a-5b7f-484e-8b5b-8d1eafb439f5/Habitat-Forever-,LLC.aspx; Morgan et al., "A State Assessment of Private Lands Wildlife Conservation in the United States."

131. Inglis.

132. Pheasants Forever, "Find a Chapter," last accessed July 27, 2019, https://pheasantsforever.org/Participate/Find-a-Chapter.aspx.

133. American Bird Conservancy, "Mission & Strategy," last accessed October 3, 2019, https://abcbirds.org/about/mission-and-strategy/.

134. Shawn Graff, interview by Christopher E. Segal, October 3, 2019.

135. American Bird Conservancy, "Birdscapes," last accessed October 3, 2019, https://abcbirds.org/birds/birdscapes/; Graff.

136. American Bird Conservancy, "Golden-Winged Warbler," last accessed October 3, 2019, https://abcbirds.org/bird/golden-winged-warbler/; American Bird Conservancy, "$10 Mil. Forest Restoration Project Will Benefit Imperiled Golden-Winged Warbler," press release, January 14, 2015, https://abcbirds.org/article/10-mil-forest-restoration-project-will-benefit-imperiled-golden-winged-warbler/; Cornell Lab of Ornithology, "Golden-Winged Warbler: Conservation Strategy and Resources," last accessed October 3, 2019, https://www.birds.cornell.edu/home/golden-winged-warbler-conservation-strategy-and-resources/; Graff.

137. Quality Deer Management Association, "Mission & History," last accessed July 27, 2019, https://www.qdma.com/about/mission-and-history/.

138. U.S. Internal Revenue Service, *990 Form Filed by Quality Deer Management Association* (2017), 1.

139. Kip Adams, interview by Christopher E. Segal, July 10, 2019. Branches are the equivalent of chapters.

140. Quality Deer Management Association, "Find a Branch," last accessed July 27, 2019, https://www.qdma.com/get-involved/find-a-branch/; Quality Deer Management Association, "Attend an Event," last accessed July 27, 2019, https://www.qdma.com/get-involved/events/. Examples of other resources on QDMA's website include information on food plots, habitat improvement, and herd management.

141. Ruffled Grouse Society & American Woodcock Society, "History of RGS," last accessed July 28, 2019, https://ruffedgrousesociety.org/history-of-rgs/.

142. U.S. Internal Revenue Service, *990 Form Filed by Ruffed Grouse Society* (2018), 1.

143. Brent A. Rudolph, interview by Christopher E. Segal, July 25, 2019.

144. For information on the Regional Conservation Partnership Program, see chapter 9.

145. Ruffled Grouse Society & American Woodcock Society, "Chapter Info," last accessed July 28, 2019, https://ruffedgrousesociety.org/chapter-info/; Rudolph.

146. Pennsylvania Land Trust Association, "What Is a Land Trust?" last accessed March 21, 2019, https://conservationtools.org/guides/150-what-is-a-land-trust.

147. Ibid.

148. Land Trust Alliance, "Find a Land Trust: The Trustees of Reservations–an Accredited Land Trust," last accessed March 21, 2019, https://www.findalandtrust.org/land_trusts/767.

149. Land trusts have grown greatly since 2005, when they protected just 36,481,797 total acres and 6,113,108 acres of conservation easements. Land Trust Alliance, *2015 National Land Trust Census Report: Our Common Ground and Collective Impact* (2016), 5.

150. Land Trust Alliance, "Find a Land Trust: The Nature Conservancy–an Accredited Land Trust," last accessed March 21, 2019, https://www.findalandtrust.org/land_trusts/263/national.

151. For example, Land Trust Alliance, "Find a Land Trust: Kittery Land Trust," last accessed March 21, 2019, https://www.findalandtrust.org/land_trusts/1138; Maine Land Trust Network, "Foothills Land Conservancy," last accessed March 21, 2019, http://www.mltn.org/trust_detail.php?t=1246.

152. Land Trust Alliance, "Find a Land Trust."

153. Annand.

154. North Carolina Coastal Land Trust, "Who We Are," last accessed March 22, 2019, https://coastallandtrust.org/who-we-are/; Land Trust Alliance, "Find a Land Trust: North Carolina Coastal Land Trust–an Accredited Land Trust," last accessed March 22, 2019, https://www.findalandtrust.org/land_trusts/10371.

155. Monitoring allows tax authorities, funders, and the public to see that the land trust is properly managing its property.

156. North Carolina Coastal Land Trust, "How We Save Land," last accessed March 22, 2019, https://coastallandtrust.org/how/.

157. U.S. Air Force, "Readiness and Environmental Protection Integration [REPI] Program Project Profile: Dare County Bombing Range: North Carolina," 2017, https://www.repi.mil/Portals/44/Documents/Current%20Year%20Fact%20Sheets/DareCounty.pdf; North Carolina Coastal Land Trust, "Salmon Creek State Natural Area," last accessed March 22, 2019, https://coastallandtrust.org/lands/salmon-creek-natural-area/; North Carolina Coastal Land Trust, "Coastal Land Trust Saves 'Site X,'" last accessed March 22, 2019, https://coastallandtrust.org/sitex/; North Carolina State Parks, "About the State Parks System: State Natural Areas," last accessed March 22, 2019, https://www.ncparks.gov/more-about-us/about-state-parks-system/state-natural-areas.

V

ACHIEVING REGULATORY CERTAINTY UNDER THE ENDANGERED SPECIES ACT FOR MANAGEMENT OF SPECIES THAT ARE LISTED OR MAY BECOME LISTED

> *"The hope of the future lies not in curbing the influence of human occupancy—it is already too late for that—but in creating a better understanding of the extent of that influence and a new ethic for its governance."*
>
> –ALDO LEOPOLD, *GAME MANAGEMENT*

12 Understanding a Landowner's Liability Under the Endangered Species Act

THIS CHAPTER EXPLAINS the relationship of the Endangered Species Act of 1973 (ESA) to private land. The ESA is frightening to many landowners, but to a large extent this is the product of political mythology and a legacy of the sagebrush rebellion of the late 1970s and 1980s, reignited by the tea party revolution of the 2010s. State- and local-rights movements have focused in part on the ESA due to high public awareness of a relatively small number of extremely controversial species, such as the snail darter and the northern spotted owl, and especially complicated situations, such as the hydrology of the Sacramento–San Joaquin River Delta and its effects on the threatened delta smelt.

Although many private landowners may have threatened or endangered species on their land, the overwhelming majority will not find themselves greatly inconvenienced as a result. This chapter explains the ESA and provides readers with an understanding of what typical landowners should expect from ESA regulation. Chapter 13 will discuss a number of voluntary programs that Congress and the U.S. Fish and Wildlife Service (FWS) have created over the decades to provide landowners with relief from, or future protection against, ESA regulations.

Federal Jurisdiction over Threatened and Endangered Species

The majority of this book has focused on "at-risk" species, meaning species listed under the federal ESA, species listed under similar state laws, candidates for listing under the ESA, Species of Greatest Conservation Need identified in State Wildlife Action Plans, species with high global ranks under NatureServe's

Conservation Status Assessment method,[1] and other species experiencing population declines. In contrast to this expansive definition of at-risk species, the ESA only applies to species that FWS or the National Marine Fisheries Service (NMFS) have designated as "threatened" or "endangered" in a Final Rule published in the Federal Register. This process is known as "listing" a species. The FWS website allows users to search by state or even county to access a database of all listed species that are known or believed to exist in that area (https://www.fws.gov/endangered/).[2] If a species is not listed, the ESA has no jurisdiction over it.

The Endangered Species Act Prohibits "Take" of Listed Species

Section 9 of the ESA states that it is unlawful for any person to take any species of fish or wildlife listed as endangered.[3] The law defines "take" to mean "harass, harm, pursue, hunt, shoot, wound, kill, trap, capture, . . . collect, or . . . attempt to engage in any such conduct."[4] According to FWS regulations "harm" includes "significant habitat modification or degradation where it actually kills or injures wildlife."[5] As a result, modification of habitat can expose landowners to civil and criminal penalties under the ESA. Activities that may run afoul of the ESA include "development, infrastructure construction, recreational activities, grazing, energy development, and land-use changes or conversions."[6] Violations of the ESA expose landowners to civil fines of up to $25,000 for a "knowing" violation and $500 for any violation regardless of knowledge. Knowing violations are also subject to criminal penalties of up to $50,000 and one year in prison. For purposes of these penalties, each animal taken is a separate violation.[7]

Note that the previous paragraph applies only to *fish or wildlife* listed as *endangered*. Plants are not subject to the same protections. Although listed plants are protected from importation and other commercial activities, it is unlawful to damage, destroy, or remove a plant only if the plant is on public land.[8] This means that private landowners do not need to worry about taking listed plants. In addition, most ESA-listed plants are very rare and do not spread easily, so the vast majority of landowners don't have any of them present. Of course, state laws may provide additional protection to plants, and these vary widely.[9] State fish and wildlife, forestry, and agricultural agencies are the best resources to turn to for information on relevant state laws. As with plants, Section 9 of the ESA is also silent with regard to fish or wildlife listed as *threatened*. Section 4(d), however, provides that for any species listed as threatened, "[t]he Secretary may by regulation prohibit . . . any act prohibited under section [9]."[10]

4(d) Rules for Threatened Species

For most of the ESA's history, FWS (but not NMFS), had in place a "blanket" 4(d) rule. First promulgated in 1975, this rule stated that (with a few narrow exceptions) the take restrictions of Section 9 were imposed on every threatened species unless the species was also the subject of a species-specific 4(d) rule that modified the blanket rule.[11] In 2019, FWS abandoned this rule for all future listings, making its application of section 4(d) consistent with NMFS practice, and protecting threatened species from take only when a species-specific 4(d) rule is issued.[12]

As of August 2019, FWS and NMFS have issued species-specific rules under section 4(d) for 148 threatened species out of a total of 303 listed threatened species of fish or wildlife.[13] In most cases, a species-specific 4(d) rule establishes take prohibitions similar to those provided for endangered species with exceptions for certain activities. For example, for a rangeland species that has coexisted with farming and ranching for many decades but cannot tolerate oil and gas development, such as the lesser prairie chicken, a 4(d) rule could exempt take in conjunction with traditional farming and ranching practices, but not take caused by energy development. Rules such as this do not necessarily incentivize conservation, but they reduce human-wildlife conflict and can make the ESA less burdensome, and conservation practices more attractive, to landowners. FWS has also written 4(d) rules that prohibit take unless it occurs in accordance with applicable state law. This provision was first used for three species of trout listed in the summer of 1975 (Lahontan cutthroat trout, Paiute cutthroat trout, and Arizona trout), and today applies to a total of fifteen threatened species.

Another approach is to exempt take that occurs due to actions carried out under a specific conservation plan, which incentivizes participation in voluntary conservation programs. For example, FWS adopted a 4(d) rule for the coastal California gnatcatcher shortly after the species was listed in 1993. The 4(d) rule exempts incidental take of the species if it results from activities "conducted pursuant to the State of California's Natural Community Conservation Planning Act of 1991 (NCCP), and in accordance with a NCCP plan for the protection of coastal sage scrub habitat, prepared consistent with the State's NCCP Conservation and Process Guideline."[14] The NCCP program is a California state initiative to identify and conserve ecosystems where several species of concern are located, streamlining landowner and industry compliance with the ESA, the California Endangered Species Act, and other environmental laws such as the National Environmental Policy Act of 1969.[15]

FWS took a similar approach with the lesser prairie chicken, listed in 2014. In addition to exempting take associated with the continuation of "routine agricultural practices," the 4(d) rule exempted take resulting from a list of approved conservation practices and also take "from activities that are conducted by a participant enrolled in, and operating in compliance with, the Lesser Prairie-Chicken Interstate Working Group's Lesser Prairie-Chicken Range-Wide Conservation Plan, as endorsed by the U.S. Fish and Wildlife Service."[16] Because the lesser prairie chicken was delisted due to a court order in 2016, this 4(d) rule is no longer in effect.[17]

What to Do for Landowners Worried About Potential Take

If a landowner is worried about running afoul of the ESA's take prohibition, there are concrete actions they can take. First, they need to figure out if any threatened or endangered species is present on their property.[18] As described earlier in this chapter, a geographic database of listed species is available on the FWS website at https://www.fws.gov/endangered. Having identified listed species that may be present on their property, landowners need to determine which of these species are actually present on their property. Many landowners, especially those active in agricultural work, will already know this. Others, perhaps seeking to develop an old family plot, may not. Conservation districts, state and federal agencies, and other conservation organizations in their area may be able to help. Another option is to hire a private biologist to survey their land. Private consulting biologists can be found through the LandCAN database or by contacting the appropriate state fish and wildlife agency for a referral.

If a landowner does have threatened or endangered species present, the next question is what take restrictions apply. For endangered wildlife species, this is easy: any take, including through habitat modification, is prohibited. For threatened species, the landowner will need to determine what 4(d) rule, if any, applies. FWS maintains an easy-to-access database of all 4(d) rules currently in effect, which can be accessed at https://ecos.fws.gov/ecp0/reports/species-fourd-report.[19] If the species the landowner is dealing with appears on this list, the linked 4(d) rule will explain what actions are regarded as taking that species and what actions are exempt from taking. As the coastal California gnatcatcher and lesser prairie chicken examples in the preceding section illustrate, routine agricultural practices or practices subject to approval by state governments may by exempt from take. Threatened species not listed on the webpage are not subject to any take restrictions *unless* listed by FWS prior to the 2019 revocation of the blanket 4(d) rule. Species listed before that change are fully protected from take as if they were listed as endangered. If a landowner has any uncertainty about this, their local FWS field office, state fish and wildlife agency, or local private conservation organization will be able to help.

If a landowner does have a protected species present on their land, they are responsible for ensuring that their land management activities do not injure or cause the death of any individual of that species. Avoiding this is something that should be brought up with anyone helping to prepare a land management plan. In most cases, it will be possible to do what the landowner wants to with their land without running afoul of this restriction, although some modifications may be necessary. For example, where ground-nesting birds such as greater sage-grouse are present, it may be necessary to reduce the intensity of grazing or adopt seasonal restrictions. In the rare situation where a land-use project cannot go forward without taking listed species, the landowner may be able to prepare a Habitat Conservation Plan or join an existing Habitat Conservation Plan and receive an Incidental Take Permit.[20] Incidental take can also be authorized through a Biological Opinion issued to a federal government agency at the conclusion of a Section 7 consultation. This process, described later in this chapter, allows most landowners enrolled in conservation programs to implement actions under those programs without any issues arising from take of listed species.

Landowners should also be aware of the possibility of state-protected species on their land. Their state fish and wildlife agency can provide information on what species are protected under state law, what restrictions state law imposes, and what avenues may be available to resolve conflicts caused by state wildlife laws.

The Endangered Species Act Prohibits "Destruction or Adverse Modification" of Critical Habitat

Critical habitat is a subject that makes many landowners uneasy, due to a perception that the federal government overdesignates and overregulates critical habitat. There have been high-profile court cases over critical habitat, including the 2018 dusky gopher frog case, where the U.S. Supreme Court overturned a critical habitat designation that included habitat unable to actually support the dusky gopher frog.[21] But what is critical habitat? And when do private landowners need to worry about it?

The history of critical habitat is complicated. The only way that critical habitat has any impact on land use activities is through the interagency consultation process under Section 7 of the ESA. Federal government agencies, not private citizens, are the only entities subject to critical habitat restrictions, but private landowners can be impacted by critical habitat if there is a federal nexus on their land. Specifically, federal agencies must "insure that any action authorized, funded, or carried out by such agency is not likely to jeopardize the continued existence of any endangered species or threatened species or result in the destruction or adverse modification of habitat of such species which is determined by the Secretary, after consultation as appropriate with the affected States, to be critical."[22] The difficulty for landowners arises from the phrase "actions authorized, funded, or carried out by" federal agencies.

While it is obvious that work done directly by land management agencies on public land is being "carried out" by a government agency, often an action on private land is either "authorized" (through a permit) or "funded" by a federal agency. Critical habitat protections apply to such actions. The connection through a federal agency is known among ESA practitioners as a "federal nexus." Common ways for a federal nexus to arise include agricultural loans, crop subsidies, crop insurance, Farm Bill conservation program funding, federal purchase of a conservation easement, disaster relief, other federal grants, water-related permits from the Army Corps of Engineers or Environmental Protection Agency, and ESA incidental take permits. For working lands, many have an established federal nexus of some sort.[23] Many of the conservation programs discussed in this book are subject to critical habitat regulations due to the involvement of a federal agency in some way.

Most Wildlife Conservation Programs Include Programmatic Consultations

The good news for private landowners is that for most of the wildlife conservation programs described in this book, the federal agency involved, known as the "action agency," has already completed the Section 7 consultation process. Section 7 states that when an agency action "may" impact critical habitat or a listed species, the agency must enter into a consultation with FWS or NMFS. Though the consultation process, FWS or NMFS determines if the proposed action will destroy or adversely modify critical habitat, or jeopardize the continued existence of the species. If it will not, the consultation is completed and the action may proceed. If it will, FWS or NMFS suggests changes to the action, the action agency responds, and the entire process continues under procedures laid out in Section 7.[24] The document in which FWS or NMFS issues its final determination is known as a Biological Opinion. A private permit or license applicant's involvement in this process is limited, but under federal regulations applicants are entitled to submit information to FWS or NMFS, be informed of delays in the process, help identify possible changes to the action, and review draft biological opinions.[25]

Contrary to popular myth, the Section 7 consultation process is highly efficient. Two recent studies demonstrate this. One looked at all 88,290 consultations completed by FWS from January 2008 through April 2015, whereas the other looked at all 24,760 consultations completed by NMFS from January

2000 through June 2017, a total of 113,050 consultations. Of these, 101,287 (89.6 percent) were shorter, informal consultations, whereas only 11,763 (10.4 percent) were longer, formal consultations resulting in biological opinions. FWS was responsible for 81,461 informal and 6,829 formal consultations; NMFS was responsible for 19,826 informal and 4,934 formal consultations. Across all of these consultations, zero projects were stopped as a result of a jeopardy or adverse modification finding and most projects were able to proceed without substantial changes. At FWS, there were only two findings of jeopardy or adverse modification, one of which was reversed by a court whereas the other project was able to proceed with amendments. At NMFS, there were ninety-two findings of jeopardy or adverse modification, but all affected projects were ultimately able to proceed.[26] The study of FWS also considered the length of time consultations took. Informal consultations were completed in a median of thirteen days, whereas formal consultations took a median of sixty-two days. Among programmatic consultations at NMFS, informal consultations were completed in a median of thirteen days, whereas formal consultations took a median of eighty-two days.[27] The actual work involved in consultations, however, can be longer than these timeframes suggest because project proponents often prepare documents, conduct surveys, and carry out other planning activities to prepare for a consultation.

The concept of a programmatic consultation is very important. Programmatic consultations are consultations on "plans, programs, or regulations that authorize future actions."[28] Many Farm Bill conservation programs and major government grant programs carry out programmatic consultations for the entire program, which can then allow them (with some limited exceptions) to carry out that program without further consultation. For example, for the Sage Grouse Initiative, FWS and NRCS carried out a single consultation covering two thousand landowners and two million acres of land.[29] In most cases, a landowner enrolling in a conservation program discussed in this book will find that the Section 7 consultation process for any listed species and critical habitat on their land has already been completed. In the case where a required Section 7 consultation has not been completed, the landowner should work with the federal action agency in order to be included in the Section 7 consultation process.

What to Do for Landowners Worried About Critical Habitat

The first threshold question for a landowner to ask is if there is any critical habitat on their property. The easiest way to do this is to use online mapping tools provided by FWS and NMFS. Unlike the FWS database of 4(d) rules referenced earlier, critical habitat maps are not all in the same place. For species listed by FWS or jointly by FWS and NMFS, FWS hosts data in a single online mapping tool available at https://fws.maps.arcgis.com/home/webmap/viewer.html?webmap=9d8de5e265ad4fe09893cf75b8dbfb77; individual species (including downloadable GIS shapefiles) are hosted in a database available at https://ecos.fws.gov/ecp/report/table/critical-habitat.html.[30] For species listed solely by NMFS, that agency hosts maps and data in a series of pages available through https://www.fisheries.noaa.gov/national/endangered-species-conservation/critical-habitat#critical-habitat-designations,-maps,-and-gis-data.[31] Another option is for landowners to ask their local FWS field office if their property includes any critical habitat. Because NMFS jurisdiction is limited to marine and anadromous species, the vast majority of landowners will find themselves dealing only with FWS.

Even if a landowner finds that there is critical habitat on their property, and they are interested in a conservation program with a federal nexus, there is a good chance that their individual conservation actions will be covered by an existing Section 7 consultation. The best way for a landowner to find out if this is the case is by asking. This applies to programs run directly by federal agencies such as NRCS, FWS, or the U.S. Forest Service, but also to programs run by state agencies, the National Fish and Wildlife Foundation, or private foundations that may have federal funding. If a landowner's proposed conservation actions are already authorized by a Section 7 consultation, they need not worry about critical habitat. Another way for landowners to find information on already-concluded consultations is by looking at the database of Section 7 Consultation Issued Biological Opinions maintained by FWS at https://ecos.fws.gov/ecp/report/biological-opinion.html and NMFS biological opinions on pages available through https://www.fisheries.noaa.gov/national/endangered-species-conservation/biological-opinions.[32]

Situations where Section 7 consultations present major obstacles are relatively uncommon and often relate to major infrastructure projects such as electrical transmission lines, oil pipelines, bridges, and highways. In these cases, where delays in the consultation process can cause major financial losses to project proponents, industry has responded by spending its own money to prepare regulatory compliance documents that are supposed to be prepared by federal agencies. These cases have also been seized upon by members of Congress and the media as evidence that the ESA is harmful. Although these cases are problematic, they are a very small minority of Section 7 consultations and are not representative of what a typical landowner should expect if they are engaged in a project that requires a Section 7 consultation.

The Endangered Species Act Demystified

Hopefully, this chapter has helped illuminate the ways that the ESA impacts private landowners and how private landowners can navigate the ESA's challenges. The actual extent of the ESA's impact on private landowners is quite small. Its psychological impact, however, is large, fed by the media's biased, sixty-second sound bites designed for dramatic impact on television that have created a dark mythology around the law. Nonetheless, there are situations where the ESA can pose very real challenges for private landowners and situations where the presence of a listed species can impede development projects, agricultural operations, and other private enterprise. In these situations, there are a variety of tools, developed by Congress, FWS, and NMFS over the last forty years, designed to help private landowners by providing relief from the ESA. These tools, most of which involve permits from FWS and/or NMFS under Section 10 of the ESA, are the subject of the next chapter.

Notes

1. NatureServe, "Conservation Status Assessment," last accessed July 29, 2019, http://www.natureserve.org/conservation-tools/conservation-status-assessment.

2. U.S. Fish and Wildlife Service, "Endangered Species," last accessed July 29, 2019, https://www.fws.gov/endangered/.

3. U.S. Fish and Wildlife Service, *Endangered Species Act of 1973, as Amended through the 108th Congress*, 25.

4. Ibid., 3.

5. This definition was upheld by the U.S. Supreme Court in 1995. *Babbitt v. Sweet Home Chapter of Communities for a Great Oregon.*

6. U.S. Fish and Wildlife Service, "Endangered Species Act Listings: Frequently Asked Questions," July 2013, https://www.fws.gov/greatersagegrouse/factsheets/ESA%20Landowner%20Fact%20Sheet_080713.pdf.

7. U.S. Fish and Wildlife Service, Endangered Species Act of 1973, as Amended through the 108th Congress, 34–35.

8. 16 U.S.C. § 1538(a)(2).

9. Carter.

10. U.S. Fish and Wildlife Service, Endangered Species Act of 1973, as Amended through the 108th Congress, 9

11. Reclassification of the American Alligator and Other Amendments, 40 Fed. Reg. 44412 (September 26, 1975), 44425.

12. Regulations for Prohibitions to Threatened Wildlife and Plants, 84 Fed. Reg. 44753 (August 27, 2019).

13. A total of 174 species of plants are also listed as threatened.

14. Special Rule Concerning Take of the Threatened Coastal California Gnatcatcher, 58 Fed. Reg. 65088 (December 10, 1993).

15. See Douglas P. Wheeler, "An Ecosystem Approach to Species Protection," *Natural Resources & Environment* 10, no. 3 (Winter 1996).

16. Special Rule for the Lesser Prairie-Chicken; Final Rule, 79 Fed. Reg. 20074 (2014), 20084–85.

17. Lesser Prairie-Chicken Removed from the List of Endangered and Threatened Wildlife, 81 Fed. Reg. 47047 (July 20, 2016).

18. This is only applicable to fish or wildlife, because for plants, there are no take restrictions on private property. 16 U.S.C. § 1538(a)(2).

19. U.S. Fish and Wildlife Service, "Environmental Conservation Online System: Species with 4d Rules," last accessed July 31, 2019, https://ecos.fws.gov/ecp0/reports/species-fourd-report.

20. These mechanisms are discussed in detail in chapter 13.

21. *Weyerhaeuser Co. v. U.S. Fish & Wildlife Serv.*, 139 S. Ct. 361 (2018). In response to this case, FWS and NMFS issued regulations stating that unoccupied habitat must contain "one or more of those physical or biological features essential to the conservation of the species" in order to be designated critical habitat. Regulations for Listing Species and Designating Critical Habitat, 84 Fed. Reg. 45020 (August 27, 2019), 45053.

22. U.S. Fish and Wildlife Service, Endangered Species Act of 1973, as Amended through the 108th Congress, 15. The current language is very similar to the original version of the ESA. Amendments in 1978 added the word "adverse" and established procedures for designating critical habitat, including citizen petitions for designation and judicial review thereof. Endangered Species Act Amendments of 1978, Pub. L. No. 95-632, 92 Stat. 3751 (1978). Section 7 applies to both actions that may jeopardize the continued existence of a listed species and actions that may destroy or adversely modify critical habitat. The discussion in this chapter focuses on critical habitat.

23. Although in theory the ESA requires consultation on all activities with a federal nexus, in practice some activities can fall through the cracks. This is most common in situations, for example certain small business loans, where it is very difficult to show a chain of causation between a federal action and an impact on a listed species.

24. U.S. Fish and Wildlife Service, Endangered Species Act of 1973, as Amended through the 108th Congress, 15–17.

25. U. S. Fish and Wildlife Service and National Marine Fisheries Service, *Endangered Species Consultation Handbook* (March 1998), 2–13.

26. Jacob W. Malcom and Ya-Wei Li, "Data Contradict Common Perceptions About a Controversial Provision of the US Endangered Species Act," *Proceedings of the National Academy of Sciences of the United States of America* 112, no. 52 (2015); Michael J. Evans, Jacob W. Malcom, and Ya-Wei Li, "Novel Data Show Expert Wildlife Agencies Are Important to Endangered Species Protection," *Nature Communications* 10 (2019).

27. Malcom and Li, "Data Contradict Common Perceptions About a Controversial Provision of the US Endangered Species Act."

28. Ibid., 15845. In some cases, the terms "umbrella" or "master" may be used in place of "programmatic," but the concept is the same.

29. Hall and Griffiths.

30. U.S. Fish and Wildlife Service, "Critical Habitat for Threatened & Endangered Species [USFWS]," last accessed August 1, 2019, https://fws.maps.arcgis.com/home/webmap/viewer.html?webmap=9d8de5e265ad4fe09893cf75b8dbfb77; U.S. Fish and Wildlife Service, "Environmental Conservation Online System: USFWS Threatened & Endangered Species Active Critical Habitat Report," last accessed August 1, 2019, https://ecos.fws.gov/ecp/report/table/critical-habitat.html.

31. National Oceanic and Atmospheric Administration Fisheries, "Critical Habitat," last accessed August 1, 2019, https://www.fisheries.noaa.gov/national/endangered-species-conservation/critical-habitat#critical-habitat-designations,-maps,-and-gis-data.

32. U.S. Fish and Wildlife Service, "Environmental Conservation Online System: Section 7 Consultation Issued Biological Opinions," last accessed August 1, 2019, https://ecos.fws.gov/ecp/report/biological-opinion.html; National Oceanic and Atmospheric Administration Fisheries, "Biological Opinions," last accessed August 1, 2019, https://www.fisheries.noaa.gov/national/endangered-species-conservation/biological-opinions.

13

Tools for Private Landowners to Navigate the Endangered Species Act

As noted in Chapter 12, there are some situations in which a project cannot avoid the Endangered Species Act's (ESA's) prohibition against take of listed species, which can occur either through directly harming a listed species or modifying its habitat such that individuals of the listed species are harmed. For these situations, Section 10(a)(1)(B) of the ESA authorizes Incidental Take Permits (ITPs), which allow some take of listed species provided that the take occurs within the context of an approved Habitat Conservation Plan (HCP). The first part of this chapter explains this mechanism in greater detail.

Another situation that can arise is one where the ESA interferes not with development, but ironically impedes efforts to conserve endangered, threatened, and other at-risk species of fish and wildlife. The profit incentive imbedded in American land ownership can act as a disincentive for wildlife conservation.[1] Whenever a private landowner is interested in improving habitat to conserve an at-risk species, they find themselves on the horns of a dilemma. If they manage their land to encourage the establishment or expansion of a population of a species listed or in danger of listing under the ESA, they expose themselves to regulation due to the presence of that species. As the U.S. Fish and Wildlife Service (FWS) put it in 1999, "[m]any property owners are willing to voluntarily manage their property to benefit fish and wildlife, provided these beneficial actions do not result in new restrictions being placed on the future use of their property. . . . some are still concerned that additional land, water, and/or natural resource use restrictions may result if listed species colonize their property or increase in numbers or distribution due to their conservation efforts."[2] Social research has borne out this hypothesis.[3] This phenomenon is colorfully referred to as the three "Ss"—"shoot, shovel, and shut up"—the implication of which is that property owners are better off if they preemptively eliminate protected species and their habitat, or prevent suitable habitat from developing. To alleviate this, Section 10(a)(1)(A) of the ESA authorizes Enhancement of Survival Permits, which allow some take of listed species provided that the take is either a consequence of a conservation effort or is being done by an individual or entity that previously carried out conservation efforts. This last mechanism reverses the "shoot, shovel, and shut up" dilemma by rewarding rather than punishing proactive conservation efforts, especially for species that are not yet listed. The second part of this chapter addresses the various ESA mechanisms that utilize Enhancement of Survival Permits.

The chapter concludes with a short discussion of other ways that ESA regulations can benefit landowners. These include excluding from critical habitat land enrolled in conservation, creating programmatic conservation plans that landowners can easily enroll in, providing for landowners to simultaneously meet the requirements of the ESA and other state or federal regulations, and properly evaluating conservation efforts to avoid unnecessary listings under the ESA. Because all of the programs in this chapter are regulatory programs run by FWS and the National Marine Fisheries Service (NMFS), landowners interested in utilizing them should contact local FWS and NMFS field offices for more information.

Habitat Conservation Plans and Incidental Take Permits

Prior to 1982, the ESA's prohibition on taking was absolute, which made the law a potential impediment to development projects. One such project led directly to a major ESA amendment in 1982. Between 1975 and 1982, a billion-dollar commercial development project on San Bruno Mountain, just south of San Francisco, had been in the planning process. The surrounding community was fiercely opposed to development, and the mountain was home to three species of imperiled butterflies, the ESA-listed endangered mission blue butterfly, the ESA-listed endangered San Bruno elfin butterfly, and the ESA listing candidate callippe silverspot. A coalition that eventually included four landowners and residential housing developers, local community members, three municipalities, San Mateo County, two state agencies, and representatives of FWS had spent four years negotiating over the scope of the development project and another three years developing a plan to save the butterflies.[4]

For the butterflies, the residential housing developers invested $1 million on two years of research and developed a plan that would destroy only 14 percent of the butterfly habitat on the mountain. They ultimately produced

Figure 13.1. View of San Bruno Mountain State & County Park, San Mateo County, California. Home to several species of imperiled butterfly, San Bruno Mountain provided the model for the development of Habitat Conservation Plans in the early 1980s.
Courtesy of San Mateo County, California.

a comprehensive biological plan that covered fifty-four different species present on the mountain, including listed and unlisted species.[5] It featured phased construction projects, perpetual protection for some habitat, an ongoing management plan, and a sustainable source of funding for habitat restoration and other management activities.[6]

Under the ESA, this was inadequate. As the stakeholders wrote in their final plan, "the Endangered Species Act prohibits killing or injuring any endangered butterfly. No development [on San Bruno Mountain] could occur on

Figure 13.2. Three species of imperiled butterflies live on San Bruno Mountain. From left to right, the endangered Mission blue butterfly (*Icaricia icarioides missionensis*), the endangered San Bruno elfin butterfly (*Callophrys mossii bayensis*), and the endangered callippe silverspot butterfly (*Speyeria callippe callippe*).
Photographs by (L-R) U.S. Fish and Wildlife Service, Donna Pomeroy, and Patrick Kobernus.

private or public land without killing or injuring some of the butterflies, and therefore, at present, no development is possible."[7] They had considered seeking an enhancement of survival permit "for scientific purposes or to enhance the propagation or survival of the affected species," but attorneys within the group felt that their argument for such a permit was weak.[8] Only an amendment to the ESA could break the impasse.

Fortunately, in 1982 Congress was already working on a major reauthorization of the ESA, and the coalition working on San Bruno Mountain had the political connections to influence Congress. As a result of their efforts, the 1982 amendment to the ESA created Section 10(a)(1)(B), which provides for ITPs that authorize take of protected species "if such taking is incidental to, and not the purpose of, the carrying out of an otherwise lawful activity." Applicants must submit, and FWS or NFMS approve, an HCP, and applicants must ensure adequate funding to implement the HCP.[9] The Conference Report on the bill specifically cited the San Bruno Mountain plan as the model for HCPs and discussed it at length, emphasizing that the San Bruno plan was based on extensive biological studies, preserved significant amounts of open space, and was adequately funded.[10]

To apply for an ITP, an applicant must submit an HCP that, among other things, specifies the impacts of the taking, outlines steps and funding to minimize and mitigate impacts, and lays out alternatives considered and why they were not adopted. To issue an ITP, the Secretary must find that under the HCP any taking will be incidental; the applicant will, to the maximum extent practicable, minimize and mitigate the impacts of such taking; the plan will be adequately funded; and the taking will not appreciably reduce the likelihood of the survival and recovery of the species in the wild. If these findings are all made, the ITP must be issued. Although HCPs primarily are intended to benefit listed species, non-listed species that occupy the same habitat may also be included under a multi-species HCP, which can expedite the issuance of ITPs for any covered species that are later listed. The San Bruno plan, for example, included dozens of non-listed species.

HCP development can be a time-consuming, collaborative, and iterative process. The project applicant is responsible for developing the HCP, including biological goals and objectives of the plan and the conservation activities needed to meet them. Many applicants hire environmental consultants or contractors to help in this process. FWS, NMFS, and state governments can also provide assistance. During plan development, FWS or NMES simultaneously begins work on other environmental compliance documents, such as National Environmental Policy Act of 1969 (NEPA) and Section 7 consultation, saving time in the review and approval phase of the HCP. State agencies can provide assistance in incorporating any independent state permitting requirements for land and wildlife management. Finally, under Section 6 of the

ESA, federal grants are available through state agencies to assist in both HCP planning and also HCP land acquisition.[11]

A special case is a "low-effect" HCP, where the ITP would have "a minor or negligible effect on the species covered in the habitat conservation plan." Usually applied to small projects such as construction of a single home, these HCPs are eligible for Categorical Exclusions under NEPA, greatly reducing the time and expense required to prepare them.[12] After a HCP is completed and submitted, FWS or NMFS reviews the plan and completes work on other environmental compliance documents. In the past there have been efforts to incorporate NEPA requirements into HCPs directly, but current practice separates the two, because the HCP is the responsibility of the applicant while NEPA compliance is the responsibility of FWS or NMFS.[13]

Expansion and Refinement of Habitat Conservation Plans

Initially, interest in developing HCPs was low, with only fourteen HCPs approved between 1982 and 1992.[14] When Bruce Babbitt, secretary of the interior in the new Clinton administration, reviewed the endangered species program, he saw potential for improvement. The new administration "recognized[d] that implementation of the ESA should be improved . . . by exercising greater administrative flexibility to minimize socio-economic effects and assure fair treatment for landowners."[15] As part of this initiative, in 1994 the administration strengthened HCPs with the issuance of the "no surprises" policy, codified as a rule in 1998.[16] The rule states that during the term of an ITP issued in conjunction with an HCP, no additional land use restrictions or financial obligations will be imposed upon landowners. The only exception is if a continuation of the permitted activity would jeopardize the existence of a listed species, in which case an approved ITP can be revoked. This situation has never arisen.[17] In sum, the rule "provides the necessary assurances to landowners who are engaged in development activities over a period of many years that their habitat conservation planning permits will remain valid for the life of the permits."[18]

The regulatory and economic certainty offered to landowners by the no surprises policy and rule greatly increased the attractiveness of HCPs, and during the Clinton administration (1993–2001) a total of 177 HCPs were approved by FWS. Between 2001 and 2009, a further 424 HCPs were approved.[19] To date, over one thousand HCPs encompassing approximately forty-six million acres have been approved.[20] The *Habitat Conservation Planning Handbook*, another Clinton administration creation, was updated in 2016 to be more user-friendly and to "streamline the HCP process and increase the overall effectiveness of the program."[21] The handbook, which is available online, is designed for FWS and NMFS staff, but "can be helpful to other HCP practitioners, such as applicants, consultants, and partners."[22]

Because the HCP development process is slow, the tool is best suited to industry, large landowners, residential housing developers, commercial real estate developers, and local governments establishing master plans for future development. Individual landowners are relatively unlikely to need to develop an HCP or to find the time and resource investment in HCP development worthwhile. However, landowners can benefit from HCPs by joining an existing HCP that that covers an entire county or other region where they live.

Multiple Species and Regional Habitat Conservation Plans

Many HCPs are not developed in isolation to benefit one species on one parcel of land. Their development is too cost- and labor-intensive for that. For example, the very first HCP, the San Bruno Mountain Area Habitat Conservation Plan, was also the first multi-species HCP. The HCP and its associated ITP covered the three species of butterflies on the mountain and also the endangered San Francisco garter snake, and have since expanded to include the threatened Bay checkerspot butterfly.[23]

A newer development is a regional HCP, in which local stakeholders develop a single HCP for an entire city or county. Regional HCPs are usually developed by county governments or state highway authorities, and are designed to protect high-value habitat for listed species in order to permit incidental take due to development elsewhere. An example is the Balcones Canyonlands Conservation Plan (BCCP), adopted in 1996 by Travis County

Figure 13.3. The black-capped vireo (*Vireo atricapilla*), listed as an endangered species in 1987, was delisted in 2018, as a result of efforts by stakeholders that included federal and state agencies, local government, NGOs, and private landowners.
Photograph by Gil Eckrich.

and the city of Austin, Texas, as well as other stakeholders, to govern taking of eight endangered species by creating a preserve system in the county to protect those species as well as twenty-seven other species of concern.[24]

Landowners in Travis County may enroll in the BCCP, although they are also free to pursue their own ESA permits. Landowner enrollment is managed by the county government and involves paying fees according to a set schedule, with reduced-rate options for small landowners. The BCCP operates as an in-lieu fee mitigation program. Travis County uses the fees to expand and maintain its network of preserve lands for listed species, whereas participating landowners can develop their land without concern over take of any listed species. Enrollment information is available online.[25]

The BCCP is typical of regional HCPs in that landowner enrollment is handled by the county, which holds the actual ESA permit. In this respect, a regional HCP is a type of programmatic conservation plan, similar to Farm Bill programs run by partner organizations and programmatic ESA permits discussed later in this chapter. Landowners who want to enroll in a regional HCP should contact the sponsoring state or local government entity. FWS field offices can also provide information. Finally, information about HCPs currently in effect is available through a database maintained by FWS at https://ecos.fws.gov/ecp0/conservationPlan/region/summary?region=9&type=HCP.[26]

Enhancement of Survival Permits

When Congress passed the ESA in 1973, it recognized that some taking of listed species was inevitable, even though incidental take of listed species in the course of commercial development was not authorized until ITPs were introduced in 1982. As first passed in 1973, Section 10(a) of the ESA explicitly authorized take "for scientific purposes or to enhance the propagation or survival of the affected species," whereas Section 10(b) provided for hardship exemptions to execute contracts entered into before the species at issue was listed. As part of the 1982 amendment to the Act, Congress moved the language of Section 10(a) into new subparagraph 10(a)(1)(A) and expanded it. Today, Section 10(a)(1)(A) authorizes take "for scientific purposes or to enhance the propagation or survival of the affected species, including, but not limited to, acts necessary for the establishment and maintenance of experimental populations pursuant to subsection (j)."

For over twenty years, 1973 to 1993, the application of this exception was largely limited to scientific research and captive breeding of listed species, which may occasionally cause inadvertent take.[27] For example, enhancement of survival permits have been issued for the removal of snail darters from the Little Tennessee River and release into the Hiwassee River[28]; for captive rearing of peregrine falcons and whooping cranes; for research on American alligators and gray wolves; and for zoo displays of exotic species such as giant pandas, lemurs, and tigers.[29]

Although enhancement of survival permits played a relatively minor role at first, their importance has grown. First came the 1982 amendments to the ESA and the addition of an explicit provision governing experimental populations, and then in the late 1990s the creation of Safe Harbor Agreements (SHAs) and Candidate Conservation Agreements with Assurances, both tied to enhancement of survival permits.

Safe Harbor Agreements

HCPs, especially coupled with "no surprises," can be valuable tools to reduce ESA conflicts when dealing with established populations of listed species and authorizing development projects. But for listed species, recovery and eventual delisting will require increasing populations, which usually means expanding or improving habitat. This presents the "shoot, shovel, and shut up" dilemma, inviting landowner hostility.

One species for which this has been an issue is the red-cockaded woodpecker. As discussed in chapter 10, woodpecker protections created obstacles to military readiness at Fort Bragg and other U.S. Department of Defense (DoD) installations in the southeast in the late 1980s and early 1990s. This created an opportunity for DoD, FWS, non-government organizations, and private landowners to collaborate on woodpecker conservation. One result of this was DoD's Readiness and Environmental Protection Integration program. Contemporaneously, FWS worked with the Environmental Defense Fund to develop the first-ever SHA in the Sandhills region of North Carolina around Fort Bragg.[30] The SHA encourages voluntary restoration and enhancement of woodpecker habitat by relieving landowners from "any additional liability under the [ESA] beyond that which exists at the time the agreement is signed."[31] "As long as a landowner carries out the agreed-upon habitat improvements and maintains his or her baseline habitat responsibilities (if any) on their property, they may develop, harvest trees upon, or make any other lawful use of the property, even if such use incidentally results in the loss of [red-cockaded woodpeckers] or their habitat."[32]

This first SHA was such a success that additional red-cockaded woodpecker SHAs were created elsewhere in North Carolina and in South Carolina, Georgia, Florida, Alabama, Mississippi, Louisiana, Texas, and Virginia, covering nearly the entire range of the species.[33] These SHAs ultimately enrolled over four hundred landowners and 2.5 million acres of land in woodpecker conservation,[34] and directly led to a 20 percent increase in the woodpecker population.[35] The SHA tool was formalized in 1999 with adoption of an official FWS/NMFS Safe Harbor Policy and accompanying implementing regulations,[36] and has since become widely used.[37]

Under an SHA, landowners agree to implement conservation measures to benefit covered species that are listed as threatened or endangered. In exchange, landowners "receive assurances from the Services that additional conservation measures will not be required and additional land, water, or resources use restrictions will not be imposed should the covered species become more numerous as a result of the property owners' actions."[38] SHAs can be attractive to owners of a variety of working lands, including farmers, ranchers, and forest landowners.[39] For species, SHAs offer benefits such as a "reduction of habitat fragmentation rates; the maintenance, restoration, or enhancement of habitats; increase in habitat connectivity; maintenance or increase of population numbers or distribution; reduction of the effects of catastrophic events; establishment of buffers for protected areas; and establishment of areas to test and develop new and innovative conservation strategies."[40] There is no fixed minimum or maximum term for an SHA.[41]

The development of an SHA is a collaborative process in which landowners and FWS/NMFS field staff work together to craft a management plan that will work for both landowners and listed species. It begins with collecting information about the wildlife and habitat conditions on the property, and how management of the property could enhance its value as habitat for one or more listed species. The second step is the evaluation of the biological "baseline conditions" of the property. "Baseline conditions" include "population estimates and distribution and/or habitat characteristics and determined area of the enrolled property that sustain seasonal or permanent use by the covered species."[42] Accurately describing baseline conditions is critical because property enrolled in an SHA may be returned to its baseline conditions at a later time, but impacts that reduce populations and/or degrade habitat to below baseline conditions would violate the ESA. Once baseline conditions are established, the landowner and agency can collaboratively develop a land management plan that can be reasonably expected to provide a net conservation benefit, which is required by the SHA policy.[43] Activities conveying a net conservation benefit "provide for an increase in a species' population and/or the enhancement, restoration, or maintenance of covered species' suitable habitat" that "contribute[s], either directly or indirectly, to the recovery of the covered species."[44]

Finally, the baseline conditions and management plan are formalized into an SHA and submitted to FWS with an application for an enhancement of survival permit. In addition to the baseline conditions and the management plan, a final SHA must include several other pieces of information. The SHA must describe the anticipated incidental take associated with the management plan. If appropriate, it must incorporate a provision to notify FWS prior to any incidental taking of the covered species, in order to give FWS biologists the opportunity to remove individual animals and relocate them. It must describe the activities that would be expected to return the property to baseline conditions, and the extent of incidental take that would result from those activities. Finally, it must identify a schedule and responsible parties for monitoring the maintenance of baseline conditions, implementation of terms and conditions of the agreement, and any incidental take as authorized in the permit.[45] After FWS reviews the application and completes the review process as required by Section 10 of the ESA, including public notice and comment, the SHA can be finalized and an enhancement of survival permit issued.[46]

Once the enhancement of survival permit is issued, the landowner is permitted to incidentally take species both in the course of implementing the SHA management plan and in order to later return their land to the baseline conditions documented in the SHA. The first authorization lasts for as long as the SHA management plan is in effect. After the term of the agreement, landowners may return their land to baselines conditions, but not below, and this assurance persists indefinitely and may be exercised by a subsequent landowner.[47] In the event that a landowner completes the terms of an SHA and wishes to continue to manage their land to benefit the covered species, such conservation actions would be regarded as additional to those required by the SHA and could be treated as credits in a compensatory mitigation program.[48]

Programmatic Safe Harbor Agreements

SHAs need not be designed and approved individually for each landowner. The SHA policy and implementing regulations permit a state agency or nongovernment organization (and in some cases a FWS employee who manages a specific conservation program) to apply for and receive a "programmatic" SHA. The concept is the same as for a programmatic Section 7 consultation, discussed in chapter 12.[49] In a programmatic SHA, the programmatic permit holder, known as the "program administrator," recruits individual landowners into the agreement and certifies that they are complying with its terms. The landowners are then covered by the permit holder's enhancement of survival permit. The advantage of this approach is that all of the work involved in developing and approving the SHA is done for the programmatic permit and need not be repeated for each individual landowner.[50] This was actually the approach used for the red-cockaded woodpecker. The original Sandhills SHA for woodpecker was a programmatic SHA that covered all of the participating landowners. Today, although there are four hundred landowners conserving 2.5 million acres of red-cockaded woodpecker habitat under SHAs, there are actually only a total of eleven active SHAs. Two of these SHAs, both in Mississippi, cover individual tracts of land. The other nine are programmatic SHAs, including the Sandhills SHA and eight state-wide SHAs.[51]

Another species for which programmatic SHAs have proven invaluable is the endangered black-footed ferret. Ferret reintroduction programs were initially conducted using Section 10(j) rules for experimental populations (discussed later in this chapter), beginning with the Shirley Basin of Wyoming in 1991. Additional 10(j) rules followed, but each one required hundreds of thousands of dollars in FWS resources and two years of review.[52] After experimenting with several individual SHAs involving a Native American tribe, a national park, and one private landowner in Kansas, in 2013 the FWS Black-footed Ferret Coordinator, Dr. Donald "Pete" Gober, hit upon the idea of using a programmatic SHA to expedite the process.[53] Black-footed ferrets' primary food source is prairie dogs, and even with management, prairie dog colonies are vulnerable to sylvatic plague. As a result, any individual black-footed ferret population is in danger of eventual loss. Therefore, Gober proposed an SHA where any landowner could withdraw from the program at any time, for any reason, with the only condition that FWS be given an opportunity to remove black-footed ferrets from the land before it is returned

Figure 13.4. The endangered black-footed ferret (*Mustela nigripes*) was thought to be extinct until a small population was discovered in Meeteetse, Wyoming, in 1981. A successful captive breeding program led to an ongoing reintroduction campaign. Delisting the species may be possible in the near future if funding can be secured for plague management by state wildlife officials and private landowners.
Kerry Hargrove/Shutterstock.

to its baseline condition from when the landowner enrolled in the program. Even where ferrets were ultimately removed, the population of the species would have had an opportunity to grow.[54]

Another question Gober wrestled with was how to handle split estates and family ranches with multiple owners. Because the purpose of the programmatic SHA was to expedite enrollment and build momentum toward ferret recovery, it would be counterproductive to bog the process down by involving every possible landowner, no matter how small their stake. The solution settled on was to accept applications from owner/operators of properties while freely allowing take by "non-participating landowner interests." The same permissive approach was applied to adjoining landowners. So if a landowner enrolls in the black-footed ferret SHA program, neither they, nor their successors, nor their minority owners, nor their neighbors, would ever be liable for incidentally taking ferrets, provided that they abided by the terms of the SHA and gave notice before intentionally returning their land to its baseline condition. The programmatic SHA for black-footed ferret effectively eliminated the regulatory burden of the ESA.[55]

The programmatic SHA also expedited enrollment, with the average time to enroll a landowner dropping from years to just a few months.[56] This was possible because FWS carried out the required Section 7 consultation and NEPA compliance as part of its approval of the programmatic SHA, eliminating that paperwork for subsequent enrollments. Being able to move quickly when working with private landowners is critical. In Gober's words, "If I find a landowner that's ready to go, and you want to put ferrets out tomorrow, and you've got a pretty good prairie dog complex, and the state is on board, and we think we can manage the disease issue, you can sign . . . and they can unload the truck, just like that."[57] De-conflicting the ESA made ferret recovery possible on a much larger scale than it ever was before. As Gober put it: "[W]ildlife is not a hard sell to these ranching families. They just don't want to have a species that's going to cause them problems. They don't want to shoot themselves in the foot. I mean, we ought to be going out there and lauding these people, you know, putting a special star on their gate. [They're] saving an endangered species for the whole country."[58]

In addition to programmatic SHAs, there is a third type of SHA known as a "template agreement." The underlying concept is the same as for a programmatic SHA, that a number of similarly situated landowners carrying out similar actions for the same species should not need redundant approvals. Where a programmatic SHA designates a program administrator to bring individual landowners into a program, in a template SHA, FWS approves each landowner individually but processes a series of similar permits in a batch, saving agency resources.[59]

Candidate Conservation Agreements with Assurances

In 1994, while the red-cockaded woodpecker stakeholders around Fort Bragg were developing what would become the first SHA, FWS and NMFS also took up the question of how to promote the conservation of species that were merely candidates for listing under the ESA. They started to develop an update to the 1983 FWS policy on Conservation Agreements for Candidate Species. In 1997, a proposed policy and implementing rule were published that spoke of unspecified "assurances" to be identified as part of the process of developing each agreement.[60] When the policy and rule were finalized in 1999, the agencies described "candidate conservation agreements with assurances" (CCAAs), that used future Section 10(a)(1)(A) enhancement of survival permits to provide landowners with regulatory predictability.[61]

CCAAs are related to SHAs, but represent an innovation beyond SHAs or HCPs, both of which apply only to listed species. CCAAs, in contrast, use enhancement of survival permits as an incentive for conservation of species that are not yet listed.[62] "Early conservation maximizes options for landowners and for the species, minimizes the cost of recovery, and reduces the potential for restrictive land use policies that may be necessary in the future."[63]

Candidate species, not formally defined by the text of the ESA, are species "being considered . . . for listing as an endangered or threatened species, but not yet the subject of a proposed rule." This includes species found to be warranted but precluded for listing as well as species "for which the Service . . . has on file sufficient information on biological vulnerability and threats to support a proposal to list as endangered or threatened under the Endangered Species Act."[64] Candidate species are maintained on a list that is published annually and can be found online.[65] Species on the candidate list are given a "listing priority number" between one and twelve, with one being the most and twelve the least urgent.[66] Species eligible for inclusion in a CCAA are those that are designated as candidate species as well as species proposed for listing under the ESA, or likely to become candidates or proposed for listing in the near future.[67] Thus, practically any at-risk species is eligible for a CCAA. Once a species is listed and is no longer a candidate species, it is no longer possible to enter into a CCAA for that species.

The dilemma facing a landowner with suitable habitat for a candidate species is the same as that facing a landowner with suitable habitat for an endangered species, albeit less pressing: if the landowner manages their land to improve its quality as habitat, they will attract the species in question to their land. If the species is then listed, they will be subject to an increased regulatory burden under the ESA, a burden that might not have fallen on them if they had not managed their land to benefit the species in question.

Like an SHA, a CCAA is an agreement between FWS and/or NMFS and one or more non-federal landowners in which the landowners agree to implement conservation measures to benefit one or more species. In exchange for their voluntary efforts on behalf of the species covered by the CCAA, the landowners receive two regulatory assurances. First, that additional conservation measures above what they have agreed to will not be required, even if the species becomes listed under the ESA. Second, that landowners may continue existing and other agreed-upon land uses that have the potential to incidentally cause take. A landowner who enters into a CCAA will receive an enhancement of survival permit, a completed Form 3-201, that becomes effective upon listing of the covered species.[68]

CCAAs must include "the population levels (if available or determinable) of the covered species existing at the time the parties negotiate the Agreement," as well as descriptions of the covered species' populations and information about existing habitat. They do not include the rigorous biological baseline conditions of an SHA.[69] As with SHAs, the heart of a CCAA is a conservation agreement negotiated between the landowner and FWS or NMFS, the scope of which is highly flexible. Depending on circumstances, suitable conservation actions may include long-term management actions such as prescribed fires, one-time construction of a habitat feature such as breeding ponds, or one-time removal of a threat such as grazing.[70] In some cases, merely agreeing to continue present land management practices into the future may be sufficient.[71] To approve a CCAA, FWS and/or NMFS must only "make a written finding that the benefits of the conservation measures to be implemented by a property owner under an Agreement would *reasonably be expected to result in a net conservation benefit to the covered species and improve its status*."[72] As with an SHA, land enrolled in a CCAA may later be used for compensatory mitigation in certain situations.[73]

In addition to describing the current population and habitat and conservation measures to be implemented, CCAAs must also include an explanation of the expected benefits to species (including increases in population, improvements in habitat, or removal of threats), an explanation of the regulatory assurances provided to the landowner, monitoring provisions, and, when appropriate, a provision for notification for the rescue of covered individuals by federal or state wildlife officials prior to any incidental take. In the event of changed circumstances requiring greater conservation efforts than those outlined in the CCAA, FWS and/or NMFS may not require "the commitment of additional land, water, or financial compensation, or additional restrictions on the use of land, water, or other natural resources available for development or use under the original terms of the CCAA without the consent of the property owner."[74] As with SHAs, the intended beneficiaries of CCAAs are "non-federal landowners," but the CCAA policy also recognizes the potential role of state regulatory agencies (or other third parties) as permit holders for programmatic CCAAs.

Like the Sandhills SHA for red-cockaded woodpecker, the first CCAA was a programmatic CCAA, an agreement with the Oregon Department of Fish and Wildlife (ODFW) to benefit the Columbian sharp-tailed grouse.[75] Completed on October 11, 2000, this CCAA covers 156,000 acres in Wallowa County, Oregon. The species had been extirpated from that part of Oregon by the late 1960s, but ODFW reintroduced a total of 179 individuals between 1991 and 1997. Landowners cooperated with ODFW and welcomed the grouse onto their lands, but there was concern that the species might be listed. To assuage these concerns, ODFW developed a CCAA.[76] The CCAA, initially valid for twenty years, required participating landowners to maintain enrolled lands as suitable Columbian sharp-tailed grouse habitat for a period of at least five years. If they did so, they would be covered for take of the species, if listed, for the twenty-year term of the permit issued to ODFW.[77] As of 2006, the most recent year for which data is available, seven landowners had enrolled a total of 2,752 acres in this CCAA.[78] The Columbian sharp-tailed grouse remains unlisted in part because of the voluntary conservation efforts under the CCAA.

CCAAs can be much more sweeping. In 2014, the Western Association of Fish and Wildlife Agencies (WAFWA) developed a programmatic CCAA

Figure 13.5. The Columbian sharp-tailed grouse (*Tympanuchus phasianellus columbianus*), seen here on a lek, was considered for listing under the Endangered Species Act several times between 1989 and 2006. It remains unlisted, in part due to conservation efforts by private landowners. franzfoto.com/Alamy Stock Photos.

to benefit lesser prairie chicken. This CCAA is open to oil and gas producers within the species' range in five states: Colorado, Kansas, New Mexico, Oklahoma, and Texas. The CCAA, valid for thirty years, requires oil and gas producers both to allow WAFWA to survey their property for chickens and to mitigate impacts from certain activities. These include seismic and land surveys; construction; drilling, completion, major repairs, and conversion of wells; routine operations and maintenance; and oil and gas remediation and restoration activities. Mitigation, as explained in chapter 5, includes avoiding impacts to the species, minimizing such impacts when unavoidable, and paying fees to offset remaining impacts. Fees from this CCAA, when required, are paid into funds used by WAFWA to procure conservation contracts on private land under its Lesser Prairie Chicken Range-wide Conservation Plan.[79] As of December 31, 2017, a total of 105 companies had entered 111 properties into this CCAA, totaling 6,889,478.3 acres of land.[80]

CCAAs are a refinement of an older conservation tool, a Candidate Conservation Agreement (CCA). A CCA is a "formal, voluntary agreement . . . between FWS and one or more parties to address the conservation needs of one or more candidate species or species likely to become candidates in the near future." CCAs carry no assurances of future treatment under the ESA, unlike CCAAs.[81] In most cases, private landowners prefer CCAAs and the regulatory predictability they provide. There are cases, however, where a CCA is used as a vehicle for a coalition of public and private entities to formalize conservation commitments, as in the case of the 2015 Sicklefin Redhorse CCA, which benefits the sicklefin redhorse, a freshwater fish found in North Carolina and Georgia. Six organizations are parties to the CCA,[82] and four more are cooperators.[83] The CCA provides a formal framework for cooperating on actions that conserve, manage, and improve sicklefin redhorse populations range-wide. It lays out specific conservation commitments that each party or cooperator has made, which include habitat improvements, fish stocking, and support for research and monitoring.[84]

Other Endangered Species Act Tools That Can Benefit Private Landowners

Up to this point, this chapter has focused on regulatory mechanisms under the ESA in which landowners can enroll in order to reduce or avoid restrictions imposed on them by the ESA. However, there are other authorities under the ESA that FWS and NMFS can use to reduce burdens on landowners. Some of these function by additionally rewarding landowners who enroll in conservation programs, whereas others may apply without any action at all on the part of landowners.

One such tool is the development and adoption of special 4(d) rules to govern the taking of threatened species, discussed in detail in chapter 12. Depending on the individual rule, this can exempt landowners from ESA take restrictions if they are participating in a conservation program, or even if they use traditional agricultural practices in their farming or ranching operations. Other tools that the two regulatory services employ include exclusions from critical habitat designations and also experimental populations, both discussed in the following.

Critical Habitat Exclusions

As discussed in chapter 12, the only explicit protection for habitat in the ESA is provided for critical habitat through interagency consultations under Section 7. In the original 1973 ESA, there were no other references to critical habitat as such. Procedures for designating critical habitat were added to Section 4 of the ESA in 1978. Included in this amendment was the authority to exclude an area from critical habitat if "the benefits of such exclusion outweigh the benefits of specifying the area as part of the critical habitat."[85] Critical habitat exclusions are a discretionary determination that the Secretary may make in conjunction with a mandatory critical habitat designation. In 2016, FWS and NMFS formalized a policy to guide this determination.[86] The policy provides for consideration of six factors. Five factors (partnerships and conservation plans, conservation plans permitted under Section 10 of the ESA, national security and homeland security impacts, tribal lands, and economic impacts) generally support excluding land from critical habitat, whereas one factor (federal lands) generally weights against excluding land from critical habitat.[87]

The policy provides that when FWS or NMFS undertakes a discretionary exclusion analysis, they always consider excluding lands covered by permitted conservation agreements such as HCPs, SHAs, and CCAAs. This involves analyzing whether the plan in question is being properly implemented, covers the relevant species, and is tailored to benefit the relevant species. To be considered, these agreements generally must be finalized and not still under development.[88] FWS or NMFS also evaluates non-permitted plans or agreements and their attendant partnerships. This is a subjective analysis, with factors to be considered including the degree to which a critical habitat designation would harm the partnership, the extent of public participation in the development of the plan, whether NEPA compliance was included, the demonstrated implementation and success of the plan, the degree to which the plan is tailored to benefit the relevant species, whether the plan will continue to be implemented, and whether the plan includes ongoing monitoring and adaptive management.[89]

Excluding land under an implemented conservation plan from critical habitat is now standard practice,[90] and it yields multiple benefits. It creates an incentive for landowners to develop conservation plans for species that are not yet listed or do not yet have critical habitat designated. Exclusions also reduce the regulatory burden for federal agencies, because critical habitat is one of the triggers for Section 7 consultation. Finally, because the areas excluded from critical habitat are subject to conservation plans that often provide benefits to the species that are greater than the benefits critical habitat would provide, there is little harm to species where these areas are excluded from critical habitat.

Reintroduction of Listed Species as Experimental Populations

One technique for conserving at-risk species is to capture individual members of a species, breed them in captivity, and reintroduce them into the wild. The whooping crane and California condor recovery programs are well-known examples of this concept, which predates the ESA. A captive breeding program for Arabian oryx was initiated by the Phoenix (Arizona) Zoo in 1962,[91] and FWS launched its famous whooping crane breeding program in Patuxent, Maryland, in 1966.[92] Ironically, passage of the 1973 ESA created a new obstacle for captive breeding. While incidental take of species in the course of their capture, breeding, and release could be authorized through enhancement of survival permits, once a captive-bred threatened or endangered species was released into the wild, it would be entitled to the full protections of the ESA just like any other member of its species. The same concerns apply to species translocated directly from one area to another. The prospect of ESA regulation can create a chilling effect on private landowner enthusiasm for hosting reintroduced or translocated populations.

In the 1982 amendment to the ESA, Congress responded to this concern by creating a new subsection of the act, Section 10(j), to govern reintroduced "experimental" populations. In the words of the House Committee on Merchant Marine and Fisheries: "To encourage efforts to establish . . . experimental populations when the conservation needs of a species would be served by doing so, this amendment relaxes certain restrictions otherwise applicable to listed species and authorizes the Secretary to relax others."[93]

Designated "experimental populations" under the ESA must be geographically separate from all other populations of the same species and can either be "essential to the continued existence" of the species ("essential") or not ("nonessential"). "Essential" experimental populations are treated as threatened species, whereas "nonessential" experimental populations are treated as proposed for listing rather than threatened (with some minor exceptions) and cannot have critical habitat designated. Because Section 10 experimental populations are treated as threatened and not endangered, even when the species as a whole is endangered, their Section 10(j) rules may make special provisions for recapture, relocation, and even lethal take, in accordance with Section 4(d) of the ESA. In theory, these provisions reduce burdens on landowners, but results have been mixed.

In the case of the black-footed ferret, Section 10(j) rulés developed for the Shirley Basin in Wyoming in 1991, and for the entire state in 2015, were flexible enough to address landowner concerns and help landowners welcome ferrets on their property. These rules provided that landowners hosting ferrets were entirely free of regulation under the ESA except that they must inform FWS before eradicating prairie dogs (the ferret's primary food source) from their property, to allow the agency to capture and relocate resident ferrets. In contrast, Section 10(j) rules for the Mexican gray wolf in Arizona and New Mexico and the red wolf in North Carolina have generated conflict and acrimony because state officials and local landowners feel that their concerns have not been fairly considered. Landowners may or may not find a particular Section 10(j) rule compatible with their individual land use practices, but Section 10(j) rules have the potential to be powerful conservation tools when designed correctly.

Figure 13.6. The endangered Mexican gray wolf (*Canis lupus baileyi*) was first reintroduced to Arizona and New Mexico in 1998, but the reintroduced population has suffered due to human-wolf and state-federal conflicts, and the species' future remains in doubt.
Nagel Photography/Shutterstock.

Simultaneous Compliance with the Endangered Species Act and Other Laws and Regulations

ESA compliance can generate a lot of regulatory paperwork. But so can any land development project or real estate transaction. One way that the marginal burden of the ESA can be reduced is by combining ESA permitting with other regulatory compliance. Examples of regulatory regimes that can be combined with ESA compliance include Clean Water Act mitigation, NEPA compliance, and state and local environmental and tax regulations.[94] Because the other regulatory burdens that fall on landowners vary by state, location, and project, it is impossible to provide a comprehensive summary in this space. But in general, when embarking on a conservation project or dealing with ESA permits, landowners should ask if they can save time and money by simultaneously meeting the requirements of any other federal, state, or local environmental and land use regulations.

One example of a mechanism for simultaneous compliance with federal and state regulatory requirements is California's Natural Community Conservation Planning (NCCP) program, briefly mentioned in chapter 12. "An NCCP identifies and provides for the regional protection of plants, animals, and their habitats, while allowing compatible and appropriate economic activity."[95] NCCPs achieve the dual goals of long-term species protection and economic development "through the establishment of permanent, legally-enforceable conservation strategies designed for specific habitat and species."[96] Through a single planning process, NCCPs allow landowners to meet the requirements of HCPs under the ESA, the California Endangered Species Act, NEPA, and the California Environmental Quality Act.[97] Today, California has fourteen NCCPs in place and twenty more in development, altogether covering seven million acres of land and nearly four hundred species of conservation concern.[98] Because NCCPs operate at the ecosystem level, interested landowners in California should contact the state fish and wildlife agency and/or their local government for information about NCCPs in their area.

The Policy for Evaluation of Conservation Efforts When Making Listing Decisions

One of the most important ESA innovations of the last twenty years is a FWS and NMFS policy called the Policy for Evaluation of Conservation Efforts When Making Listing Decisions (PECE). With the creation of CCAAs in 1999, FWS and NMFS for the first time established ESA regulatory permits for actions that benefited species not yet listed under the ESA. One reason to incentivize such efforts is the hope of avoiding the need to list species in the future. But FWS and NMFS needed a tool to evaluate the impact of voluntary conservation efforts on species. For conservation efforts that had already yielded results, the results would be reflected in the scientific assessment of the species' status. To supplement this, in 2000, FWS and NMFS developed PECE, finalized with only slight revisions in 2003.[99] PECE "identifies criteria we will use in determining whether formalized conservation efforts that have yet to be implemented or to show effectiveness contribute to making listing a species as threatened or endangered unnecessary."[100] PECE evaluates planned conservation actions and asks whether or not there is sufficient certainty a) that the conservation actions will be implemented, and b) that the conservation actions will be effective.

The first determination, *certainty of implementation*, includes identification of nine criteria: 1) all necessary parties to the agreement, including those that will provide funding and staff support; 2) legal authority for the actions; 3) legal procedural requirements (such as NEPA or other environmental review) needed to implement the effort; 4) authorizations needed to implement the effort, such as from landowners; 5) the incentives included for voluntary participants in the effort; 6) regulatory or procedural mechanisms in place for the effort; 7) available resources for the conservation actions; 8) an implementation schedule for actions; and 9) the agreement of all parties.[101] The second determination, *certainty of effectiveness*, includes six criteria, whether the conservation effort: 1) properly describes threats to the species and how they will be addressed; 2) includes specific, incremental conservation objectives; 3) includes detailed steps that will to be taken to address the threats to the species; 4) includes quantifiable parameters that will show achievement of objectives, as well as standards for the parameters to be measured; 5) includes performance measures to evaluate compliance with and effectiveness of the plan; and 6) incorporates principles of adaptive management.[102]

PECE can apply when a species is being considered for listing, and allows FWS or NMFS to evaluate conservation efforts that have not yet yielded results. To evaluate such conservation efforts in the first place, however, FWS or NMFS needs to be aware of their existence. Accordingly, when working on an ESA listing FWS or NMFS often requests information that includes regulatory mechanisms and other conservation efforts affecting the species. The accurate gathering of this information relies on close communication between the biologists evaluating the species' status, FWS or NMFS field staff, and state wildlife agencies within the species' range.

Assembling this information was particularly difficult for the greater sage-grouse, because the species' range sprawls over 165 million acres in eleven states. To facilitate a comprehensive review of the many state, federal, and private conservation efforts undertaken on behalf of sage-grouse, FWS created the Sage Grouse Conservation Efforts Database, a single web portal through which all conservation programs could be submitted for review. More than one hundred partners entered information about sixty-two hundred projects into the database. After evaluation, 44 percent of these, covering three million acres, were deemed complete and effective at addressing the primary threats to the species.[103] A similar database was subsequently created for the monarch butterfly, and use of such databases will expand in the future.

FWS and NMFS developed PECE in order to both encourage more voluntary conservation efforts and to provide a formal mechanism to determine whether those voluntary conservation efforts can avoid the need to list species under the ESA. Through PECE, voluntary conservation, including by private landowners, can be fully considered in the ESA listing process. By voluntarily participating in conservation programs with certain implementation and effectiveness, private landowners can help keep species off of the threatened and endangered species lists.

Notes

1. Morgan et al., "A State Assessment of Private Lands Wildlife Conservation in the United States."

2. Announcement of Final Safe Harbor Policy, 64 Fed. Reg. 32717 (June 17, 1999), 32717; Michael Harris, "Federal Tools for Conservation of at-Risk Species on Private Working Forests in the Southeast" (paper presented at the Southeastern Association of Fish and Wildlife Agencies Annual Conference, Mobile, AL, October 22, 2018).

3. For example, Drake and Jones, "Forest Management Decisions of North Carolina Landowners Relative to the Red-Cockaded Woodpecker"; Hansen et al., *Cooperative Conservation: Determinants of Landowner Engagement in Conserving Endangered Species.*

4. San Bruno Mountain Habitat Conservation Plan Steering Committee, *San Bruno Mountain Area Habitat Conservation Plan–Final* (November 1982); Lindell L. Marsh and Robert D. Thornton, "San Bruno Mountain Habitat Conservation Plan," in *Managing Land-Use Conflicts: Case Studies in Special Area Management*, ed. David J. Brower and Daniel S. Carol (Durham, NC: Duke University Press, 1987); Lindell L. Marsh, interview by Christopher E. Segal, November 30, 2018; Joshua Chin, "Fighting to Save, and Popularize, San Bruno Mountain," *Bay Nature*, February 26, 2015; San Bruno Mountain Watch, "San Bruno Mountain Watch Oral History Project: The Story of the Breakup of the CSSBM and the Evolution of the HCP," February 25, 2011, http://www.mountainwatch.org/sbmw-oral-history-project/2011/2/25/the-story-of-the-breakup-of-the-cssbm-and-the-evolution-of-t.html; San Bruno Mountain Watch, "San Bruno Mountain Watch Oral History Project: Bette Higgins: The Queen of the Environmental Movement on San Bruno Mountain," February 25, 2011, http://www.mountainwatch.org/sbmw-oral-history-project/2011/2/25/bette-higgins.html; San Bruno Mountain Watch, "San Bruno Mountain Watch Oral History Project: Who Is Bette Higgins?" February 25, 2011, http://www.mountainwatch.org/sbmw-oral-history-project/2011/2/25/who-is-bette

-higgins.html; Leon Lindsay, "A Tale of Butterflies and a Billion-Dollar Project," *Christian Science Monitor*, March 14, 1983.

5. Marsh and Thornton, "San Bruno Mountain Habitat Conservation Plan," 122; Marsh.

6. Laura H. Watchman, Martha Groom, and John D. Perrine, "Science and Uncertainty in Habitat Conservation Planning: A Review of 43 Habitat Conservation Plans Reveals Numerous Ways to Reduce Uncertainty for Landowners and Imperiled Species Alike," *American Scientist* 89, no. 4 (2001): 352; Marsh and Thornton, "San Bruno Mountain Habitat Conservation Plan," 124–25. The plan provided for developers to pay startup fees and residents to pay annual fees thereafter.

7. San Bruno Mountain Habitat Conservation Plan Steering Committee, *San Bruno Mountain Area Habitat Conservation Plan–Final*, I-1.

8. Marsh.

9. U.S. Fish and Wildlife Service, Endangered Species Act of 1973, as Amended through the 108th Congress, 28-29; Committee on Conference, Endangered Species Act Amendments of 1982, Conference Report to Accompany H.R. 6133, H.R. Rep. No. 97-835 (1982), 13.

10. Committee on Conference, Endangered Species Act Amendments of 1982, Conference Report to Accompany H.R. 6133, H.R. Rep. No. 97-835, 30–32; Marsh and Thornton, "San Bruno Mountain Habitat Conservation Plan," 126–27.

11. See chapter 10.

12. U.S. Fish and Wildlife Service and National Marine Fisheries Service, *Habitat Conservation Planning and Incidental Take Permit Processing Handbook* (2016).

13. Ibid.

14. Douglas P. Wheeler and Ryan M. Rowberry, "Habitat Conservation Plans and the Endangered Species Act," in *Endangered Species Act: Law, Policy, and Perspectives*, ed. Donald C. Baur and William Robert Irvin (Chicago, IL: American Bar Association, 2010), 224.

15. U.S. Fish and Wildlife Service, *Protecting America's Living Heritage: A Fair, Cooperative and Scientifically Sound Approach to Improving the Endangered Species Act* (March 6, 1995), 2.

16. Habitat Conservation Plan Assurances ("No Surprises") Rule, 63 Federal Register 8859 (February 23, 1998).

17. Michael J. Bean, "Landowner Assurances under the Endangered Species Act," Working Paper (Madison, WI: Sand County Foundation/Environmental Policy Innovation Center, 2017), 4.

18. U.S. Fish and Wildlife Service, *Protecting America's Living Heritage: A Fair, Cooperative and Scientifically Sound Approach to Improving the Endangered Species Act*, 7.

19. Wheeler and Rowberry, "Habitat Conservation Plans and the Endangered Species Act," 224–25.

20. Bean, "Landowner Assurances under the Endangered Species Act," 9.

21. U.S. Fish and Wildlife Service and National Marine Fisheries Service, *Habitat Conservation Planning and Incidental Take Permit Processing Handbook*, i.

22. Ibid., 1–13.

23. Friends of Endangered Species, Inc. v. Jantzen, 596 F.Supp 518 (N.D. Cal. 1984), 521; U.S. Fish and Wildlife Service, Federal Fish and Wildlife Permit, San Bruno Mountain HCP Permittees (March 29, 2013). The San Bruno Mountain HCP was amended in 2009, and its associated ITP was extended in 2013.

24. Travis County Texas, "The Balcones Canyonlands Conservation Plan (BCCP)," last accessed August 3, 2019, https://www.traviscountytx.gov/tnr/bccp; Travis County Texas, "Species Covered under the Balcones Canyonlands Conservation Plan," last accessed September 20, 2019, https://www.traviscountytx.gov/images/tnr/Docs/species_covered.pdf. The listed species were

the black-capped vireo, golden-cheeked warbler, Tooth Cave pseudoscorpion, Tooth Cave spider, Bee Creek Cave harvestman, Bone Cave harvestman, Tooth Cave ground beetle, and Kretschmarr Cave mold beetle. The black-capped vireo was subsequently delisted due to recovery.

25. Travis County Texas, "The Balcones Canyon Conservation Plan: Participating," last accessed August 3, 2019, https://www.traviscountytx.gov/tnr/bccp/participating. There are still some restrictions on actions in golden-cheeked warbler habitat and on disturbing newly discovered caves that have not yet been surveyed for listed species.

26. U.S. Fish and Wildlife Service, "Environmental Conservation Online System: Habitat Conservation Plans," last accessed August 3, 2019, https://ecos.fws.gov/ecp0/conservationPlan/region/summary?region=9&type=HCP. Unfortunately, the information in this database is not well organized or easy to understand. The database can be searched by species, but the locations of HCPs and whether they are open to additional participants are not always clear.

27. Bean, "Landowner Assurances under the Endangered Species Act," 5.

28. For example, Endangered Species Permit: Receipt of Addendum to Application, 41 Fed. Reg. 8515 (February 27, 1976).

29. For example, Endangered Species Permits: Official Action, 41 Fed. Reg. 9576 (March 5, 1976); Endangered Species Permit: Notice of Receipt of Application, 39 Fed. Reg. 18483 (May 28, 1974).

30. U.S. Fish and Wildlife Service, A Habitat Conservation Plan to Encourage the Voluntary Restoration and Enhancement of Habitat for the Red-Cockaded Woodpecker on Private and Certain Other Land in the Sandhills Region of North Carolina by Providing "Safe Harbor" to Participating Landowners (1995); Drake and Jones, "Forest Management Decisions of North Carolina Landowners Relative to the Red-Cockaded Woodpecker"; Beard.

31. U.S. Fish and Wildlife Service, A Habitat Conservation Plan to Encourage the Voluntary Restoration and Enhancement of Habitat for the Red-Cockaded Woodpecker on Private and Certain Other Land in the Sandhills Region of North Carolina by Providing "Safe Harbor" to Participating Landowners, 2.

32. Ibid.

33. U.S. Fish and Wildlife Service, "Environmental Conservation Online System: Red-Cockaded Woodpecker (Picoides Borealis)," last accessed October 3, 2018, https://ecos.fws.gov/ecp0/profile/speciesProfile?spcode=B04F. As of 2017, the Virginia programmatic SHA had only one enrolled landowner, The Nature Conservancy, which is also the permit holder. Bean, "Endangered Species Act Safe Harbor Agreements An Assessment," 8.

34. Bean, "Endangered Species Act Safe Harbor Agreements: An Assessment," 4; Harris, "Federal Tools for Conservation of at-Risk Species on Private Working Forests in the Southeast."

35. Between 1993 and 2006 (the last year in which a five-year review was completed, although a new review is now underway), the woodpecker's population increased by 30 percent. Two-thirds of this increase was attributable to SHA-enrolled lands. Costa and U.S. Fish and Wildlife Service, 5-Year Review: Red-Cockaded Woodpecker (*Picoides Borealis*), 10; Harris, "Federal Tools for Conservation of at-Risk Species on Private Working Forests in the Southeast." It should be noted that red-cockaded woodpecker populations are counted in terms of "active clusters"—locations occupied by at least one family group of woodpeckers—rather than in terms of individuals.

36. Announcement of Final Safe Harbor Policy, 64 Fed. Reg. 32717; Safe Harbor Agreements and Candidate Conservation Agreements with Assurances, 64 Fed. Reg. 32706 (June 17, 1999). Minor corrections to the regulations were made by Safe Harbor Agreements and

Candidate Conservation Agreements with Assurances (Correction), 64 Fed. Reg. 52676 (September 30, 1999).

37. By June 2005, "more than 325 landowners had enrolled over 3.6 million acres in 32 Safe Harbor Agreements" to benefit thirty-six listed species. U.S. Fish and Wildlife Service, "Working Together: Tools for Helping Imperiled Wildlife on Private Lands," 2005, https://digitalmedia.fws.gov/digital/collection/document/id/887/. By 2017, one hundred agreements had been created. Bean, "Endangered Species Act Safe Harbor Agreements: An Assessment," iii–iv.

38. Announcement of Final Safe Harbor Policy, 64 Fed. Reg. 32717, 32717.

39. U.S. Fish and Wildlife Service, "20 Years of the Safe Harbor Program: An Interview with Michael Bean," 2015, https://www.youtube.com/watch?time_continue=622&v=htd4xCljx9I.

40. Announcement of Final Safe Harbor Policy, 64 Fed. Reg. 32717, 32723.

41. The SHA policy simply states that SHAs "must be of sufficient duration to reasonably allow enough time to achieve the expected 'net conservation benefit' for the listed species." Ibid., 32719.

42. Ibid., 32722.

43. "Net conservation benefit" means "cumulative benefits of . . . management activities . . . that provide for an increase in a species' population and/or the enhancement, restoration, or maintenance of covered species' . . . habitat . . . taking into account the length of the [Safe Harbor] Agreement and any off-setting adverse effects attributable to the incidental taking allowed by the enhancement of survival permit. Net conservation benefits must be sufficient to contribute, either directly or indirectly, to the recovery of the covered species." Ibid.

The phrase "reasonably expected" was added to implementing regulations in 2004. This was done "to address confusion regarding the word 'will' in the issuance criteria, which could have been interpreted as suggesting that we must determine with complete certainty that a net conservation benefit will occur before a permit can be issued. This unrealistic standard was not the intent of the Safe Harbor Policy or the regulation." Safe Harbor Agreements and Candidate Conservation Agreements with Assurances; Revisions to the Regulations, 69 Fed. Reg. 24084 (May 3, 2004), 24086–87.

44. Announcement of Final Safe Harbor Policy, 64 Fed. Reg. 32717, 32722. When the term "net conservation benefit" is used in the context of a prelisting action such as a Candidate Conservation Agreement with Assurances (see discussion later in this chapter), rather than an SHA, it is defined to contribute to the "conservation" of a species rather than "recovery," because the concept of recovery under the ESA does not apply to unlisted candidate species. U.S. Fish and Wildlife Service, Director's Order No. 218: Policy Regarding Voluntary Prelisting Conservation Actions (January 18, 2017), at Attachment, 34–35.

45. Announcement of Final Safe Harbor Policy, 64 Fed. Reg. 32717, 32723.

46. U.S. Fish and Wildlife Service, "Safe Harbor Agreements for Private Landowners," October, 2017, https://www.fws.gov/endangered/esa-library/pdf/harborqa.pdf.

47. On a case-by-case basis, landowners may return to baseline conditions prior to completion of an SHA agreement. The SHA policy allows early termination prior to the realization of net conservation benefits for covered species due to "circumstances out of the landowner's control." Announcement of Final Safe Harbor Policy, 64 Fed. Reg. 32717, 32725. Some SHAs, such as the DuPage County (Illinois) Forest Preserve District SHA for the Hine's Emerald Dragonfly, simply adopt this provision. Forest Preserve District of DuPage County and U.S. Fish and Wildlife Service, Final Draft Safe Harbor Agreement for Voluntary Protection, Restoration, and Enhancement of Hine's Emerald Dragonfly Habitat within Waterfall Glen Forest Preserve, Dupage County, Illinois (2008), 13. Others, such as the nationwide Black-Footed Ferret Programmatic Safe Harbor Agreement, allow a return to baseline conditions at any time. U.S. Fish

and Wildlife Service Black-footed Ferret Recovery Program, Black-Footed Ferret Programmatic Safe Harbor Agreement (October 23, 2013), 24.

48. See chapter 5.

49. See chapter 12.

50. This makes the process of enrolling land much more efficient and less intimidating for landowners. It also leverages the resources of the program administrator and takes advantage of any preexisting relationships the program administrator may have with landowners. Bean, "Endangered Species Act Safe Harbor Agreements: An Assessment," 4–5; Hansen et al., *Cooperative Conservation: Determinants of Landowner Engagement in Conserving Endangered Species.*

51. U.S. Fish and Wildlife Service, "Environmental Conservation Online System: Red-Cockaded Woodpecker (Picoides Borealis)."

52. Donald "Pete" Gober, interview by Lowell E. Baier and Christopher E. Segal, May 25, 2018, Washington, DC.

53. Ibid.

54. U.S. Fish and Wildlife Service Black-footed Ferret Recovery Program, Black-Footed Ferret Programmatic Safe Harbor Agreement; Gober.

55. Gober.

56. In the first year of the SHA, fifty-six or fifty-eight landowners expressed interest, and within two years, six new reintroductions had been carried out.

57. Gober.

58. Ibid.

59. Bean, "Endangered Species Act Safe Harbor Agreements: An Assessment," 5–6.

60. Announcement of Draft Policy for Candidate Conservation Agreements, 62 Fed. Reg. 32183 (June 12, 1997); Safe Harbor Agreements and Candidate Conservation Agreements, 62 Fed. Reg. 32189 (June 12, 1997).

61. Announcement of Final Policy for Candidate Conservation Agreements with Assurances, 64 Fed. Reg. 32726 (June 17, 1999); Safe Harbor Agreements and Candidate Conservation Agreements with Assurances, 64 Fed. Reg. 32706.

62. Announcement of Final Policy for Candidate Conservation Agreements with Assurances, 64 Fed. Reg. 32726, 32735.

63. U.S. Fish and Wildlife Service, "Candidate Species: Section 4 of the Endangered Species Act," October 2017, https://www.fws.gov/endangered/esa-library/pdf/candidate_species.pdf.

64. Endangered Species Act Compensatory Mitigation Policy, 81 Fed. Reg. 95316, 95346-47.

65. Candidate Notices of Review are available for the last several years at https://www.fws.gov/endangered/what-we-do/cnor.html.

66. Endangered and Threatened Species Listing and Recovery Priority Guidelines, 48 Fed. Reg. 43098 (September 21, 1983).

67. Announcement of Final Policy for Candidate Conservation Agreements with Assurances, 64 Fed. Reg. 32726, 32734.

68. U.S. Fish and Wildlife Service, *Draft Candidate Conservation Agreements with Assurances Handbook* (June, 2003), 29; Announcement of Final Policy for Candidate Conservation Agreements with Assurances, 64 Fed. Reg. 32726, 32735.

69. U.S. Fish and Wildlife Service, *Draft Candidate Conservation Agreements with Assurances Handbook*, 12.

70. Ibid., 14.

71. Ibid., 9–10.

72. Candidate Conservation Agreements with Assurances Policy, 81 Fed. Reg. 95164 (December 27, 2016), 95,172. (Emphasis added.)

73. Endangered Species Act Compensatory Mitigation Policy, 81 Fed. Reg. 95316, 95337, 41 (withdrawn).

74. Candidate Conservation Agreements with Assurances Policy, 81 Fed. Reg. 95164, 95172.

75. U.S. Fish and Wildlife Service, *Draft Candidate Conservation Agreements with Assurances Handbook*, 5.

76. Oregon Department of Fish and Wildlife, Candidate Conservation Agreement with Assurances for Columbian Sharp-Tailed Grouse between the Oregon Department of Fish and Wildlife and the U.S. Fish and Wildlife Service (2000), 4–5.

77. Ibid., 6.

78. U.S. Fish and Wildlife Service, "ODFW Columbian-Sharp Tailed Grouse Candidate Conservation Agreement with Assurances," last accessed October 17, 2018, https://ecos.fws.gov/ecp0/conservationPlan/plan?plan_id=854.

79. Range-Wide Oil and Gas Candidate Conservation Agreement with Assurances for the Lesser Prairie-Chicken (*Tympanuchus Pallidicinctus*) in Colorado, Kansas, New Mexico, Oklahoma and Texas between the U.S. Fish and Wildlife Service and the Western Association of Fish and Wildlife Agencies/Foundation for Western Fish and Wildlife (February 28, 2014), 9–10.

80. Roger L. Wolfe et al., *The 2017 Lesser Prairie-Chicken Range-Wide Conservation Plan Annual Progress Report* (March 2018), 27–32. The enrollment fees paid for these properties were not reported because some of the properties had been transferred to the CCAA from a mechanism within the WAFWA Range-wide Plan known as a WAFWA Conservation Agreement. These Conservation Agreements pay into the same in-lieu fee mitigation fund as the CCAA, and WAFWA annual reports combine them for financial purposes. Land in the CCAA and in WAFWA Conservation Agreements combined generated $42,719,342 in enrollment fees through 2017. Ibid., 100. However, compared to the 6,889,478.3 acres enrolled in the CCAA, there were only 673,538 acres enrolled in WAFWA Conservation Agreements, meaning 91 percent of the total land was enrolled in the CCAA. Ibid., 21. This does not necessarily mean that the CCAA generated 91 percent of the fees, because the enrollment fee calculation varies by project. See Van Pelt et al., *The Lesser Prairie-Chicken Range-Wide Conservation Plan*, 233.

81. The concept of a CCA dates back to at least 1983, when FWS first published a policy on "Conservation Agreements for Candidate Species." The policy was designed to allow a conservation agreement to preclude listing a species if the agreement "removes all known threats to the species that would otherwise warrant listing." This policy was discontinued in 1985 and reimplemented in 1992, although there is some evidence that it was used during the period 1985 to 1992 as well. A revised draft guidance was published in 1994, but the tool was soon overshadowed by the development of CCAAs. CCAs are used today, but their application is limited, with most use of CCAs by state governments or federal agencies meeting Section 7 obligations. U.S. Fish and Wildlife Service, "Candidate Conservation Agreements," last accessed May 30, 2018, https://www.fws.gov/endangered/esa-library/pdf/CCAs.pdf.

82. The Fish and Wildlife Service, North Carolina Wildlife Resources Commission, Duke Energy Carolinas, LLC, Tennessee Valley Authority, Eastern Band of Cherokee Indians, and Georgia Department of Natural Resources.

83. Brookfield Smoky Mountain Hydro, Conservation Fisheries, Inc., Warm Springs National Fish Hatchery, and the North Carolina Department of Transportation.

84. Candidate Conservation Agreement for the Sicklefin Redhorse (*Moxostoma* Sp.) between the United States Fish and Wildlife Service, North Carolina Wildlife Resources Commission, Duke Energy Carolinas, LLC, Tennessee Valley Authority, Eastern Band of Cherokee Indians, and Georgia Department of Natural Resources (December 2015).

85. Endangered Species Act Amendments of 1978. An additional condition is that excluding the area from critical habitat must not result in the extinction of the species.

86. Policy Regarding Implementation of Section 4(B)(2) of the Endangered Species Act, 81 Fed. Reg. 7226 (February 11, 2016).

87. Harris, "Federal Tools for Conservation of at-Risk Species on Private Working Forests in the Southeast"; Policy Regarding Implementation of Section 4(B)(2) of the Endangered Species Act, 81 Fed. Reg. 7226 (February 11, 2016).

88. Policy Regarding Implementation of Section 4(B)(2) of the Endangered Species Act, 81 Fed. Reg. 7226 (February 11, 2016), 7247.

89. Ibid.

90. Mike Harris, At-Risk Species Coordinator, Southeast Regional Office, U.S. Fish and Wildlife Service, conversation with Christopher Segal, October 22, 2018.

91. PBS, "The Loneliest Animals: Captive Breeding Success Stories," April 1, 2009, http://www.pbs.org/wnet/nature/the-loneliest-animals-captive-breeding-success-stories/4920/.

92. Karin Brulliard, "A 50-Year Effort to Raise Endangered Whooping Cranes Comes to an End," *Washington Post*, Spetember 18, 2017.

93. Committee on Merchant Marine and Fisheries, Endangered Species Act Amendments, Report to Accompany H.R. 6133, H.R. Rep. No. 97-567, Part 1 (1982), 33.

94. For discussion of the Clean Water Act and NEPA, see chapter 5. For discussion of state and local environmental and tax regulations, see chapters 4 and 11.

95. California Department of Fish and Wildlife, "Natural Community Conservation Planning (NCCP)," last accessed October 5, 2019, https://www.wildlife.ca.gov/Conservation/Planning/NCCP.

96. Douglas P. Wheeler, "A Strategy for the Future," *Stanford Environemental Law Journal* 12 (1993): ix.

97. Douglas P. Wheeler, Interview by Lowell E. Baier and Christopher E. Segal, March 24, 2017.

98. California Department of Fish and Wildlife.

99. Announcement of Draft Policy for Evaluation of Conservation Efforts When Making Listing Decisions, 65 Fed. Reg. 37102 (June 13, 2000); Policy for Evaluation of Conservation Efforts When Making Listing Decisions, 68 Fed. Reg. 15100 (March 28, 2003).

100. Policy for Evaluation of Conservation Efforts When Making Listing Decisions, 68 Fed. Reg. 15100, 15100.

101. Ibid., 15114–15.

102. Ibid., 15115.

103. 12-Month Finding on a Petition to List Greater Sage-Grouse (*Centrocercus Urophasianus*) as an Endangered or Threatened Species, 80 Fed. Reg. 59858, 58871–72.

Conclusion

One of the great strengths of voluntary conservation is that the tools available are constantly evolving, a testament to the ingenuity and creativity of private landowners, the conservation community, and America as a whole. The tools, programs, and organizations described in this book will change and evolve over time, and new opportunities will emerge. The conservation programs and organizations that work with private landowners in every state and county in the nation provide starting points for identification of contemporary conservation opportunities in their communities.

The federal Farm Bill and its programs are amended every five years, and other federal government initiatives evolve as more or less funding is available, as congressional, administration, and agency priorities change, and as the science of wildlife conservation evolves and matures. Among the nongovernmental organization community, each organization brings its own unique mission, culture, experience, and expertise to its work, whether that is providing funding for conservation, in-kind contributions, or technical support to make delivery of federal funds more efficient. Finally, partnerships play a central role in successful conservation programs. From the state-federal relationship that defines our federalist system of government to public-private partnerships that increasingly shape conservation programs, these coalitions are ever-multiplying and ever-changing. By working with both governments and landowners to make conservation accessible and affordable, partnerships generate win-win solutions for wildlife and for humans and the working landscapes upon which we rely.

The case studies in this book show how diverse tools can all be used to benefit species. In all likelihood, there is no piece of land and no planned development or land use that is not compatible with some conservation program or creative conservation solution. It is simply a matter of identifying the correct tool and utilizing it. If you love wildlife and want to work to improve wildlife and its habitats, there is an organization out there that will work with you, and there are government programs, foundations, and organizations willing to help develop and fund conservation on your land. American landowners have a long and proud legacy of conserving and caring for their land, and history has shown that when they do, all benefit. As Aldo Leopold teaches us, private landowners are the key stewards of our natural resources, our rich legacy and heritage, and our future.

Appendix A
Contacts for Landowners

THIS BOOK PRESENTS a variety of resources for landowners seeking assistance in designing, implementing, and paying for conservation on their land. This appendix summarizes the resources described elsewhere in the book, providing contact information for each.

U.S. Department of Agriculture Service Centers

U.S. Department of Agriculture Service Centers hosting Natural Resources Conservation Service and Farm Service Agency field offices/staff can be found in nearly every one of the more than three thousand counties in the United States. Landowners can find their local Service Center through the following website: https://offices.sc.egov.usda.gov/locator/app

U.S. Fish and Wildlife Service

Partners for Fish and Wildlife

Contact Information for Partners for Fish and Wildlife in each region and state is available here: https://www.fws.gov/partners/contactUs.html

In addition, most U.S. Fish and Wildlife Service regions maintains webpages about Partners for Fish and Wildlife with additional information:

Midwest Region: https://www.fws.gov/midwest/partners/
(Illinois, Indiana, Iowa, Michigan, Minnesota, Missouri, Ohio, Wisconsin)

Mountain-Prairie Region: https://www.fws.gov/mountain-prairie/refuges/partners.php
(Colorado, Kansas, Montana, Nebraska, North Dakota, South Dakota, Utah, Wyoming)

Northeast Region: https://www.fws.gov/northeast/EcologicalServices/partners.html
(Connecticut, Delaware, Maine, Maryland, Massachusetts, New Hampshire, New Jersey, New York, Pennsylvania, Rhode Island, Vermont, Virginia, West Virginia)

Pacific Region: https://www.fws.gov/pacific/ecoservices/habcon/partners/index.html
(Hawaii and Pacific Islands, Idaho, Oregon, Washington)

Pacific Southwest Region: https://www.fws.gov/cno/conservation/Partners.html
(California, Nevada, and Klamath Basin, Oregon)

Southeast Region: https://www.fws.gov/southeast/our-services/partners-program/
(Alabama, Arkansas, Florida, Georgia, Kentucky, Louisiana, Mississippi, North Carolina, South Carolina, Tennessee, Puerto Rico, and U.S. Virgin Islands)

Southwest Region: https://www.fws.gov/southwest/es/Partners_Main.html
(Arizona, New Mexico, Oklahoma, Texas)

Migratory Bird Joint Ventures

The national Migratory Bird Joint Ventures website includes a directory: https://mbjv.org/who-we-are/directory/

Other Landowner Programs

https://www.fws.gov/endangered/landowners/index.html

U.S. Forest Service

Most U.S. Forest Service services for private landowners are delivered through state forestry agencies (see state listings later in this appendix). However, there are several U.S. Forest Service–affiliated websites that provide nationwide information.

U.S. Forest Service Forest Taxation Website: https://www.fs.fed.us/spf/coop/programs/loa/tax.shtml
National Timber Tax Website: https://www.timbertax.org/
National Agroforestry Center (with the Natural Resources Conservation Service): https://www.fs.usda.gov/nac/index.shtml

Bureau of Land Management

The Bureau of Land Management's funding for private land conservation is currently allocated at the District Office level. A directory of offices is available here: https://www.blm.gov/office/national-office

Landowners may also find the national Threatened and Endangered Species program office helpful: https://www.blm.gov/programs/fish-and-wildlife/threatened-and-endangered

Conservation Districts

The National Association of Conservation Districts maintains a national directory of conservation districts: https://www.nacdnet.org/general-resources/conservation-district-directory/

Land Trusts

The Land Trust Alliance maintains a national directory of land trusts: https://www.findalandtrust.org

The Land Trust Alliance provides other landowner resources as well: https://www.conservationeasement.us/resources

Agricultural Extensions

The U.S. Department of Agriculture maintains a directory of agricultural extensions and other services provided by land-grant universities: https://nifa.usda.gov/land-grant-colleges-and-universities-partner-website-directory

Private Websites

Land Conservation Assistance Network: https://www.landcan.org/
Conservation Finance Center: http://www.conservationfinancecenter.org/

State Agencies

Alabama

Alabama Department of Conservation and Natural Resources Wildlife and Freshwater Fisheries Division

Main Office
https://www.outdooralabama.com/about-us/wildlife-and-freshwater-fisheries-division
334-242-3465

Alabama Forestry Commission

Main Office
http://www.forestry.alabama.gov/
334-240-9300
Online Request for Assistance
http://www.forestry.alabama.gov/Pages/Other/Assistance.aspx
Employee Directory (Search by County)
http://www.forestry.alabama.gov/Pages/Other/Employee_Directory.aspx

Alaska

Alaska Department of Fish and Game

Main Office
https://www.adfg.alaska.gov/
Contact Directory by Location
https://www.adfg.alaska.gov/index.cfm?adfg=contacts.main
Landowner Assistance
https://www.adfg.alaska.gov/index.cfm?adfg=habitatrestoration.landowner

Alaska Department of Natural Resources, Division of Forestry

Main Office
http://forestry.alaska.gov/
Public Information Centers
http://dnr.alaska.gov/commis/pic/
800-770-8973

Arizona

Arizona Game & Fish Department

Main Office
https://www.azgfd.com/

Landowner Relations & Habitat Enhancement Program
https://www.azgfd.com/wildlife/landowners/
Contact Directory
https://s3-us-west-2.amazonaws.com/azgfd-portal-wordpress-pantheon/wp-content/uploads/archive/finalPromo2019_08_181.pdf

Arizona Department of Forestry and Fire Management
Main Office
https://dffm.az.gov/
602-771-1400
Landowner Assistance Programs
https://dffm.az.gov/forestry-community-forestry

Arkansas

Arkansas Game and Fish Commission
Main Office
https://www.agfc.com/en/
800-364-4263
Private Lands Program
https://www.agfc.com/en/wildlife-management/private-lands-program/
Find A Private Lands Biologist
https://www.agfc.com/en/wildlife-management/private-lands-program/private-lands-biologist/

Arkansas Department of Agriculture Forestry Division
Main Office
https://www.aad.arkansas.gov/arkansas-forestry-commission
501-225-1598
Landowner Assistance
https://www.aad.arkansas.gov/arkansas-famous-and-historic-trees

California

California Department of Fish and Wildlife
Main Office
https://www.wildlife.ca.gov/
Regional Offices
https://www.wildlife.ca.gov/Regions
Grants
https://www.wildlife.ca.gov/Grants

Landscape Conservation Planning
https://www.wildlife.ca.gov/Conservation/Planning

California Department of Forestry and Fire Protection
Main Office
https://www.fire.ca.gov/
Grants
https://www.fire.ca.gov/grants/
Resource Management
https://www.fire.ca.gov/programs/resource-management/

Colorado

Colorado Parks and Wildlife
Main Office
https://cpw.state.co.us/
303-297-1192
Private Land Program
https://cpw.state.co.us/aboutus/Pages/PrivateLand.aspx

Colorado State Forest Service
Main Office
https://csfs.colostate.edu/
970-491-6303
Resources for Homeowners & Landowners
https://csfs.colostate.edu/homeowners-landowners/

Connecticut

Connecticut Department of Energy & Environmental Protection—Wildlife Division
Main Office
https://www.ct.gov/deep/cwp/view.asp?a=2723&q=325726&deepNav_GID=1655
Wildlife and Habitat Management
https://www.ct.gov/deep/cwp/view.asp?a=2723&q=325728&deepNav_GID=1655

Connecticut Department of Energy & Environmental Protection—Forestry Division
Main Office
https://www.ct.gov/deep/cwp/view.asp?a=2697&q=322792&deepNav_GID=1631

Forest Landowner Assistance
https://www.ct.gov/deep/cwp/view.asp?a=2697&q=322796&deepNav_GID=1631

Delaware

Delaware Department of Natural Resources and Environmental Control—Division of Fish & Wildlife
Main Office
https://dnrec.alpha.delaware.gov/fish-wildlife/
302-739-9910
Delaware Private Lands Assistance Program
http://www.dnrec.delaware.gov/fw/dplap/Pages/default.aspx

Delaware Forest Service
Main Office
https://agriculture.delaware.gov/forest-service/
302-698-4547
Landowner Assistance
https://agriculture.delaware.gov/forest-service/landowner-assistance/

Florida

Florida Fish and Wildlife Conservation Commission
Main Office
https://myfwc.com/
850-488-4676
Working with Private Landowners
https://myfwc.com/conservation/special-initiatives/lap/

Florida Forest Service
Main Office
https://www.freshfromflorida.com/Divisions-Offices/Florida-Forest-Service
For Landowners
https://www.freshfromflorida.com/Divisions-Offices/Florida-Forest-Service/For-Landowners

Georgia

Georgia Department of Natural Resources—Wildlife Resources Division
Main Office
https://georgiawildlife.com/
706-557-3333

Landowner Programs
https://georgiawildlife.com/landowners

Georgia Forestry Commission
Main Office
http://www.gfc.state.ga.us/
800- 428-7337
Landowner Services
http://www.gfc.state.ga.us/forest-management/private-forest-management/landowner-programs/

Hawaii

Hawaii Department of Land and Natural Resources—Division of Forestry and Wildlife
Main Office
https://dlnr.hawaii.gov/dofaw/
808-587-0166
Wildlife Programs Main Office
https://dlnr.hawaii.gov/wildlife/
Forestry Programs Main Office
https://dlnr.hawaii.gov/forestry/
Forestry Programs Landowner Assistance Programs
https://dlnr.hawaii.gov/forestry/lap/

Idaho

Idaho Department of Fish & Game
Main Office
https://idfg.idaho.gov/
208-334-3700
Lands/Landowner Programs
https://idfg.idaho.gov/wildlife/lands

Idaho Department of Lands—Forestry
Main Office
https://www.idl.idaho.gov/forestry/index.html
Assistance for Forest Landowners
https://www.idl.idaho.gov/forestry/service/index.html

Illinois

Illinois Department of Natural Resources

Main Office

https://www.dnr.illinois.gov/Pages/default.aspx

Conservation Programs

https://www.dnr.illinois.gov/conservation/Pages/default.aspx

Forestry Resources

https://www.dnr.illinois.gov/conservation/Forestry/Pages/default.aspx

Grant Program

https://www.dnr.illinois.gov/grants/Pages/default.aspx

University of Illinois Extension–Illinois Forestry

Main Office

https://web.extension.illinois.edu/forestry/home.html

Landowner Assistance Programs

https://web.extension.illinois.edu/forestry/forest_mgmt/land_assistance.html

Indiana

Indiana Department of Natural Resources—Division of Fish & Wildlife

Main Office

https://www.in.gov/dnr/fishwild/

Habitat and Wildlife Landowner Assistance

https://www.in.gov/dnr/fishwild/2352.htm

Indiana Department of Natural Resources—Division of Forestry

Main Office

https://www.in.gov/dnr/forestry/

Private Forestland Management

https://www.in.gov/dnr/forestry/2845.htm

Iowa

Iowa Department of Natural Resources—Wildlife

Main Office

https://www.iowadnr.gov/Conservation/Iowas-Wildlife

Wildlife Landowner Assistance

https://www.iowadnr.gov/Conservation/Wildlife-Landowner-Assistance

Iowa Department of Natural Resources—Forestry
Main Office
https://www.iowadnr.gov/Conservation/Forestry
Forestry Landowner Assistance
https://www.iowadnr.gov/Conservation/Forestry

Kansas

Kansas Department of Wildlife, Parks and Tourism
Main Office
https://ksoutdoors.com/
Private Landowner Assistance
https://ksoutdoors.com/Services/Private-Landowner-Assistance

Kansas Forest Service
Main Office
https://www.kansasforests.org/
785-532-3300
Services & Landowner Assistance
https://www.kansasforests.org/rural_forestry/servicesandlandownerassistance.html

Kentucky

Kentucky Department of Fish & Wildlife
Main Office
https://fw.ky.gov/Pages/default.aspx
800-858-1549
Landowners Services
https://fw.ky.gov/Wildlife/Pages/Landowner-Services.aspx

Kentucky Energy and Environment Cabinet—Division of Forestry
Main Office
https://eec.ky.gov/Natural-Resources/Forestry/Pages/default.aspx
Forest Stewardship Program and Landowner Services
https://eec.ky.gov/Natural-Resources/Forestry/forest-stewardship-program-and-landowner-services/Pages/default.aspx

Louisiana

Louisiana Department of Wildlife and Fisheries

Main Office
http://www.wlf.louisiana.gov/
800-256-2749
Private Land Management Assistance
http://www.wlf.louisiana.gov/assistance-private-landowners-and-m

Louisiana Department of Agriculture and Forestry–Forestry

Main Office
http://www.ldaf.state.la.us/forestry/
866-927-2476
Forestry Management
http://www.ldaf.state.la.us/forestry/management/

Maine

Maine Department of Inland Fisheries and Wildlife

Main Office
https://www.maine.gov/ifw/
207-287-8000
Private Lands
https://www.maine.gov/ifw/fish-wildlife/wildlife/lands/private-lands.html

Maine Forest Service

Main Office
https://www.maine.gov/dacf/mfs/
Forest Policy and Management
https://www.maine.gov/dacf/mfs/policy_management/index.html

Maryland

Maryland Department of Natural Resources

Main Office
http://dnr.maryland.gov/Pages/default.aspx
877-620-8367
Habitat for Wildlife
http://dnr.maryland.gov/Wildlife/Pages/habitat/home.aspx

Maryland Department of Natural Resources—Forest Service
Main Office
http://dnr.maryland.gov/forests/Pages/default.aspx
Forestry Assistance to Landowners
http://dnr.maryland.gov/forests/Pages/programapps/faland.aspx

Massachusetts

Massachusetts Department of Fish and Game
Main Office
https://www.mass.gov/orgs/department-of-fish-and-game
617-626-1500
Conservation Resources for Private Landowners
https://www.mass.gov/conservation-resources-for-private-landowners

Massachusetts Department of Conservation and Recreation—Bureau of Forest Fire Control and Forestry
Main Office
https://www.mass.gov/orgs/bureau-of-forest-fire-control-and-forestry
Service/Private Lands Forestry
https://www.mass.gov/service-details/service-forestry
Programs and Services for Forest Land Owners
https://www.mass.gov/programs-and-services-for-forest-land-owners

Michigan

Michigan Department of Natural Resources
Main Office
https://www.michigan.gov/dnr/
Private Lands Program
https://www.michigan.gov/dnr/0,4570,7-350-79136_79238_83065—-,00.html

Michigan Department of Natural Resources—Forestry
Main Office
https://www.michigan.gov/dnr/0,4570,7-350-79136_79237—-,00.html
Assistance for Private Forest Land Owners
https://www.michigan.gov/dnr/0,4570,7-350-79136_79237_80945—-,00.html

Minnesota

Minnesota Department of Natural Resources—Fish and Wildlife Division

Main Office

https://www.dnr.state.mn.us/fishwildlife/index.html

Private Land Habitat

https://www.dnr.state.mn.us/privatelandhabitat/index.html

Minnesota Department of Natural Resources—Division of Forestry

Main Office

https://www.dnr.state.mn.us/forestry/index.html

Forest Legacy Program

https://www.dnr.state.mn.us/forestlegacy/index.html

Forest Stewardship

https://www.dnr.state.mn.us/foreststewardship/index.html

Mississippi

Mississippi Wildlife, Fisheries, & Parks

Main Office

https://www.mdwfp.com/

601-432-2400

Private Lands Program

https://www.mdwfp.com/wildlife-hunting/private-lands-program/private-lands-program/

Mississippi Forestry Commission

Main Office

https://www.mfc.ms.gov/

Private Landowner Services

https://www.mfc.ms.gov/private-lands

Missouri

Missouri Department of Conservation

Main Office

https://mdc.mo.gov/

Your Property

https://mdc.mo.gov/property

Missouri Department of Conservation—Forestry Division
Main Office
https://mdc.mo.gov/trees-plants
Forest Care
https://mdc.mo.gov/trees-plants/forest-care
Timber Management Assistance
https://mdc.mo.gov/trees-plants/forest-care/timber-management-assistance

Montana

Montana Fish, Wildlife & Parks
Main Office
http://fwp.mt.gov/
406-444-2535
Montana Landowners
http://fwp.mt.gov/fishAndWildlife/landowners/

Montana Department of Natural Resources and Conservation—Forestry Division
Main Office
http://dnrc.mt.gov/divisions/forestry
406-542-4300
Forestry Assistance
http://dnrc.mt.gov/divisions/forestry/forestry-assistance

Nebraska

Nebraska Game and Parks Commission
Main Office
http://outdoornebraska.gov/
402-471-0641
Landowner Programs
http://outdoornebraska.gov/landownerprograms/

Nebraska Forest Service
Main Office
https://nfs.unl.edu/
Financial Assistance
https://nfs.unl.edu/financial-assistance

Nevada

Nevada Department of Wildlife

Main Office

http://www.ndow.org/

Private Lands for Wildlife Program

http://www.ndow.org/Nevada_Wildlife/Conservation/Nevada_Landowner_Incentive_Program/

Nevada Division of Forestry

Main Office

http://forestry.nv.gov/

775-684-2500

Grants

http://forestry.nv.gov/grants/

Natural Resources Programs

http://forestry.nv.gov/forestry-resources/

New Hampshire

New Hampshire Fish and Game Department

Main Office

https://www.wildlife.state.nh.us/

603-271-3421

Landowner Relations Program

https://www.wildlife.state.nh.us/landshare/

New Hampshire Division of Forests and Lands

Main Office

https://www.nh.gov/nhdfl/

Landowner Resources

https://www.nh.gov/nhdfl/land-conservation/index.htm

New Jersey

New Jersey Department of Environmental Protection—Division of Fish & Wildlife

Main Office

https://www.njfishandwildlife.com/

Wildlife Habitat Management

https://www.nj.gov/dep/fgw/habitat_mgt.htm

New Jersey Forest Service
Main Office
https://www.state.nj.us/dep/parksandforests/forest/
Your Property
https://www.state.nj.us/dep/parksandforests/forest/njfs_private_lands_mgt.html

New Mexico

New Mexico Department of Game & Fish
Main Office
http://www.wildlife.state.nm.us/
505-476-8000

New Mexico Energy, Minerals and Natural Resources Department—State Forestry Division
Main Office
http://www.emnrd.state.nm.us/SFD/
505-476-3325
Forest Legacy Program
http://www.emnrd.state.nm.us/SFD/ForestMgt/ForestLegacy.html
Land Conservation Tax Credits
http://www.emnrd.state.nm.us/SFD/LandConservationTaxCredit.html
Forest Stewardship Program
http://www.emnrd.state.nm.us/SFD/ForestMgt/ForestStewardship.html

New York

New York State Department of Environmental Conservation—Fish and Wildlife
Main Office
https://www.dec.ny.gov/23.html
518-402-8883
Landowner Incentive Program
https://www.dec.ny.gov/animals/32722.html

New York State Department of Environmental Conservation—Lands and Forests
Main Office
https://www.dec.ny.gov/lands/309.html
518-402-9405
Private Forest Management
https://www.dec.ny.gov/lands/4972.html

North Carolina

North Carolina Wildlife Resources Commission

Main Office

https://www.ncwildlife.org/

Land Management

https://www.ncwildlife.org/Conserving/Land-Management

North Carolina Forest Service

Main Office

https://www.ncforestservice.gov/

(919) 857-4801

Managing Your Forest

https://www.ncforestservice.gov/Managing_your_forest/managing_your_forest.htm

North Dakota

North Dakota Game and Fish Department

Main Office

https://gf.nd.gov/

701-328-6300

Landowner Programs

https://gf.nd.gov/plots/landowner

North Dakota Forest Service

Main Office

https://www.ag.ndsu.edu/ndfs

701-228-5422

Forest Stewardship

https://www.ag.ndsu.edu/ndfs/programs-and-services/forest-stewardship-1

Ohio

Ohio Department of Natural Resources—Division of Wildlife

Main Office

http://wildlife.ohiodnr.gov/

800-945-3543

Private Lands Management

http://wildlife.ohiodnr.gov/species-and-habitats/private-lands-management

Ohio Department of Natural Resources—Division of Forestry

Main Office

http://forestry.ohiodnr.gov/

877-247-8733

Landowner Assistance
http://forestry.ohiodnr.gov/landownerassistance
Ohio Forest Legacy Program
http://forestry.ohiodnr.gov/legacyprogram
Ohio Forest Tax Law
http://forestry.ohiodnr.gov/oftl

Oklahoma

Oklahoma Department of Wildlife Conservation

Main Office
https://www.wildlifedepartment.com/
405-521-3851
Landowner Programs
https://www.wildlifedepartment.com/lands-and-minerals/landowner-programs
Your Side of the Fence (Publication)
https://www.wildlifedepartment.com/lands-and-minerals/landowner-programs

Oklahoma Forestry Services

Main Office
http://www.forestry.ok.gov/
405-522-6158
Forest and Woodland Management
http://www.forestry.ok.gov/management

Oregon

Oregon Department of Fish and Wildlife

Main Office
https://www.dfw.state.or.us/
503-947-6000
Wildlife Habitat
https://www.dfw.state.or.us/lands/index.asp
Wildlife Grant Opportunities
https://www.dfw.state.or.us/wildlife/grants/index.asp

Oregon Department of Forestry

Main Office
https://www.oregon.gov/odf
503-945-7200

Helping Landowners
https://www.oregon.gov/ODF/Working/Pages/HelpingLandowners.aspx

Pennsylvania

Pennsylvania Fish and Boat Commission
Main Office
https://www.fishandboat.com/Pages/default.aspx
Private Landowners Stewardship
https://www.fishandboat.com/Resource/PrivateLandownersStewardship/Pages/d efault.aspx

Pennsylvania Game Commission
Main Office
https://www.pgc.pa.gov/Pages/default.aspx
717-787-4250

Pennsylvania Department of Conservation and Natural Resources—Bureau of Forestry
Main Office
https://www.dcnr.pa.gov/about/Pages/Forestry.aspx
Managing Your Woods
https://www.dcnr.pa.gov/Conservation/ForestsAndTrees/ManagingYourWoods/Pa ges/default.aspx

Rhode Island

Rhode Island Department of Environmental Management—Division of Fish and Wildlife
Main Office
http://www.dem.ri.gov/programs/fish-wildlife/

Rhode Island Department of Environmental Management—Division of Forest Environment
Main Office
http://www.dem.ri.gov/programs/forestry/
Forest Legacy Program
http://www.dem.ri.gov/programs/forestry/forestlegacy/
Forest Stewardship Program
http://www.dem.ri.gov/programs/forestry/stewardship/index.php

South Carolina

South Carolina Department of Natural Resources
Main Office
http://www.dnr.sc.gov/

South Carolina Forestry Commission
Main Office
https://www.state.sc.us/forest/
Landowner Services
https://www.state.sc.us/forest/refserv.htm

South Dakota

South Dakota Game, Fish, and Parks
Main Office
https://gfp.sd.gov/landowner-programs/
605-223-7660
Landowner Programs
https://gfp.sd.gov/landowner-programs/

South Dakota Department of Agriculture—Division of Resource Conservation & Forestry
Main Office
https://sdda.sd.gov/conservation-forestry/
Forest Legacy Program
https://sdda.sd.gov/conservation-forestry/forest-legacy-program/
Forest Stewardship Program
https://sdda.sd.gov/conservation-forestry/forest-stewardship/forest-stewardship-program/
Grants & Loans
https://sdda.sd.gov/conservation-forestry/grants-loans/

South Dakota Habitat Pays
https://habitat.sd.gov/

Tennessee

Tennessee Wildlife Resources Agency
Main Office
https://www.tn.gov/twra.html
615-781-6500

Private Lands Management & Assistance
https://www.tn.gov/content/tn/twra/wildlife/habitat.html#assistance

Tennessee Department of Agriculture–Division of Forestry
Main Office
https://www.tn.gov/agriculture/forests.html
Landowner Assistance
https://www.tn.gov/agriculture/forests/landowners.html

Texas

Texas Parks & Wildlife Department
Main Office
https://tpwd.texas.gov/
800-792-1112
Private Lands and Habitat Program
https://tpwd.texas.gov/landwater/land/private/
Private Lands Assistance and Technical Guidance Program
https://tpwd.texas.gov/landwater/land/technical_guidance/
Texas Farm and Ranch Lands Conservation Program
https://tpwd.texas.gov/landwater/land/private/farm-and-ranch/

Texas A&M Forest Service
Main Office
https://tfsweb.tamu.edu/
Landowner Assistance
https://tfsweb.tamu.edu/LandownerAssistance/

Utah

Utah Department of Natural Resources—Division of Wildlife Resources
Main Office
https://wildlife.utah.gov/index.html
801-538-4700

Utah Department of Natural Resources—Division of Forestry, Fire & State Lands
Main Office
https://ffsl.utah.gov/
Private Landowner Forestry Assistance
https://ffsl.utah.gov/forestry/private-landowner-forestry-assistance/

Vermont

Vermont Fish & Wildlife Department
Main Office
https://vtfishandwildlife.com/
802-828-1000
Landowner Resources
https://vtfishandwildlife.com/node/276

Vermont Department of Forest, Parks and Recreation
Main Office
https://fpr.vermont.gov/
802-828-1531
Managing Your Woodlands
https://fpr.vermont.gov/forest/managing-your-woodlands

Virginia

Virginia Department of Game and Inland Fisheries
Main Office
https://www.dgif.virginia.gov/
Habitat for Wildlife
https://www.dgif.virginia.gov/wildlife/habitat/

Virginia Department of Forestry
Main Office
http://www.dof.virginia.gov/
434-977-6555
Programs and Services
http://www.dof.virginia.gov/services/index.htm

Washington

Washington Department of Fish & Wildlife
Main Office
https://wdfw.wa.gov/
360-902-2200
Habitat Recovery and Protection
https://wdfw.wa.gov/species-habitats/habitat-recovery

Washington State Department of Natural Resources
Main Office
https://www.dnr.wa.gov/

Small Forest Landowner Office
https://www.dnr.wa.gov/sflo
Forest Health Assistance for Small Forest Landowners
https://www.dnr.wa.gov/cost-share
Forest Legacy Program
https://www.dnr.wa.gov/leaving-legacy-forests
Forest Practices
https://www.dnr.wa.gov/programs-and-services/forest-practices

West Virginia

West Virginia Division of Natural Resources
Main Office
http://www.wvdnr.gov/

West Virginia Division of Forestry
Main Office
http://wvforestry.com/
Landowner Assistance
https://wvforestry.com/landowner-assitance/

Wisconsin

Wisconsin Department of Natural Resources—Division of Fish, Wildlife, and Parks
Main Office
https://dnr.wi.gov/about/divisions/FishWildlifeParks/
Wildlife and Your Land
https://dnr.wi.gov/topic/WildlifeHabitat/yourland.html

Wisconsin Department of Natural Resources—Forestry Division
Main Office
https://dnr.wi.gov/about/divisions/Forestry/
Woodland Owners
https://dnr.wi.gov/topic/ForestLandowners/

Wyoming

Wyoming Game and Fish Department
Main Office
https://wgfd.wyo.gov/
307-777-4600

Wyoming State Forestry Division

Main Office
https://wsfd.wyo.gov/
Forestry Assistance Programs
https://wsfd.wyo.gov/forestry-assistance-programs

Washington, DC

District of Columbia Department of Energy & Environment—Wildlife Management Branch

Main Office
https://doee.dc.gov/service/wildlife-management

District of Columbia Department of Transportation—Urban Forestry Division

Main Office
https://ddot.dc.gov/page/ddot-urban-forestry

Non-Governmental Organizations

American Bird Conservancy

National Office
https://abcbirds.org/
BirdScapes
https://abcbirds.org/birds/birdscapes/

The Conservation Fund

National Office
https://www.conservationfund.org/

Ducks Unlimited

National Office
https://www.ducks.org/
Regional Offices
https://www.ducks.org/about-ducks-unlimited/regional-offices

Mule Deer Foundation

National Office
https://muledeer.org/
Chapters
https://muledeer.org/state

National Audubon Society
National Office
https://www.audubon.org/
Chapters
https://www.audubon.org/about/audubon-near-you
Working Lands
https://www.audubon.org/conservation/working-lands

National Wild Turkey Federation
National Office
http://www.nwtf.org/
Chapters
http://www.nwtf.org/about/nation

National Wildlife Federation
National Office
https://www.nwf.org/
State Affiliates
https://www.nwf.org/About-Us/Regional-Centers-and-Affiliates

The Nature Conservancy
National Office
https://www.nature.org/en-us/
State Offices
https://www.nature.org/en-us/connect/contact-us/

Pheasants Forever
National Office
https://pheasantsforever.org/
Chapters
https://pheasantsforever.org/Participate/Find-a-Chapter.aspx

Quality Deer Management Association
National Office
https://www.qdma.com/
Branches
https://www.qdma.com/get-involved/find-a-branch
Events
https://www.qdma.com/get-involved/events

Rocky Mountain Elk Foundation
National Office
https://www.rmef.org/
Chapters
http://www.rmef.org/NewsandMedia/PressRoom/AboutRMEF/Chapters/RMEFChapterMap.aspx

Ruffed Grouse Society & American Woodcock Society
National Office
https://ruffedgrousesociety.org/
Chapters
https://ruffedgrousesociety.org/chapter-info

Trout Unlimited
National Office
https://www.tu.org/
Chapters
https://www.tu.org/find-your-chapter/

Appendix B
Annual Impacts of Farm Bill Programs

THE FEDERAL FARM BILL, discussed in chapters 6 through 9, is the single largest source of funding for conservation on private land. This appendix provides historical data on the financial investment and conservation impact of several Farm Bill programs, in order to supplement the overall discussion of the programs found earlier in this book. Data are derived from the several Farm Bills, from presidential budgets, and from Natural Resources Conservation Service (NRCS) databases. Blanks indicate years for which data were not available. Programs are presented in the same order as in the main text.

Environmental Quality Incentives Program

Table 1 provides statistics on the Environmental Quality Incentives Program (EQIP). Included are Farm Bill appropriations from 1996 to 2023 and funds actually expended and acres receiving conservation in each year from 2005 to 2017. Data is not available for funds expended and acres receiving conservation prior to 2005. At times, NRCS staffing levels have been a bottleneck that prevented some EQIP dollars from being put on the ground before their appropriation expired.[1] It is important to read the financial and acreage columns separately for EQIP and for all Farm Bill conservation programs. Farm Bill payments may be available for any foregone income, lost land value, and the cost of implementing conservation practices. Because the relationship between these expenses varies from program to program, year to year, and property to property, the data presented here is not sufficiently granular to be used to calculate a cost per acre for any program. The aggregate totals presented are intended only to convey a sense of the scale of Farm Bill conservation programs.

Table 1. Environmental Quality Incentives Program (EQIP) Annual Expenditures and Acres Enrolled

Year	*Funds Appropriated by the Farm Bill*	*Funds Expended*	*Acres Receiving Conservation*
1996	$130,000,000	*	*
1997	$200,000,000	*	*
1998	$200,000,000	*	*
1999	$200,000,000	*	*
2000	$200,000,000	*	*
2001	$200,000,000	*	*
2002	$400,000,000	*	*
2003	$700,000,000	*	*
2004	$1,000,000,000	*	*
2005	$1,200,000,000	$1,024,800,000	13,713,789
2006	$1,200,000,000	$1,074,800,000	18,731,617
2007	$1,300,000,000	$1,070,900,000	23,226,022
2008	$1,200,000,000	$1,252,300,000	23,834,264
2009	$1,337,000,000	$1,054,581,600	23,175,167
2010	$1,450,000,000	$1,184,252,200	24,114,797
2011	$1,588,000,000	$1,244,659,400	22,458,802
2012	$1,750,000,000	$1,391,561,700	24,301,639
2013	$1,750,000,000	$1,391,160,000	24,418,414
2014	$1,350,000,000	$1,313,933,500	19,465,088
2015	$1,600,000,000	$1,248,419,000	18,604,761
2016	$1,650,000,000	$1,454,101,200	15,734,247
2017	$1,650,000,000	$1,671,911,600	17,074,562
2018	$1,750,000,000	$1,877,884,300	17,734,084
2019	$1,750,000,000	N/A	N/A
2020	$1,750,000,000	N/A	N/A
2021	$1,800,000,000	N/A	N/A
2022	$1,850,000,000	N/A	N/A
2023	$2,025,000,000	N/A	N/A

*Indicates data unavailable.

Sources: Federal Agriculture Improvement and Reform Act of 1996, 1007; Farm Security and Rural Investment Act of 2002, 278; Food, Conservation, and Energy Act of 2008, 1072; American Taxpayer Relief Act of 2012, 2363; Agricultural Act of 2014, 757; Agricultural Improvement Act of 2018, 90; Natural Resources Conservation Service, "Environmental Quality Incentives Program (EQIP)"; Natural Resources Conservation Service, "NRCS Conservation Programs."

Conservation Stewardship Program

Table 2 provides the authorized appropriations for the Conservation Stewardship Program (CSP) and its predecessor the Conservation Security Program (CSeP) from 2004 through 2023. The table includes columns for funds actually expended under CSeP and CSP for 2005 to 2017 and acres enrolled in CSP at the end of each year for 2010 to 2017.

Table 2. Conservation Stewardship Program (CSP) Annual Expenditures and Acres Enrolled

Year	*CSeP Appropriations*	*CSP Appropriations*	*Total CSeP and CSP Appropriations*	*CSeP Funds Expended*	*CSP Funds Expended*	*Total CSP and CSeP Funds Expended*	*CSP Acres Enrolled (Year End)*
2004	$40,000,000	N/A	$40,000,000	$40,400,000	N/A	$40,400,000	N/A
2005	$202,000,000	N/A	$202,000,000	$201,600,000	N/A	$201,600,000	N/A
2006	$257,000,000	N/A	$257,000,000	$257,200,000	N/A	$257,200,000	N/A
2007	$294,000,000	N/A	$294,000,000	$294,400,000	N/A	$294,400,000	N/A
2008	$317,000,000	N/A	$317,000,000	$317,100,000	N/A	$317,100,000	N/A
2009	$276,000,000	*$9,000,000	$285,000,000	$276,000,000	*$9,378,200	$285,378,200	N/A
2010	$222,000,000	$390,000,000	$612,000,000	$222,200,000	$389,813,000	$612,013,000	25,164,326.9
2011	$204,000,000	$601,000,000	$805,000,000	$198,900,000	$577,803,700	$776,703,700	12,750,675.9
2012	$188,000,000	$742,000,000	$930,000,000	$188,000,000	$741,619,900	$929,619,900	12,109,876.0
2013	$159,000,000	$883,000,000	$1,042,000,000	$158,900,000	$882,547,800	$1,041,447,800	9,519,370.5
2014	$125,000,000	$1,079,000,000	$1,204,000,000	$120,400,000	$1,030,870,700	$1,151,270,700	9,598,222.1
2015	$28,000,000	$1,164,000,000	$1,192,000,000	$30,900,000	$1,095,878,800	$1,126,778,800	24,059,377.9
2016	$5,000,000	$1,225,000,000	$1,230,000,000	$2,000,000	$1,129,295,300	$1,131,295,300	17,410,782.7
2017	$5,000,000	$1,149,000,000	$1,154,000,000	$1,800,000	$1,134,534,400	$1,136,334,400	16,287,852.1
2018	N/A	$1,345,000,000	$1,345,000,000	N/A	$1,197,294,000	$1,197,294,000	11,941,694.9
2019	N/A	$700,000,000	$700,000,000	N/A	N/A	N/A	N/A
2020	N/A	$725,000,000	$725,000,000	N/A	N/A	N/A	N/A
2021	N/A	$750,000,000	$750,000,000	N/A	N/A	N/A	N/A
2022	N/A	$800,000,000	$800,000,000	N/A	N/A	N/A	N/A
2023	N/A	$1,000,000,000	$1,000,000,000	N/A	N/A	N/A	N/A

Note: CSeP: Conservation Security Program; CSP: Conservation Stewardship Program.
*CSP was created by the 2008 Farm Bill, and so the only payments in 2009 were technical assistance payments for the first enrollees. Normal payments began in 2010.

Sources: U.S. Government Accountability Office, *Conservation Security Program: Despite Cost Controls, Improved USDA Management Is Needed to Ensure Proper Payments and Reduce Duplication with Other Programs, No. GAO-06-312* (April 2006), 67, 72; U.S. Department of Agriculture, *FY 2004 Budget Summary*; U.S. Department of Agriculture, *FY 2005 Budget Summary*; U.S. Department of Agriculture, *FY 2006 Budget Summary and Annual Performance Plan*; U.S. Department of Agriculture, *FY 2007 Budget Summary and Annual Performance Plan* (2006); U.S. Department of Agriculture, *FY 2008 Budget Summary and Annual Performance Plan* (2007), 27; U.S. Department of Agriculture, *FY 2009 Budget Summary and Annual Performance Plan* (2008), 56; U.S. Department of Agriculture, *FY 2010 Budget Summary and Annual Performance Plan* (2009), 72; U.S. Department of Agriculture, *FY 2011 Budget Summary and Annual Performance Plan* (2010), 87; U.S. Department of Agriculture, *FY 2012 Budget Summary and Annual Performance Plan* (2011), 71; U.S. Department of Agriculture, *FY 2013 Budget Summary and Annual Performance Plan* (2012), 68; U.S. Department of Agriculture, *FY 2014 Budget Summary and Annual Performance Plan* (2013), 65; U.S. Department of Agriculture, *FY 2015 Budget Summary and Annual Performance Plan* (2014), 61; U.S. Department of Agriculture, *FY 2016 Budget Summary and Annual Performance Plan* (2015), 68; U.S. Department of Agriculture, *FY 2017 Budget Summary* (2016), 63; U.S. Department of Agriculture, *FY 2018 Budget Summary* (2017), 52; U.S. Department of Agriculture, *FY 2019 Budget Summary* (February 16, 2018 [Revised]), 23; U.S. Department of Agriculture, *FY 2020 Budget Summary* (2019); Agricultural Improvement Act of 2018, 90; Natural Resources Conservation Service, "NRCS Conservation Programs"; Natural Resources Conservation Service, "Conservation Stewardship Program (CSP)."

Conservation Reserve Program

The Farm Bill does not specify a funding level for the Conservation Reserve Program (CRP), instead imposing an acreage cap that indirectly limits the amount of money expended under the program. The cap applies to total acres enrolled, meaning that if the cap does not increase, new acres can be enrolled only when existing acres leave the program. Table 3 sets forth the statutory cap and the actual acres enrolled for each year from 1986 to 2023, as well as the total rental payments made in each year and the average dollars per acre paid in each year. Because payments are annual, payments in a given fiscal year reflect the acres enrolled as of the previous fiscal year. In Table 3, the figures in the "Total Rental Payments" and "Average Payment Per Acre" columns correspond to the year above, whereas "Authorized Acreage" and "Enrolled Acreage" correspond to the year with which they are aligned.

Table 3. Conservation Reserve Program Acres Enrolled

Year	*Authorized Acreage*	*Enrolled Acreage*	*Total Rental Payments*	*Average Payment Per Acre*
1986	45,000,000	1,929,064	N/A	N/A
1987	45,000,000	15,348,819	$82,931,664	$42.99
1988	45,000,000	23,870,635	$754,869,926	$49.18
1989	45,000,000	28,876,200	$1,158,201,981	$48.52
1990	45,000,000	32,522,380	$1,410,518,826	$48.85
1991	45,000,000	32,995,796	$1,595,003,251	$49.04
1992	45,000,000	33,993,313	$1,620,411,445	$49.11
1993	45,000,000	35,015,042	$1,679,566,765	$49.41
1994	45,000,000	35,015,042	$1,743,948,674	$49.81
1995	45,000,000	34,977,682	$1,743,948,674	$49.81
1996	36,400,000	34,503,815	$1,741,734,327	$49.80
1997	36,400,000	32,817,134	$1,718,006,124	$49.79
1998	36,400,000	30,134,648	$1,620,237,552	$49.37
1999	36,400,000	29,823,429	$1,376,893,866	$45.69
2000	36,400,000	31,425,584	$1,346,567,729	$45.15
2001	36,400,000	33,608,344	$1,420,310,010	$45.20
2002	36,400,000	33,964,386	$1,557,667,405	$46.35
2003	39,200,000	34,111,135	$1,602,690,826	$47.19
2004	39,200,000	34,707,287	$1,625,819,554	$47.66
2005	39,200,000	34,902,300	$1,665,612,834	$47.99
2006	39,200,000	36,003,300	$1,690,207,900	$48.43
2007	39,200,000	36,770,984	$1,762,529,086	$48.95
2008	32,000,000	34,612,696	$1,819,216,652	$49.47
2009	32,000,000	33,721,252	$1,756,904,966	$50.76

Year	*Authorized Acreage*	*Enrolled Acreage*	*Total Rental Payments*	*Average Payment Per Acre*
2010	32,000,000	31,298,245	$1,736,504,541	$51.50
2011	32,000,000	31,124,517	$1,666,815,708	$53.26
2012	32,000,000	29,525,599	$1,717,101,362	$55.17
2013	32,000,000	26,838,728	$1,693,828,736	$57.37
2014	27,500,000	25,448,835	$1,632,892,901	$60.84
2015	26,000,000	24,186,715	$1,620,167,838	$63.66
2016	25,000,000	23,884,000	$1,615,296,020	$66.78
2017	24,000,000	23,433,452	$1,734,235,093	$72.61
2018	24,000,000	22,609,492	$1,798,059,135	$76.73
2019	24,000,000	N/A	$1,844,889,495	$81.60
2020	24,500,000	N/A	N/A	N/A
2021	25,000,000	N/A	N/A	N/A
2022	25,000,000	N/A	N/A	N/A
2023	27,000,000	N/A	N/A	N/A

Sources: Food Security Act of 1985, Pub. L. No. 99-198, 1509; Food, Agriculture, Conservation, and Trade Act of 1990, 3577; Federal Agriculture Improvement and Reform Act of 1996, 994; Farm Security and Rural Investment Act of 2002, 239; Food, Conservation, and Energy Act of 2008, 1028–29; American Taxpayer Relief Act of 2012, 2363; Agricultural Act of 2014, 714; Agricultural Improvement Act of 2018, 44; Farm Service Agency, "CRP Enrollment and Rental Payment History by State," 2018, https://www.fsa.usda.gov/Assets/USDA-FSA-Public/usdafiles/Conservation/Excel/CRP%20Enrollment%20and%20Rental%20Payment%20History%20by%20County%2019862017.xlsx?CRP+Enrollment+and+Rental+Payments+History+by+County+2017.

Agricultural Conservation Easement Program

Table 4 sets forth the acres enrolled in the Agricultural Conservation Easement Program (ACEP) and its predecessors from 1992 to 2017. Acres enrolled in ACEP programs during the years 1992 to 2008 are presented on the NRCS website as lump sums, and so they are combined here as well. Table 5 sets forth the dollars expended in each fiscal year, beginning with 1992. Complete information is only available from NRCS for the years 2005 to 2017. For most programs, data for previous years (1992 to 2004) is based on Congressional appropriations. Because agencies are not necessarily able to spend all of the funds appropriated in a given year, the actual expenditures may have been lower in these years. One exception to the lack of detailed financial information is that more precise figures are available for funds expended through the Wetlands Reserve Program during the years 1992 to 2007. Note that older programs often continue to have expenditures after they are no longer enrolling new acres because cost-sharing programs continue to operate for acres already enrolled.

Table 4. Agricultural Conservation Easement Program (and Predecessors) Acres Enrolled

	Farm and Ranch Lands Protection Program (and Predecessor Farmland Protection Program)	*Grassland Reserve Program*	*Wetlands Reserve Program*	*ACEP-ALE*	*ACEP-WRE*	*Total*
1992–2008	N/A	N/A	1,821,044.5	N/A	N/A	1,821,044.5
1996–2008	550,100.7	N/A	N/A	N/A	N/A	550,100.7
2003–2008	N/A	117,318.1	N/A	N/A	N/A	117,318.1
2009	99,843.3	27,744.0	167,197.1	N/A	N/A	294,784.4
2010	101,080.7	67,402.2	244,147.4	N/A	N/A	412,630.3
2011	231,362.3	74,162.0	171,918.6	N/A	N/A	477,442.9
2012	38,493.7	39,791.4	144,728.2	N/A	N/A	223,013.3
2013	30,461.1	56,799.0	75,701.8	N/A	N/A	162,961.9
2014	N/A	N/A	N/A	70,641.9	45,751.7	116,393.6
2015	N/A	N/A	N/A	59,504.0	43,217.1	102,721.1
2016	N/A	N/A	N/A	123,065.0	36,454.7	159,519.7
2017	N/A	N/A	N/A	205,793.1	90,573.1	296,366.2
2018	N/A	N/A	N/A	88,429.1	34,593.1	123,022.2
Total (all-time)	1,051,341.8	383,216.7	2,624,737.6	547,433.1	250,589.7	4,857,318.9

Note: ACEP-ALE: Agricultural Conservation Easement Program–Agricultural Land Easements; ACEP-WRE: Agricultural Conservation Easement Program–Wetland Reserve Easements.

Sources: Natural Resources Conservation Service, "NRCS Conservation Programs: Farm and Ranch Lands Protection Program (FRPP)"; Natural Resources Conservation Service, "NRCS Conservation Programs: Grassland Reserve Program (GRP)"; Natural Resources Conservation Service, "NRCS Conservation Programs: Wetlands Reserve Program (WRP);" Natural Resources Conservation Service, "NRCS Conservation Programs: Agricultural Conservation Easement Program (ACEP)."

Table 5. Agricultural Conservation Easement Program (and Predecessors) Annual Expenditures

	Farm and Ranch Lands Protection Program (and Predecessor Farmland Protection Program)	*Grassland Reserve Program*	*Wetlands Reserve Program*	*ACEP (Total ACEP-ALE and ACEP-WRE)†*	*Total*
1992	N/A	N/A	$0	N/A	$0
1993	N/A	N/A	$0	N/A	$0
1994	N/A	N/A	$6,773,079	N/A	$6,773,079
1995	N/A	N/A	$67,128,240	N/A	$67,128,240
1996	*$14,000,000	N/A	$79,138,885	N/A	$93,138,885
1997	*$2,000,000	N/A	$122,012,740	N/A	$124,012,740
1998	*$17,000,000	N/A	$184,453,310	N/A	$201,453,310
1999	*$0	N/A	$119,069,890	N/A	$119,069,890
2000	*$0	N/A	$145,334,800	N/A	$145,334,800
2001	*$17,000,000	N/A	$136,685,230	N/A	$153,685,230
2002	*$51,000,000	N/A	$222,369,960	N/A	$273,369,960
2003	*$98,000,000	*$54,000,000	$241,736,390	N/A	$393,736,390
2004	*$91,000,000	*$57,000,000	$231,979,860	N/A	$379,979,860
2005	$111,800,000	$71,300,000	$267,300,000	N/A	$450,400,000
2006	$73,600,000	$35,500,000	$191,100,000	N/A	$300,200,000
2007	$73,100,000	$13,000,000	$247,900,000	N/A	$334,000,000
2008	$96,200,000	$2,800,000	$182,900,000	N/A	$281,900,000
2009	$118,766,200	$47,658,100	$435,711,300	N/A	$602,135,600
2010	$152,098,900	$100,108,400	$630,169,300	N/A	$882,376,600

(*continued*)

Table 5. *Continued*

	Farm and Ranch Lands Protection Program (and Predecessor Farmland Protection Program)	*Grassland Reserve Program*	*Wetlands Reserve Program*	*ACEP (Total ACEP-ALE and ACEP-WRE)†*	*Total*
2011	$170,908,700	$77,944,700	$569,085,800	N/A	$817,939,200
2012	$144,914,900	$65,264,400	$587,986,200	N/A	$798,165,500
2013	$118,129,900	$62,857,900	$400,218,500	N/A	$581,206,300
2014	$2,876,700	$1,452,400	$61,852,800	$316,900,300	$383,082,200
2015	$4,827,900	$6,950,800	$81,922,600	$297,347,800	$391,049,100
2016	$2,827,500	$1,758,600	$54,326,600	$345,677,200	$404,589,900
2017	$7,654,600	$4,955,200	$127,278,800	$536,032,500	$675,921,100
2018	$2,694,100	$1,461,900	$152,974,800	$337,539,300	$494,670,100
Total (all-time)	$1,228,399,400	$493,012,400	$5,547,409,084	$1,833,497,100	$9,355,317,984

Note: ACEP-ALE: Agricultural Conservation Easement Program–Agricultural Land Easements; ACEP-WRE: Agricultural Conservation Easement Program–Wetland Reserve Easements.

*Indicates appropriated, rather than expended, funds.

† Note that the Farm Bill authorizes expenditures for ACEP as a whole, and the Natural Resources Conservation Service subsequently divides the ACEP budget into ACEP-ALE and ACEP-WRE funding based on demand.

Sources: Natural Resources Conservation Service, "NRCS Conservation Programs;" Natural Resources Conservation Service, "NRCS Conservation Programs: Farm and Ranch Lands Protection Program (FRPP)"; Natural Resources Conservation Service, "NRCS Conservation Programs: Grassland Reserve Program (GRP)"; Natural Resources Conservation Service, "NRCS Conservation Programs: Wetlands Reserve Program (WRP)"; United States Department of Agriculture Economic Research Service, "Wetland Reserve Program Expenditures"; Natural Resources Conservation Service, "NRCS Conservation Programs: Agricultural Conservation Easement Program (ACEP)"; Office of Management and Budget, *The Budget for Fiscal Year 1998: Department of Agriculture*, 168; Office of Management and Budget, *The Budget for Fiscal Year 1999: Department of Agriculture*, 99; Office of Management and Budget, *The Budget for Fiscal Year 2000: Department of Agriculture*, 104; Office of Management and Budget, *The Budget for Fiscal Year 2001: Department of Agriculture*, 110; Office of Management and Budget, *The Budget for Fiscal Year 2002: Department of Agriculture*, 108; Office of Management and Budget, *The Budget for Fiscal Year 2003: Department of Agriculture*, 111; U.S. Department of Agriculture, *FY 2004 Budget Summary*, 24; U.S. Department of Agriculture, *FY 2005 Budget Summary*, 34; U.S. Department of Agriculture, *FY 2006 Budget Summary and Annual Performance Plan*, 29.

Healthy Forests Reserve Program

Table 6 shows the acres enrolled and funds expended under the Healthy Forests Reserve Program for 2006 to 2017. Detailed NRCS data begins in 2009. For 2006 to 2008, acres enrolled are available only as a lump sum, and the expenditures column uses appropriated, rather than expended, dollars.

Table 6. Healthy Forests Reserve Program Annual Expenditures and Acres Enrolled

	Annual Expenditures on HFRP	*Acres Enrolled in HFRP*
2006	*$2,000,000	N/A
2007	*$2,000,000	N/A
2008	*$0	N/A
Total 2006–2008	*$4,000,000	139,311.1
2009	$2,621,000	645.9
2010	$7,946,100	5,724.8
2011	$17,910,800	20,849.8
2012	$9,843,100	500,333.2
2013	$6,443,300	8,057.4
2014	$576,700	0
2015	$1,284,200	1,258.6
2016	$703,900	0
2017	$153,100	0
2018	$70,500	0
Total (all-time)	$53,552,700	676,180.80

Note: HFRP: Healthy Forests Reserve Program.
*Indicates appropriated funds, whereas other years show expended funds.

Sources: Natural Resources Conservation Service, "NRCS Conservation Programs: Healthy Forests Reserve Program (HFRP)"; U.S. Department of Agriculture, *FY 2008 Budget Summary and Annual Performance Plan*, 63; U.S. Department of Agriculture, *FY 2009 Budget Summary and Annual Performance Plan*, 56; U.S. Department of Agriculture, *FY 2010 Budget Summary and Annual Performance Plan*, 72.

Note

1. Pedersen and Hatfield.

Acknowledgments

IN 2016, A BOOK that is still in production was started entitled *Playing God with Nature: The Codex of the Endangered Species Act.* This book will be an omnibus treatment of the history, application, and challenges of the 1973 Endangered Species Act (ESA) reflecting its fifty-year history come 2023. The book you are reading on private landowner incentives was initially inspired during a conversation with Dr. Lee M. Talbot back in 2017. Dr. Talbot was the principal author of the 1973 ESA and is celebrated for many distinctions throughout his incredible career. This book started as a single chapter in the original book, but developed its own identify as it grew. We recognized early on its unique importance to private landowners who shared with us their frustration when encountering endangered species challenges without any reliable single source for information, consultation, advice, or funding. This will come as no surprise to the working landowner.

The ESA was twelve pages long when enacted in 1973. Today it is thirty-six pages long. The accompanying regulations initially were thirty pages. Today the regulations are five thousand pages. Most states, moreover, have their own separate endangered species acts. Separate state and federal service centers are found in virtually every county in America. Which one do you turn to for the right kind of help or advice? Understanding the confusing laws and regulations that govern wildlife conservation, and especially at-risk and endangered species, is like trying to piece together a giant jigsaw puzzle. The pieces just don't always seem to fit, and the picture is never quite complete. This book is an attempt to provide such guidance that will in turn provide incentives to the working landowner for wildlife conservation and species preservation.

Early on in the development of this book, I was incapacitated with a major medical challenge for well over eight months. My stalwart colleague and associate counsel since 2011, Christopher E. Segal, picked up the gauntlet and assumed the burden for the majority of the tedious research involved in writing the manuscript. His intellectual keenness, persevering diligence, and expertise in research is remarkable, and his manuscript preparation, editing, and fact checking superb. Hence his name joins mine on the cover. My executive assistant since 2005, Janice Shoyooee, has been dedicated to this project and a tremendous support, especially during my medical tribulations. She scrambles to anticipate my relenting need for perfection, information, facts, figures, and deciphering legions of handwritten manuscript text, letters, notes, etc. Finally and most importantly, I thank my wife of fifty-one years, who gracefully and patiently tolerates my exhausting work schedule, whims, and Germanic temperament. These are the three people to whom I'm blessed to share this journey. They are my anchors. Moreover my thanks to my editor at Rowman & Littlefield, Jon Sisk, who recognized the importance of publishing this reference guide and has had continuing faith in my research and writing skills since the publication of my last book in 2016 that became an award winner, *Inside the Equal Access to Justice Act*.

My deepest thanks to the many people who so generously gave their time and expertise as we pieced together the puzzle in order to make this project a reality. The world of voluntary conservation has its own language, and many people provided invaluable assistance in disseminating critical information. For insights into the conservation importance of privately owned land, thanks to Daniel M. Ashe, Dale Hall and Steven Williams, all former Directors of the U.S. Fish and Wildlife Service; Robert Bonnie, Under Secretary of Agriculture for Natural Resources and the Environment at the U.S. Department of Agriculture during the Obama administration; Dr. Terry Z. Riley, Director of Policy at the North American Grouse Partnership; Dr. Edward B. Arnett, Chief Scientist at the Theodore Roosevelt Conservation Partnership; and Douglas Wheeler, who served as Deputy Assistant Secretary for Fish, Wildlife, and Parks at the U.S. Department of the Interior during the Nixon administration and later as California's Secretary for Resources.

Special thanks to Mark Rey, who served as Under Secretary for Natural Resources and Environment at the U.S. Department of Agriculture in the George W. Bush administration and to James R. Lyons, who held the same office during the Clinton administration, and was later Deputy Assistant Secretary for Land and Minerals Management at the U.S. Department of the Interior during the Obama administration. Each helped me better understand the role of federal agencies in voluntary conservation and gave insight into diverse subjects such as compensatory mitigation, conservation banking, and the challenges involved in developing effective conservation programs. Thanks also to Wayne Walker with Common Ground Capital for assistance in understanding conservation banking and to Lori Faeth with the Land Trust Alliance for information on land trusts.

Many individuals provided information about the Florida panther, including Paul Souza, David Shindle, and Constance Cassler with the U.S. Fish and Wildlife Service; Margarita Carey with The Conservation Fund; and Nick Wiley with Ducks Unlimited, formerly with the Florida Fish and Wildlife Conservation Commission. Dr. Donald "Pete" Gober with the U.S. Fish and Wildlife Service, John Emmerich with the Wyoming Game and Fish Department, and the former Director of the Department Scott Talbott provided information on black-footed ferret, and Wendy Jackson with the Land Trust Alliance, Fred Annand with The Nature Conservancy, and Jay Carter with Dr. J. H. Carter & Associates, Inc., educated me about red-cockaded woodpecker. It is impossible to thank all the people who have worked tirelessly on greater sage-grouse conservation and included us in their meetings and discussions over the last decade, but we must acknowledge San Stiver and Kenneth Mayer with the Western Association of Fish and Wildlife Agencies, Virgil Moore formerly with the Idaho Department of Fish and Game, Larry Voyles formerly with the Arizona Game and Fish Department, Tony Wasley with the Nevada Department of Wildlife, and once again James R. Lyons.

The federal Farm Bill is an astonishingly complicated subject, one that required months of dedicated research and many, many conversations with expert policymakers and practitioners. Particular thanks to James Cummins with Wildlife Mississippi and the Boone & Crockett Club, for his patience and endless willingness to answer questions by phone and email. Additional thanks to Mark Rey, Jimmy Bullock with Resource Management Service, LLC, David Nomsen and James Inglis with Pheasants Forever, Joel Pedersen and Mark Hatfield with the National Wild Turkey Federation, Andrew Schmidt with Ducks Unlimited, and Galon Hall and Tim Griffiths with the Natural Resources Conservation Service.

Many of our federal public servants offered their time to help us understand and explain their work. These include Gary Frazer and Matt Filsinger with the U.S. Fish and Wildlife Service; Galon Hall and Tim Griffiths with Natural Resources Conservation Service; Bruce Beard, formerly with the Office of the Secretary of Defense and now at Texas A&M University; Jean Mansavage with the Air Force Historical Studies Office; Joseph Burns, Laurie Schoonhoven, Linda Wang, Susan Stein, and Steven Koehn with the U.S. Forest Service; Karen Prentice and Laura Van Riper with the Bureau of Land Management; David Gagner, Greg Knadle, and Rob Blumenthal with the National Fish and Wildlife Foundation; Andy Bishop with the Rainwater Basin Joint Venture; Dave Smith and Ali Duvall with the Intermountain West

Joint Venture; and Dr. John Organ, formerly Director of the U.S. Geological Survey's Cooperative Wildlife Research Units program.

State government agencies are critical partners for every program discussed in this book, and many state employees provided us with information on their programs. Special thanks to Lane Kisonak and Gordon Batcheller at the Association of Fish and Wildlife Agencies for connecting us to our diverse community of state agencies. For sharing information about individual state programs, thanks to John Marks with the Arkansas Game and Fish Commission; Don Larsen with the Arizona Game and Fish Department; Chris Breidenbach with the Colorado Attorney General's Office; Tamara McIntosh with the Iowa Department of Natural Resources; Chris Tymeson with the Kansas Department of Wildlife, Parks, and Tourism; Jason A. Raup with the Pennsylvania Game Commission; Jon Kotilnek with the South Dakota Department of Game, Fish, and Parks; Robert Sweeney, Meredith Longoria, Justin Dreibelbis, and Christopher Abernathy with the Texas Parks & Wildlife Department; Jeff Davis with the Washington Department of Fish and Wildlife; and Amy Dona with the Washington State Office of the Attorney General.

Equally helpful were the men and women who work for the nongovernmental organizations that partner with and support our private landowners. Thanks to Shawn Graff with the American Bird Conservancy; Kelly Reed with The Conservation Fund; John Devney with Delta Waterfowl; Nick Wiley and Johann Walker with Ducks Unlimited; Miles Moretti with the Mule Deer Foundation; Lynn Scarlett and Brent Keith with The Nature Conservancy; Sarah Greenberger and Marshall Johnson with the National Audubon Society; Joel Pedersen and Mark Hatfield with the National Wild Turkey Federation; Mike Leahy and Julie Sibbing with the National Wildlife Federation; David Nomsen and James Inglis with Pheasants Forever; Kip Adams with the Quality Deer Management Association; Blake Henning with the Rocky Mountain Elk Foundation; Brent Rudolph with the Ruffed Grouse Society & American Woodcock Society; Christy Plumer with the Theodore Roosevelt Conservation Partnership; Chris Wood and Robert Masonis with Trout Unlimited; Russ Austad with Whitetails Unlimited; and Kevin Hurley and Clay Brewer with the Wild Sheep Foundation.

The research for this project started where the finished product ends: with the 1973 ESA. Many people too numerous to list have educated us about the ESA over the years, but special mention must be made of Daniel M. Ashe, Dale Hall, James R. Lyons, Mark Rey, Steven Williams, and the extraordinary Michael Bean. It has been my privilege to know, work with, and interview many of the people who stewarded the ESA through Congress and later implemented it, including the late Representative John D. Dingell, Jr., Dr. Lee M. Talbot, the late Nathanial P. Reed, E.U. Curtis "Buff' Bohlen, Douglas

Wheeler, and Lynn Greenwalt. For this book specifically, Donald Barry, Lynn Greenwalt, Dale Hall, and Lindell Marsh provided us with recollections and information on the development of the San Bruno Mountain Habitat Conservation Plan and the 1982 amendment to the ESA.

Finally, a deep thank you to the farmers, ranchers, miners, tree farmers, and others that make up our national community of private landowners. I know them, because I grew up on a family farm in Indiana and grandad's ranch in Montana, and was raised by them. There are no people more dedicated to fish and wildlife conservation, and no effort on behalf of at-risk species that can succeed without their help. This book captures their efforts at just a single moment of time—the present—but their long legacy of conservation and the challenges that lie before them are truly humbling. A final thank you must be made to Steve Jester, Executive Director of Partners for Conservation, for his tireless work to amplify the voice of our landowner conservationists, and for his guidance in the scope and direction of this book.

Lowell E. Baier
Bethesda, Maryland
October 1, 2019

Bibliography

Books and Book Chapters

Ahner, Mark D. *Can the United States Army Adjust to the Endangared Species Act of 1973?* Carlisle Barracks, PA: U.S. Army War College, 1992.

American Forests. *Partnering for Forests: A Look at the U.S. Forest Service's Cooperative Forestry Program and Its Partners*. 2017.

Bean, Michael J., and Melanie J. Rowland. *The Evolution of National Wildlife Law.* Third edition. Westport, CT: Praeger, 1997.

Bureau of Land Management. *Public Land Statistics 2017.* June 2018.

Coppess, Jonathan. *The Fault Lines of Farm Policy: A Legislative and Political History of the Farm Bill.* Lincoln: University of Nebraska Press, 2018.

Costa, Ralph, and U.S. Fish and Wildlife Service. *5-Year Review: Red-Cockaded Woodpecker* (Picoides Borealis). 2006.

Cummins, James L. "Private Lands Conservation Programs." In *North American Wildlife Policy and Law*, edited by Bruce D. Leopold, Winifred B. Kessler, and James L. Cummins, 231–46. Missoula, MT: Boone and Crockett Club, 2018.

Gartner, Todd, C. Josh Donlan, Michael G. Sorice, James Mulligan, Mary Snieckus, and Rhett Johnson. "The Gopher Tortoise, Military Readiness, and Pre-Listing Conservation." In *Proactive Strategies for Protecting Species: Pre-Listing Conservation and the Endangered Species Act*, edited by C. Josh Donlan. Oakland, CA: University of California Press, 2015.

Gustanski, Julie Ann, and Roderick H. Squires. *Protecting the Land: Conservation Easements Past, Present, and Future.* Washington, DC: Island Press, 2000.

Hansen, Megan E., Sarah M. Bennett, Jennifer Morales, and Rebekah M. Yeagley. *Cooperative Conservation: Determinants of Landowner Engagement in Conserving Endangered Species*. Center for Growth and Opportunity, Utah State University, November 2018.

Hornaday, William Temple. *Save the Sage Grouse from Extinction: A Demand from Civilization to the Western States*. Permanent Wild Life Protection Fund Bulletin 5. New York, NY: New York Zoological Park, 1916.

Intermountain West Joint Venture. *2013 Implementation Plan—Strengthening Science and Partnerships*. Missoula, MT: Intermountain West Joint Venture, 2013.

Kiernan, Sean. "Perspective: Renewable Energy and Endangered Species." In *Proactive Strategies for Protecting Species: Pre-Listing Conservation and the Endangered Species Act*, edited by C. Josh Donlan, 36–42. Oakland, CA: University of California Press, 2015.

Land Trust Alliance. *2015 National Land Trust Census Report: Our Common Ground and Collective Impact*. 2016.

Leopold, Aldo. "Conservation Economics." In *The River of the Mother of God and Other Essays by Aldo Leopold*, edited by Susan L. Flader and J. Baird Callicott, 193–202. Madison, WI: University of Wisconsin Press, 1991.

———. "The Conservation Ethic." In *The River of the Mother of God and Other Essays by Aldo Leopold*, edited by Susan L. Flader and J. Baird Callicott, 191–92. Madison, WI: University of Wisconsin Press, 1991.

Lindstrom, Timothy. *A Tax Guide to Conservation Easements*. Second edition. Washington, DC: Land Trust Alliance, 2016.

Male, Timothy. *Headstarting Conservation: Reviewing New Policy to Incentivize Early Investments in Declining Species*. Madison, WI: Sand County Foundation/Mission: Wildlife, 2014.

Male, Timothy, and C. Josh Donlan. "The Future of Pre-Listing Conservation Programs for Wildlife Conservation." In *Proactive Strategies for Protecting Species: Pre-Listing Conservation and the Endangered Species Act*, edited by C. Josh Donlan, 219–29. Oakland, CA: University of California Press, 2015.

Mansavage, Jean A. *Natural Defense: U.S. Air Force Origins of the Department of Defense Natural Resources Conservation Program*. Air Force Historical Studies. Washington, DC: Air Force History and Museums Program, 2014.

Marsh, Lindell L., and Robert D. Thornton. "San Bruno Mountain Habitat Conservation Plan." In *Managing Land-Use Conflicts: Case Studies in Special Area Management*, edited by David J. Brower and Daniel S. Carol, 114–39. Durham, NC: Duke University Press, 1987.

Monke, Jim. *U.S. Library of Congress, Congressional Research Service, Farm Bill Programs without a Budget Baseline Beyond FY2018, CRS Report No. R44758*. 2018.

North American Bird Conservation Initiative Private Lands Subcommittee. *Delivering Fish and Wildlife Conservation: Building Farm Bill Capacity*. August 2009.

North American Bird Conservation Initiative U.S. Committee. *2014 Farm Bill Field Guide to Fish and Wildlife Conservation*. 2015.

North American Waterfowl Management Plan. *1994 Update to the North American Waterfowl Management Plan: Expanding the Commitment*. 1994.

———. *2018 North American Waterfowl Management Plan (NAWMP) Update: Connecting People, Waterfowl, and Wetlands*. 2018.

———. *Expanding the Vision: 1998 Update North American Waterfowl Management Plan*. 1998.

———. *North American Waterfowl Management Plan 2012: People Conserving Waterfowl and Wetlands*. 2012.

———. *Strengthening the Biological Foundation: 2004 Strategic Guidance*. 2004.

Office of Management and Budget. *The Budget for Fiscal Year 1998: Department of Agriculture*. 1997.

———. *The Budget for Fiscal Year 1999: Department of Agriculture*. 1998.

———. *The Budget for Fiscal Year 2000: Department of Agriculture*. 1999.

———. *The Budget for Fiscal Year 2001: Department of Agriculture*. 2000.

———. *The Budget for Fiscal Year 2002: Department of Agriculture*. 2001.

———. *The Budget for Fiscal Year 2003: Department of Agriculture*. 2002.

Parkhurst, Gregory M., and Jason F. Shogren. "An Economic Review of Incentive Mechanisms to Protect Species on Private Lands." In *Species at Risk: Using Economic Incentives to Shelter*

Endangered Species on Private Lands, edited by Jason F. Shogren. Austin, TX: University of Texas Press, 2005.

———. "Evaluating the Incentive Tools." In *Species at Risk: Using Economic Inceentives to Shelter Endangered Species on Private Lands*, edited by Jason F. Shogren. Austin, TX: University of Texas Press, 2005.

San Bruno Mountain Habitat Conservation Plan Steering Committee. *San Bruno Mountain Area Habitat Conservation Plan–Final*. November, 1982.

Stubbs, Megan. *U.S. Library of Congress Congressional Research Service: Agricultural Conservation: A Guide to Programs, CRS Report No. R40763.* April 17, 2018.

———. *U.S. Library of Congress Congressional Research Service: Conservation Compliance and U.S. Farm Policy, CRS Report No. R42459.* October 6, 2016.

U.S. Department of Agriculture. *FY 2004 Budget Summary*. 2003.

———. *FY 2005 Budget Summary*. 2004.

———. *FY 2006 Budget Summary and Annual Performance Plan*. 2005.

———. *FY 2007 Budget Summary and Annual Performance Plan*. 2006.

———. *FY 2008 Budget Summary and Annual Performance Plan*. 2007.

———. *FY 2009 Budget Summary and Annual Performance Plan*. 2008.

———. *FY 2010 Budget Summary and Annual Performance Plan*. 2009.

———. *FY 2011 Budget Summary and Annual Performance Plan*. 2010.

———. *FY 2012 Budget Summary and Annual Performance Plan*. 2011.

———. *FY 2013 Budget Summary and Annual Performance Plan*. 2012.

———. *FY 2014 Budget Summary and Annual Performance Plan*. 2013.

———. *FY 2015 Budget Summary and Annual Performance Plan*. 2014.

———. *FY 2016 Budget Summary and Annual Performance Plan*. 2015.

———. *FY 2017 Budget Summary*. 2016.

———. *FY 2018 Budget Summary*. 2017.

———. *FY 2019 Budget Summary*. February 16, 2018 (Revised).

———. *FY 2020 Budget Summary*. 2019.

U.S. Department of Agriculture, U.S. Department of Defense, and U.S. Department of the Interior. *Sentinel Landscapes 2016 Report*. 2016.

U.S. Department of Defense Readiness and Environmental Protection Integration Program. *13th Annual Report to Congress*. 2019.

U.S. Department of the Interior. *Budget Justifications and Performance Information Fiscal Year 2019: Fish and Wildlife Service*. 2018.

———. *Budget Justifications and Performance Information Fiscal Year 2020: Fish and Wildlife Service*. 2019.

———. *Budget Justifications and Performance Information, Fiscal Year 2018: U.S. Fish and Wildlife Service*. 2017.

U.S. Fish and Wildlife Service. *Protecting America's Living Heritage: A Fair, Cooperative and Scientifically Sound Approach to Improving the Endangered Species Act*. 1995.

U.S. Fish and Wildlife Service and Canadian Wildlife Service. *North American Waterfowl Management Plan*. May 1986.

U.S. Fish and Wildlife Service Partners for Fish and Wildlife Program. *Mountain-Prairie Region Annual Narrative FY 2017*. 2018.

U.S. Fish and Wildlife Service–Southeast Region, U.S. Department of Defense, Alabama Department of Conservation and Natural Resources, Florida Fish and Wildlife Conservation Commission, Georgia Department of Natural Resources, and South Carolina Department of

Natural Resources. *Department of Defense Gopher Tortoise (Gopherus Polyphemus) Conservation and Crediting Strategy*. March 2017.
U.S. Forest Service. *FY 2018 Budget Justification*. 2017.
———. *FY 2020 Budget Justification*. 2019.
U.S. Government Accountability Office. *Conservation Security Program: Despite Cost Controls, Improved USDA Management Is Needed to Ensure Proper Payments and Reduce Duplication with Other Programs, No. GAO-06-312*. April, 2006.
Van Pelt, William E., Sean Kyle, Jim Pitman, David Klute, Grant Beauprez, Doug Schoeling, Allan Janus, and Jonathan B. Haufler. *The Lesser Prairie-Chicken Range-Wide Conservation Plan*. Cheyenne, WY: Western Association of Fish and Wildlife Agencies, 2013.
Vincent, Carol Hardy, Laura A. Hanson, and Carla N. Argueta. *Federal Land Ownership: Overview and Data, Congressional Research Service Report R42346*. March 3, 2017.
Wheeler, Douglas P., and Ryan M. Rowberry. "Habitat Conservation Plans and the Endangered Species Act." In *Endangered Species Act: Law, Policy, and Perspectives*, edited by Donald C. Baur and Wm Robert Irvin, 220–45. Chicago, Ill.: American Bar Association, 2010.
Wilkins, R. Neal. "Improving the ESA's Performance on Private Land." In *Rebuilding the Ark: New Perspectives on Endangered Species Act Reform*, edited by Jonathan H. Adler. Washington, DC: American Enterprise Institute for Public Policy Research, 2011.
Wisconsin's Managed Forest Law: A Program Summary, PUB-FR-295. November 2017.
Wolfe, Roger L., Sean C. Kyle, Jim C. Pitman, Deb M. VonDeBur, and Michael E. Houts. *The 2017 Lesser Prairie-Chicken Range-Wide Conservation Plan Annual Progress Report*. March, 2018.

Journal Articles

Airey, Shea B. "Conservation Easements in Private Practice." *Real Property, Trust and Estate Law Journal* 44 (2010): 745–822.
Bauer, Marybeth, Jessica Fox, and Michael J. Bean. "Landowners Bank on Conservation: The U.S. Fish and Wildlife Service's Guidance on Conservation Banking." *Environmental Law Reporter News & Analysis* 34 (2004).
Brasher, Michael G., James J. Giocomo, David A. Azure, Anne M. Bartuszevige, Mark E. Flaspohler, Dean E. Harrigal, Brian W. Olson, et al. "The History and Importance of Private Lands for North American Waterfowl Conservation." *Wildlife Society Bulletin* 43, no. 3 (2019): 338–54.
Burger, L. Wes Jr., Kristine O. Evans, Mark D. McConnell, and Leslie M. Burger. "Private Lands Conservation: A Vision for the Future." *Wildlife Society Bulletin* 43, no. 3 (2019): 398–407.
Carr, Sunni L., Steven P. Riley, Kristal Stoner, Gerry Steinauer, and Thomas L. Walker Jr. "Influence of the Federal Landowner Incentive Program." *Wildlife Society Bulletin* 43, no. 3 (2019): 355–62.
Cohn, Jeffrey P. "Joint Ventures: A Different Approach to Conservation." *BioScience* 55, no. 10 (October 2005): 824–27.
Conner, Richard N., Daniel Saenz, and D. Craig Rudolph. "The Red-Cockaded Woodpecker: Interactions with Fire, Snags, Fungi, Rat Snakes, and Pileated Woodpeckers." *Texas Journal of Science* 56, no. 4 (2004): 415–26.
Conover, Michael R., and Anthony J. Roberts. "Declining Populations of Greater Sage-Grouse: Where and Why." *Human–Wildlife Interactions* 10, no. 2 (2016): 217–29.
Drake, David, and Edwin J. Jones. "Forest Management Decisions of North Carolina Landowners Relative to the Red-Cockaded Woodpecker." *Wildlife Society Bulletin* 30, no. 1 (2002): 121–30.
Echols, Alex, Alan Front, and James Cummins. "Broadening Conservation Funding." *Wildlife Society Bulletin* 43 (2019): 372–81.

Evans, Michael J., Jacob W. Malcom, and Ya-Wei Li. "Novel Data Show Expert Wildlife Agencies Are Important to Endangered Species Protection." *Nature Communications* 10 (2019): 3467.

Farmer, James R., Doug Knapp, Vicky J. Meretsky, Charles Chancellor, and Burnell C. Fisher. "Motivations Influencing the Adoption of Conservation Easements." *Conservation Biology* 25 (2011): 827–34.

Fox, Jessica, and Anamaria Nino-Murcia. "Status of Species Conservation Banking in the United States." *Conservation Biology* 19, no. 4 (2005): 996–1007.

Galik, Christopher S., Todd K. BenDor, Julie DeMeester, and David Wolfe. "Improving Habitat Exchange Planning through Theory, Application, and Lessons from Other Fields." *Environmental Science & Policy* 73 (2017): 45–51.

Gardner, Royal C., and Jessica Fox. "The Legal Status of Environmental Credit Stacking." *Ecology Law Quarterly* 4, no. 4 (2013): 713–57.

Germano, Jennifer M., Kimberleigh J. Field, Richard A. Griffiths, Simon Clulow, Jim Foster, Gemma Harding, and Ronald R. Swaisgood. "Mitigation-Driven Translocations: Are We Moving Wildlife in the Right Direction?" *Frontiers in Ecology and the Environment* 13, no. 2 (2015): 100–05.

Huff, Julia LeMense. "Protecting Ecosystems Using Conservation Tax Incentives: How Much Bang Do We Get for Our Buck?". *Missouri Environmental Law & Policy Review* 11 (2004): 138–58.

Lindstrom, Timothy. "State Tax Incentives for Conservation Easements Can Benefit Everyone." *Journal of Multistate Taxation and Incentives* 12-NOV (2002): 20–48.

Lzgon, J. David, Peter B. Stacey Richard N. Conner, Carl E. Bock, and Curtis S. Adkisson. "Report of the American Ornithologists' Union Committee for the Conservation of the Red-Cockaded Woodpecker." *The Auk* 103 (October, 1986): 848–55.

Malcom, Jacob W., and Ya-Wei Li. "Data Contradict Common Perceptions About a Controversial Provision of the US Endangered Species Act." *Proceedings of the National Academy of Sciences of the United States of America* 112, no. 52 (2015): 15844–49.

Martinez, Michael. "Working Lands for Wildlife: Targeted Landscape-Scale Wildlife Habitat Conservation." *Natural Resources & Environment* 29, no. 3 (2015): 36–39.

McConnell, Mark D. "Bridging the Gap between Conservation Delivery and Economics with Precision Agriculture." *Wildlife Society Bulletin* 43, no. 3 (2019): 391–97.

McLaughlin, Nancy A. "Increasing the Tax Incentives for Conservation Easement Donations–a Responsible Approach." *Ecology Law Quarterly* 31 (2004): 1–112.

Merenlender, A. M., L. Huntsinger, G. Guthey, and S. K. Fairfax. "Land Trusts and Conservation Easements: Who Is Conserving What for Whom?" *Conservation Biology* 18 (2004): 65–75.

Morgan, John J., Cody M. Rhoden, Bill White, and Steven P. Riley. "A State Assessment of Private Lands Wildlife Conservation in the United States." *Wildlife Society Bulletin* 43, no. 3 (2019): 328–37.

Riley, Steven P., Robert Abernethy, and Donald F. McKenzie. "Case Studies of Scalable Wildlife Conservation in Working Landscapes." *Wildlife Society Bulletin* 43, no. 3 (2019): 363–71.

Riley, Terry Z., Noe Marymor, Nancy D. Riley, and Bill White. "Delivery of Habitat Conservation Programs on Private Land." *Wildlife Society Bulletin* 43, no. 3 (2019): 408–13.

Rudolph, D. Craig, Richard N. Conner, and Jeffrey R. Walters. "Red-Cockaded Woodpecker Recovery: An Integrated Strategy." *U.S. Forest Service Miscellaneous Publications* (2004).

Ruhl, J.B., Alan Glen, and David Hartman. "Practical Guide to Habitat Conservation Banking Law and Policy." *Natural Resources and Environment* 20 (2005): 26–32.

Watchman, Laura H., Martha Groom, and John D. Perrine. "Science and Uncertainty in Habitat Conservation Planning: A Review of 43 Habitat Conservation Plans Reveals Numerous

Ways to Reduce Uncertainty for Landowners and Imperiled Species Alike." *American Scientist* 89, no. 4 (2001): 351–59.
Wheeler, Douglas P. "A Strategy for the Future." *Stanford Environemental Law Journal* 12 (1993): ix–xv.
———. "An Ecosystem Approach to Species Protection." *Natural Resources & Environment* 10, no. 3 (Winter 1996): 7–9.

Newspapers and Periodicals

Associated Press. "Nature Conservancy to Preserve Historic Ranch." *Daily Journal of Commerce*, January 20, 2000.
Barker, Rocky. "Sage Grouse Numbers Plummeted after 2007, Study Shows." *Idaho Statesman*, April 24, 2015.
Bodin, Madeline. "The Farm Bill: Celebrating Passage of a Key Land Conservation Resource." *Saving Land*, Spring 2019.
Brulliard, Karin. "A 50-Year Effort to Raise Endangered Whooping Cranes Comes to an End." *Washington Post*, Spetember 18, 2017.
Chin, Joshua. "Fighting to Save, and Popularize, San Bruno Mountain." *Bay Nature*, February 26, 2015.
Lindsay, Leon. "A Tale of Butterflies and a Billion-Dollar Project." *Christian Science Monitor*, March 14, 1983.
Lindstrom, Timothy. "The Syndication of Conservation Easement Tax Deductions." *Saving Land*, Summer 2015.
Maixner, Ed. "On Tap: First Ever under-10-Years CRP Contracts." *Agri-Pulse*, February 20, 2019.
Martin, Brian. "Guest Opinion: TNC's Vision for Montana Prairie Conservation." *Billings Gazette*, February 23, 2019.
Oregon Department of Transportation. "Conservation Bank Helps ODOT." *Moving Ahead*, April 2015.
Pittman, Craig. "Saga of Florida Panther Is 'Sordid Story.'" *St. Petersburg Times*, April 19, 2010.
Watling, Eve. "Monarch Butterflies Are Going Extinct: How Can We Save Them?" *Newsweek*, January 26, 2019.
Wiley, Nick, and Paul Souza. "Conservation Banks Help Protect Panthers." *St. Petersburg Times*, March 1, 2010.
Wright, John B., and Anthony Anella. "Saving the Ranch: Fresh Eyes on Taxes, Development, and Conservation Easements." *Rangelands*, June 2007.

Legislative Materials

16 U.S.C. § 669e.
16 U.S.C. § 777e.
16 U.S.C. § 1538(a)(2).
26 U.S.C. § 126.
26 U.S.C. § 126(a)(8).
26 U.S.C. § 170.
An Act Making Omnibus Consolidated Appropriations for the Fiscal Year Ending September 30, 1997, and for Other Purposes, Pub. L. No. 104-308 § 124, 110 Stat. 3009. 1996.

An Act to Amend the Act of March 10, 1934 Entitled "an Act to Promote the Conservation of Wildlife, Fish, and Game, and for Other Purposes," Pub. L. No. 79-132, 60 Stat. 1080. 1946.
An Act to Amend the Act of March 10, 1934, to Provide for More Effective Integration of a Fish and Wildlife Conservation Program with Federal Water-Resource Developments, and for Other Purposes, Pub. L. No. 85-624, 72 Stat. 568. 1958.
An Act to Amend the Migratory Bird Hunting Stamp Act of March 16,1934, as Amended, Pub. L. No. 85-585, 72 Stat. 486. 1958.
An Act to Promote the Conservation of Migratory Waterfowl by the Acquisition of Wetlands and Other Essential Waterfowl Habitat, and for Other Purposes, Pub. L. No. 87-383, 75 Stat. 813. 1961.
An Act to Promote the Conservation of Wild Life, Fish, and Game, and for Other Purposes, Pub. L. No. 73-121, 48 Stat. 401. 1934.
Agricultural Act of 2014, Pub. L. No. 113-79, 128 Stat. 649. 2014.
Agricultural Credit Act of 1978–Title IV (Emergency Conservation Program), Public Law 95-334, 92 Stat. 420, 16 U.S.C. 2201-2205, as Amended through P.L. 115-334, Enacted December 20, 2018.
Agricultural Credit Act of 1978, Pub. L. No. 95-334, 92 Stat. 420. 1978.
Agricultural Improvement Act of 2018, Pub. L. No. 115-334. 2018.
American Taxpayer Relief Act of 2012, Pub. L. No. 112-240, 126 Stat. 2313. 2012.
Bob Stump National Defense Authorization Act for Fiscal Year 2003, Pub. L. No. 107-314, 116 Stat. 2458. 2002.
Committee on Conference, Agriculture Improvement Act of 2018, Conference Report to Accompany H.R. 2, H.R. Rep. No. 115-1072. 2018.
Committee on Conference, Endangered Species Act Amendments of 1982, Conference Report to Accompany H.R. 6133, H.R. Rep. No. 97-835. 1982.
Committee on Conference, Farm Security and Rural Investment Act of 2002, Conference Report to Accompany H.R. 2646, H.R. Rep. No. 107-424. 2002.
Committee on Conference, Food, Conservation, and Energy Act of 2008, Conference Report to Accompany H.R. 2419, H.R. Rep. No. 110-627. 2008.
Committee on Interstate and Foreign Commerce, Amending the Coordination Act, Report to Accompany H.R. 13138, S. Rep. No. 85-1981. 1958.
Committee on Merchant Marine and Fisheries, Endangered Species Act Amendments, Report to Accompany H.R. 6133, H.R. Rep. N. 97-567, Part 1. 1982.
Deficit Reduction Act of 1984, Pub. L. No. 98-369, 98 Stat. 494.
Department of Defense, Emergency Supplemental Appropriations to Address Hurricanes in the Gulf of Mexico, and Pandemic Influenza Act, 2006, Pub. L. No. 109-148, 119 Stat. 2680.
Department of the Interior and Related Agencies Appropriations Act, Pub. L. No. 107-63, 115 Stat. 414, 422. 2002.
Disaster Relief Appropriations, Pub. L. No. 113-2, 127 Stat. 4. 2013.
Endangered Species Act Amendments of 1978, Pub. L. No. 95-632, 92 Stat. 3751. 1978.
Farm Security and Rural Investment Act of 2002, Pub. L. No. 107-171, 116 Stat. 134. 2002.
Federal Agriculture Improvement and Reform Act of 1996, Pub. L. No. 104-127, 110 Stat. 888. 1996.
Federal Duck Stamp Act of 2014, Pub. L. 113-264, 128 Stat. 2939. 2014.
Fish and Wildlife Coordination Act (as Amended), Pub. L. No. 73-121, 48 Stat. 401. 1934.
Food Security Act of 1985, as Amended through P.L. 115–334, Enacted December 20, 2018.
Food Security Act of 1985, Pub. L. No. 99-198, 99 Stat. 1354. 1985.

Food, Agriculture, Conservation, and Trade Act of 1990, Pub. L. No. 101-624, 104 Stat. 3359. 1990.

Food, Conservation, and Energy Act of 2008, Pub. L. No. 110-234, 122 Stat. 923. 2008.

Healthy Forests Restoration Act of 2003, Pub. L. No. 108-148, 117 Stat. 1887. 2003.

John D. Dingell, Jr. Conservation, Management, and Recreation Act, Public L. No. 116-9. 2019.

National Defense Authorization Act for Fiscal Year 2004, § 318, Pub. L. No. 108-136, 117 Stat. 1392, 1433. 2003.

National Environmental Policy Act of 1969, Pub. L. No. 91-190, 83 Stat. 852. 1970.

Partners for Fish and Wildlife Act, Public Law No. 109-294, 120 Stat. 1351 (2006).

To Review USDA Farm Bill Conservation Programs: Hearing before the Subcommittee on Conservation and Forestry of the Committee on Agriculture, U.S. House of Representatives, 116th Congress. 2019.

U.S. Fish and Wildlife Service. Endangered Species Act of 1973, as Amended through the 108th Congress. 2005.

Administrative and Regulatory Materials

7 CFR § 1466.7(C).

12-Month Finding on a Petition to List Greater Sage-Grouse (Centrocercus Urophasianus) as an Endangered or Threatened Species, 80 Fed. Reg. 59858. Ootober 2, 2015.

50 CFR Part 80.

Final Redefinition of "Harm," 46 Fed. Reg. 54748. November 4, 1981.

Announcement of Draft Policy for Candidate Conservation Agreements, 62 Fed. Reg. 32183. June 12, 1997.

Announcement of Draft Policy for Evaluation of Conservation Efforts When Making Listing Decisions, 65 Fed. Reg. 37102. June 13, 2000.

Announcement of Final Policy for Candidate Conservation Agreements with Assurances, 64 Fed. Reg. 32726. June 17, 1999.

Announcement of Final Safe Harbor Policy, 64 Fed. Reg. 32717. June 17, 1999.

Bureau of Land Management. Instruction Memorandum 2018-093: Compensatory Mitigation. July 24, 2018.

Candidate Conservation Agreement for the Sicklefin Redhorse (*Moxostoma* Sp.) between the United States Fish and Wildlife Service, North Carolina Wildlife Resources Commission, Duke Energy Carolinas, LLC, Tennessee Valley Authority, Eastern Band of Cherokee Indians, and Georgia Department of Natural Resources. December, 2015.

Candidate Conservation Agreements with Assurances Policy, 81 Fed. Reg. 95164. December 27, 2016.

Compensatory Mitigation for Losses of Aquatic Resources, 73 Fed. Reg. 19594. April 10, 2008.

Endangered and Threatened Species Listing and Recovery Priority Guidelines, 48 Fed. Reg. 43098. September 21, 1983.

Endangered Species Act Compensatory Mitigation Policy, 81 Fed. Reg. 95316. December 27, 2016.

Endangered Species Act Compensatory Mitigation Policy, 83 Fed. Reg. 36469. July 30, 2018.

Endangered Species Permit: Notice of Receipt of Application, 39 Fed. Reg. 18483. May 28, 1974.

Endangered Species Permit: Receipt of Addendum to Application, 41 Fed. Reg. 8515. February 27, 1976.

Endangered Species Permits: Official Action, 41 Fed. Reg. 9576. March 5, 1976.

Executive Order No. 13783: Promoting Energy Independence and Economic Growth. March 28, 2017.

Federal Guidance for the Establishment, Use and Operation of Mitigation Banks, 60 Fed. Reg. 58605. November 28, 1995.

Federal Guidance on the Use of in Lieu-Fee Arrangements for Compensatory Mitigation under Section 404 of the Clean Water Act and Section 10 of the Rivers and Harbors Act, 65 Fed. Reg. 66914. November 7, 2000.

Finding on a Petition to List the Fluvial Population of the Arctic Grayling as Endangered, 59 Fed. Reg. 37738. July 25, 1994.

Forest Preserve District of DuPage County and U.S. Fish and Wildlife Service. Final Draft Safe Harbor Agreement for Voluntary Protection, Restoration, and Enhancement of Hine's Emerald Dragonfly Habitat within Waterfall Glen Forest Preserve, Dupage County, Illinois. 2008.

Habitat Conservation Plan Assurances ("No Surprises") Rule, 63 Federal Register 8859. February 23, 1998.

Implementation of Procedural Provisions, 43 Fed. Reg. 55978. November 29, 1978.

King, R. Andrew. Revised Programmatic Biological Opinion on the Proposed Construction, Operation, and Maintenance of Alternative 3C of Interstate 69 (I-69) from Evansville to Indianapolis for the Federally Endangered Indiana Bat (*Myotis sodalis*) and the Federally Threatened Bald Eagle (*Haliaeetus leucocephalus*) Traversing Portions of Gibson, Warrick, Pike, Daviess, Greene, Monroe, Morgan, Johnson, and Marion Counties, Indiana Submitted to the Federal Highway Administration. August 24, 2006.

Lesser Prairie-Chicken Removed from the List of Endangered and Threatened Wildlife, 81 Fed. Reg. 47047. July 20, 2016.

Oregon Department of Fish and Wildlife. Candidate Conservation Agreement with Assurances for Columbian Sharp-Tailed Grouse between the Oregon Department of Fish and Wildlife and the U.S. Fish and Wildlife Service. 2000.

Policy for Evaluation of Conservation Efforts When Making Listing Decisions, 68 Fed. Reg. 15100. March 28, 2003.

Policy Regarding Implementation of Section 4(B)(2) of the Endangered Species Act, 81 Fed. Reg. 7226. February 11, 2016.

Presidential Memorandum: Mitigating Impacts on Natural Resources from Development and Encouraging Related Private Investment. November 3, 2015.

Range-Wide Oil and Gas Candidate Conservation Agreement with Assurances for the Lesser Prairie-Chicken (*Tympanuchus pallidicinctus*) in Colorado, Kansas, New Mexico, Oklahoma and Texas between the U.S. Fish and Wildlife Service and the Western Association of Fish and Wildlife Agencies/Foundation for Western Fish and Wildlife February 28, 2014.

Reclassification of the American Alligator and Other Amendments, 40 Fed. Reg. 44412. September 26, 1975.

Recovery Crediting Guidance, 73 Fed. Reg. 44761. July 31, 2008.

Regulations for Listing Species and Designating Critical Habitat, 84 Fed. Reg. 45020. August 27, 2019.

Regulations for Prohibitions to Threatened Wildlife and Plants, 84 Fed. Reg. 44753. August 27, 2019.

Revised 12-Month Finding on a Petition to List the Upper Missouri River Distinct Population Segment of Arctic Grayling as an Endangered or Threatened Species, 79 Fed. Reg. 49384. August 20, 2014.

Safe Harbor Agreements and Candidate Conservation Agreements with Assurances (Correction), 64 Fed. Reg. 52676. September 30, 1999.

Safe Harbor Agreements and Candidate Conservation Agreements with Assurances, 64 Fed. Reg. 32706. June 17, 1999.

Safe Harbor Agreements and Candidate Conservation Agreements with Assurances; Revisions to the Regulations, 69 Fed. Reg. 24084. May 3, 2004.

Safe Harbor Agreements and Candidate Conservation Agreements, 62 Fed. Reg. 32189. June 12, 1997.

Special Rule Concerning Take of the Threatened Coastal California Gnatcatcher, 58 Fed. Reg. 65088. December 10, 1993.

Special Rule for the Lesser Prairie-Chicken; Final Rule, 79 Fed. Reg. 20074. 2014.

U.S. Department of the Interior. Departmental Manual Release, Landscape-Scale Mitigation Policy (600 DM 6). October 23, 2015.

———. Secretarial Order No. 3330: Improving Mitigation Policies and Practices of the Department of the Interior. October 31, 2013.

———. Secretarial Order No. 3349: American Energy Independence. March 29, 2017.

———. Secretarial Order No. 3360: Rescinding Authorities Inconsistent with Secretary's Order 3349, "American Energy Independence" December 22, 2017.

U.S. Department of the Interior Office of Inspector General. U.S. Fish and Wildlife Service Conservation Easement Enforcement and Monitoring 2012.

U.S. Department of the Interior Office of the Solicitor. Solicitor's Memorandum M-37037: The Bureau of Land Management's Authority to Address Impacts of Its Land Use Authorizations through Mitigation. December 21, 2016.

U.S. Fish and Wildlife Service. A Habitat Conservation Plan to Encourage the Voluntary Restoration and Enhancement of Habitat for the Red-Cockaded Woodpecker on Private and Certain Other Land in the Sandhills Region of North Carolina by Providing "Safe Harbor" to Participating Landowners. 1995.

———. Director's Order No. 217: Collaboration with the U.S. Department of Agriculture's Natural Resources Conservation Service (NRCS) to Conserve Listed, Candidate, and Other at-Risk Species. 2016.

———. Director's Order No. 218: Policy Regarding Voluntary Prelisting Conservation Actions. January 18, 2017.

———. Draft Candidate Conservation Agreements with Assurances Handbook. June, 2003.

———. Federal Fish and Wildlife Permit, San Bruno Mountain HCP Permittees. March 29, 2013.

———. Final Biological Opinion on the Effects to Grizzly Bears from the Implementation of Proposed Actions Associated with Plan of Operations for the Montanore Minerals Corporation Copper/Silver Mine as Proposed by the U.S. Forest Service, Kootenai National Forest. March 31, 2014.

———. Florida Panther (Puma Concolor Coryi) 5-Year Review. 2009.

———. Florida Panther Recovery Plan: 3rd Revision. 2008.

———. Guidance for the Establishment, Use, and Operation of Conservation Banks. May 2, 2003.

———. Panther Habitat Assessment Methodology. September 24, 2012.

———. Policy Regarding Voluntary Prelisting Conservation Actions, Appendix 1, 735 FW 1.

———. Recovery Plan for the Red-Cockaded Woodpecker (Picoides Borealis) Second Revision. 2003.

U.S. Fish and Wildlife Service Black-footed Ferret Recovery Program. Black-Footed Ferret Programmatic Safe Harbor Agreement. October 23, 2013.

U. S. Fish and Wildlife Service and National Marine Fisheries Service. Endangered Species Consultation Handbook. March, 1998.

U.S. Fish and Wildlife Service and National Marine Fisheries Service. Habitat Conservation Planning and Incidental Take Permit Processing Handbook. 2016.
U.S. Fish and Wildlife Service Mitigation Policy, 81 Fed. Reg. 83440. November 21, 2016.
U.S. Fish and Wildlife Service Mitigation Policy, 83 Fed. Reg. 36472. July 30, 2018.
U.S. Forest Service. Fiscal Year 2020 Landscape Scale Restoration Competitive Process National Guidance. May 29, 2019.
———. "Greater Sage-Grouse Proposed Land Management Plan Amendments (LMPA) and Draft Environmental Impact Statement (DEIS) for the Intermountain and Rocky Mountain Regions." (October 2018).
———. Greater Sage-Grouse Record of Decision: Idaho and Southwest Montana, Nevada, Utah. September, 2015.

Court Cases

Babbitt v. Sweet Home Chapter of Communities for a Great Oregon, 515 U.S. 687 (1995).
Center for Biological Diversity v. Zinke, 900 F.3d 1053 (9th Cir. 2018).
Friends of Endangered Species, Inc. v. Jantzen, 596 F.Supp 518 (N.D. Cal. 1984).
Robertson v. Methow Valley Citizens Council, 490 U.S. 332 (1989).
Weyerhaeuser Co. v. U.S. Fish & Wildlife Serv., 139 S. Ct. 361 (2018).

Interviews and Personal Communications

Abernathy, Christopher. Email to Christopher E. Segal. July 26, 2019.
Adams, Kip. Interview by Christopher E. Segal. July 10, 2019.
Annand, Fred. Interview by Christopher E. Segal. February 22, 2019.
Barry, Donald J. Interview by Christopher E. Segal and Lowell E. Baier. February 27, 2019.
Beard, Bruce. Interview by Christopher E. Segal. July 10, 2019.
Bishop, Andy. Interview by Christopher E. Segal. July 23, 2019.
Blumenthal, Rob. Email to Christopher E. Segal. August 20, 2019.
Bullock, Jimmy. Interview by Christopher E. Segal. March 7, 2019.
Carter, Jay H. Interview by Christopher E. Segal. September 30, 2019.
Cummins, James L. Interview by Christopher E. Segal. March 6, 2019.
Devney, John. Interview by Christopher E. Segal. July 2, 2019.
Dreibelbis, Justin. Email to Christopher E. Segal. July 26, 2019.
Faeth, Lori. Interview by Christopher E. Segal. January 28, 2019.
Gagner, David. Interview by Christopher E. Segal. May 28, 2019.
Gober, Donald "Pete." Interview by Lowell E. Baier and Christopher E. Segal. May 25, 2018.
Graff, Shawn. Interview by Christopher E. Segal. October 3, 2019.
Hall, Galon, and Tim Griffiths. Interview by Christopher E. Segal. May 29, 2019.
Henning, Blake. Interview by Christopher E. Segal. July 17, 2019.
Hurley, Kevin. Interview by Lowell E. Baier and Christopher E. Segal. June 25, 2019.
Inglis, Jim. Interview by Christopher E. Segal. June 7 and October 2, 2019.
Jackson, Wendy. Interview by Christopher E. Segal. January 28, 2019.
Jester, Steve. Email to Christopher E. Segal. July 17 and August 21, 2019.
———. Interview by Lowell E. Baier and Christopher E. Segal. July 19, 2019.
Johnson, Marshall. Interview by Christopher E. Segal. October 1, 2019.
Keith, Brent. Interview by Christopher E. Segal. June 28, 2019.
Longoria, Meredith. Email to Christopher E. Segal. July 29, 2019.

Lyons, James R. Interview by Lowell E. Baier and Christopher E. Segal. October 9, 2018.
Marsh, Lindell L. Interview by Christopher E. Segal. November 30, 2018.
Masonis, Robert. Interview by Christopher E. Segal. September 13, 2019.
Moretti, Miles. Interview by Lowell E. Baier and Christopher E. Segal. July 29, 2019.
Organ, John. Interview by Christopher E. Segal. April 5, 2019.
Pedersen, Joel, and Mark Hatfield. Interview by Christopher E. Segal. June 18, 2019.
Prentice, Karen. Interview by Christopher E. Segal. September 30, 2019.
Reed, Kelly. Interview by Christopher E. Segal. July 12, 2019.
Rey, Mark. Interview by Christopher E. Segal. January 30, 2019.
Rudolph, Brent A. Interview by Christopher E. Segal. July 25, 2019.
Schoonhoven, Laurie. Email to Christopher E. Segal. July 2, 2019.
Sibbing, Julie. Interview by Christopher E. Segal. July 8, 2019.
Smith, Dave, and Ali Duvall. Interview by Christopher E. Segal. August 12, 2019.
Walker, Johann. Interview by Christopher E. Segal. August 2, 2019.
Walker, Wayne. Interview by Christopher E. Segal. October 5, 2018.
Wheeler, Douglas P. Interview by Lowell E. Baier and Christopher E. Segal. March 24, 2017.

Press Releases

American Bird Conservancy. "$10 Mil. Forest Restoration Project Will Benefit Imperiled Golden-Winged Warbler." Press release, January 14, 2015, https://abcbirds.org/article/10-mil-forest-restoration-project-will-benefit-imperiled-golden-winged-warbler/.

Delta Waterfowl. "Delta Waterfowl Applauds Inclusion of 'Working Wetlands' in 2018 Farm Bill." Press release, https://deltawaterfowl.org/delta-waterfowl-applauds-inclusion-of-working-wetlands-in-2018-farm-bill/.

Ducks Unlimited. "Ducks Unlimited Recognizes Top Fundraising Chapters across the Nation." Press release, March 22, 2016, https://www.ducks.org/press-room/news-releases/ducks-unlimited-recognizes-top-fundraising-chapters-across-the-nation.

National Fish and Wildlife Foundation. "National Fish and Wildlife Foundation Partners with Permian Basin Oil and Natural Gas Companies to Conserve Pecos Watershed in Texas and New Mexico." Press release, November 8, 2017, https://www.nfwf.org/swrivers/pecos/Documents/nfwf-partners-with-permian-basin-oil-natural-gas-companies-to-conserve-pecos-watershed-in-texas-new-mexico-2017-1109.pdf.

———. "NFWF's Partnership with Permian Basin Oil and Natural Gas Companies and Federal Agencies Leads to $3.7 Million in Conservation Impact in First Year." Press release, March 22, 2018, https://www.nfwf.org/swrivers/pecos/Documents/partnership-with-permian-basin-oil-natural-gas-companies-federal-agencies-leads-to-3-7-million-in-conservation-2018-0322.pdf.

The Conservation Fund. "The Conservation Fund and Resource Management Service, LLC Announce Partnership and Initial Funding for Largest, Single Longleaf Pine Landscape Conservation and Restoration Effort on Private Lands." Press release, January 14, 2015, https://www.conservationfund.org/news/press-releases/585-the-conservation-fund-and-resource-management-service-llc-announce-partnership-and-initial-funding-for-largest-single-longleaf-pine-landscape-conservation-and-restoration-effort-on-private-lands.

———. "Conservation Working for Ranchers and Sage Grouse in the West." Press release, February 17, 2016. https://www.conservationfund.org/news/press-releases/1220-conservation-working-for-ranchers-and-sage-grouse-in-the-west.

The Nature Conservancy. "Another Vision for Conservation on the Prairie: The Nature Conservancy Buys Land for Partnership with Neighboring Ranchers." Press release, February 5, 2019, https://www.nature.org/en-us/explore/newsroom/montana-rancher-land-partnership/.

U.S. Fish and Wildlife Service. "Interior Secretary Awards States More Than $20 Million in Grants Supporting Coastal Wetlands Conservation, Recreation and Communities." Press release, May 8, 2019, https://www.fws.gov/news/ShowNews.cfm?ref=interior-secretary-awards-states-more-than-$20-million-in-grants-&_ID=36408.

U.S. Geological Survey. "New Database Available: USGS Releases 'Species of Greatest Conservation Need' Lists." Press release, October 25, 2017, https://www.usgs.gov/news/new-database-available-usgs-releases-species-greatest-conservation-need-lists.

Wild Sheep Foundation. "Wild Sheep Foundation Honors Missoula National Wildlife Federation Leaders and BC Hunter-Conservationist." Press release, January 25, 2017, https://www.wildsheepfoundation.org/blog/wild-sheep-foundation-honors-missoula-national-wildlife-federation-leaders-and-bc-hunter-conservationist.

Online Resources

America's Longleaf Restoration Initiative. "America's Longleaf Restoration Initiative." Last accessed October 3, 2019. http://www.americaslongleaf.org/.

America's Longleaf Restoration Initiative. "American's Longleaf Restoration Initiative." November 2015. https://www.nwf.org/~/media/PDFs/Misc/ALRI-General-Messaging_FINAL.ashx.

American Bird Conservancy. "Birdscapes." Last accessed October 3, 2019. https://abcbirds.org/birds/birdscapes/.

———. "Golden-Winged Warbler." Last accessed October 3, 2019. https://abcbirds.org/bird/golden-winged-warbler/.

———. "Mission & Strategy." Last accessed October 3, 2019. https://abcbirds.org/about/mission-and-strategy/.

American Farmland Trust. "Farm Policy Update: The 2018 Farm Bill." December 21, 2018. https://farmland.org/farm-policy-update-the-2018-farm-bill/.

Arkansas Game and Fish Commission. "Acres for Wildlife Program." Last accessed July 24, 2019. https://www.agfc.com/en/wildlife-management/private-lands-program/acres-for-wildlife/.

———. "Arkansas Waterfowl Rice Incentive Conservation Enhancement." 2019. https://drive.google.com/file/d/1n9Bimc7FaQ_7Y26FCcP_IupemrHe6iQA/view.

———. "Find a Private Lands Biologist." Last accessed July 27, 2019. https://www.agfc.com/en/wildlife-management/private-lands-program/private-lands-biologist/.

———. "Private Lands Program." Last accessed July 24, 2019. https://www.agfc.com/en/wildlife-management/private-lands-program/.

Association of Fish and Wildlife Agencies. "State Wildlife Action Plans." Last accessed December 13, 2018. https://www.fishwildlife.org/afwa-informs/state-wildlife-action-plans

———. "Multistate Conservation Grant Program Overview." Last accessed July 10, 2019. https://www.fishwildlife.org/afwa-informs/multi-state-conservation-grants-program.

Audubon Dakota. "Audubon Dakota." Last accessed October 3, 2019. https://dakota.audubon.org/.

Bowman, Andrew. "Pass the Farm Bill." *Land Trust Alliance Blog*, December 11, 2018. https://www.landtrustalliance.org/blog/pass-farm-bill

Brandywine Conservancy. "Do You Need a New Conservation Plan?" Last accessed June 18, 2019. https://www.brandywine.org/conservancy/landowner-resources/do-you-need-new-conservation-plan.

Bureau of Land Management. "Coal Data, National Coal Statistics Table." 2018. https://www.blm.gov/programs/energy-and-minerals/coal/coal-data.

———. "National Office." Last accessed September 19, 2019. https://www.blm.gov/office/national-office.

———. "Oil and Gas Statistics, Table 2: Acreage in Effect." 2018. https://www.blm.gov/sites/blm.gov/files/Table2_Number_of_Acres_Leased.xlsx.

———. "Threatened and Endangered Species." Last accessed September 19, 2019. https://www.blm.gov/programs/fish-and-wildlife/threatened-and-endangered.

California Department of Fish and Wildlife. "Natural Community Conservation Planning (NCCP)." Last accessed October 5, 2019. https://www.wildlife.ca.gov/Conservation/Planning/NCCP.

Calvert Soil Conservation District. "Farm Planning." Last accessed June 19, 2019. http://www.calvertsoil.org/c/262/farm-planning.

Center for Biological Diversity. "Red-Cockaded Woodpecker." Last accessed April 25, 2019. https://www.biologicaldiversity.org/campaigns/esa_works/profile_pages/RedcockadedWoodpecker.html.

Chapman, Dan. "Coastal Headwaters Project in Florida Is a Major Step for Longleaf Pine Restoration." April 24, 2019. https://www.fws.gov/southeast/articles/coastal-headwaters-project-in-florida-is-a-major-step-for-longleaf-pine-restoration/.

———. "A Harmonious Future for Profits, Pine and at-Risk Species Along the Florida-Alabama Line." May 9, 2017. https://www.fws.gov/southeast/articles/a-harmonious-future-for-profits-pine-and-at-risk-species-along-the-florida-alabama-line/.

City of Missoula. "2019 Open Space Bond." Last accessed July 26, 2019. https://www.ci.missoula.mt.us/2429/2018-Open-Space-Bond.

Colorado Parks and Wildlife. "Colorado Wildlife Habitat Program." Last accessed July 24, 2019. https://cpw.state.co.us/aboutus/Pages/LandWaterCWHP.aspx.

———. "Colorado Wildlife Habitat Program 2019 Request for Proposals (RFP)." 2019. https://cpw.state.co.us/Documents/LandWater/COWildlifeHabitatProtectionProgram/2019-CWHP-RFP_ProposalInstructions.pdf.

———. "Habitat Partnership Program (HPP)." Last accessed July 24, 2019. https://cpw.state.co.us/aboutus/Pages/HabitatPartnershipProgram.aspx.

———. "Pheasant Habitat Improvement Program." Last accessed July 24, 2019. https://cpw.state.co.us/aboutus/Pages/PheasantHabitatImprovementProgram.aspx.

———. "Private Land Program." Last accessed July 24, 2019. https://cpw.state.co.us/aboutus/Pages/PrivateLand.aspx.

Colorado State Forest Service. "Colorado's Forest Stewardship Program." 2016. https://csfs.colostate.edu/media/sites/22/2016/05/stewardship-brochure-web.pdf.

———. "Forest Ag Program." Last accessed July 26, 2019. https://csfs.colostate.edu/forest-ag-program/.

———. "Forest Stewardship Program." Last accessed July 26, 2019. https://csfs.colostate.edu/forest-stewardship-program/.

———. "Grants & Funding Assistance." Last accessed July 27, 2019. https://csfs.colostate.edu/funding-assistance.

Conservation Finance Center. "Conservation Finance Center." Last accessed June 19, 2019. http://conservationfinancecenter.org.

Cooley, Jordan. "Conservation Expense Federal Tax Deduction." April 10, 2015. https://valuewetlands.tamu.edu/2015/04/10/conservation-expense-federal-tax-deduction/.

Cornell Lab of Ornithology. "All About Birds: Red-Cockaded Woodpecker Identification." Last accessed April 25, 2019. https://www.allaboutbirds.org/guide/Red-cockaded_Woodpecker/id#.

———. "Golden-Winged Warbler: Conservation Strategy and Resources." Last accessed October 3, 2019. https://www.birds.cornell.edu/home/golden-winged-warbler-conservation-strategy-and-resources/.

Council on Environmental Quality. "A Citizen's Guide to the National Environmental Policy Act." Last accessed November 5, 2018. https://www.fs.usda.gov/Internet/FSE_DOCUMENTS/stelprdb5310825.pdf.

Ducks Unlimited. "Conserving Wetlands & Waterfowl." Last accessed August 6, 2019. https://www.ducks.org/conservation/how-du-conserves-wetlands-and-waterfowl.

———. "DU National & Regional Offices." Last accessed August 6, 2019. https://www.ducks.org/about-ducks-unlimited/regional-offices.

———. "Fact Sheet." January 1, 2019. https://www.ducks.org/media/_global/_documents/stateFactSheets/NationalFactSheet.pdf.

———. "Where Ducks Unlimited Works." Last accessed August 6, 2019. https://www.ducks.org/conservation/where-ducks-unlimited-works.

Ducks Unlimited Canada. "Ducks Unlimited Canada." Last accessed September 18, 2019. https://www.ducks.ca/.

Ducks Unlimited De México. "Ducks Unlimited De México." Last accessed September 18, 2019. http://www.dumac.org/dumac/habitat/ing/index.htm.

Earth Tech Environmental LLC. "Red-Cockaded Woodpecker Cavity Installation." Last accessed September 16, 2019. http://eteflorida.com/red-cockaded-woodpecker-cavity-installation/.

Environmental Defense Fund. "Habitat Exchanges Transform Conservation." Last accessed November 20, 2018. https://www.edf.org/ecosystems/habitat-exchanges-transform-conservation.

———. "Habitat Exchanges: How Do They Work?" Last accessed July 12, 2018. https://www.edf.org/ecosystems/habitat-exchanges-how-do-they-work.

———. "How the Farm Bill Changes the Future of Climate and Water Conservation." Deember 12, 2018. http://blogs.edf.org/growingreturns/2018/12/12/farm-bill-climate-water-conservation/.

Environmental Protection Agency. "EPA Grants." Last accessed September 17, 2019. https://www.epa.gov/grants.

Espenshade, Jessica. "Become a Cover Crop Champion!" *National Wildlife Federation Blog*, September 21, 2017. https://blog.nwf.org/2017/09/become-a-cover-crop-champion/.

Farm Service Agency. "Commodity Credit Corporation." Last accessed February 20, 2019. https://www.fsa.usda.gov/about-fsa/structure-and-organization/commodity-credit-corporation/index.

———. "Conservation Reserve Enhancement Program." Last accessed September 6, 2018. https://www.fsa.usda.gov/programs-and-services/conservation-programs/conservation-reserve-enhancement/index.

———. "Conservation Reserve Program." Last accessed January 7, 2019. https://www.fsa.usda.gov/programs-and-services/conservation-programs/conservation-reserve-program/index.

———. "Conservation Reserve Program State Acres for Wildlife Enhancement (SAFE)." January 2008. https://www.fsa.usda.gov/Internet/FSA_File/safe08.pdf.

———. "CRP Enrollment and Rental Payment History by State." 2018. https://www.fsa.usda.gov/Assets/USDA-FSA-Public/usdafiles/Conservation/Excel/CRP%20Enrollment%20and%20Rental%20Payment%20History%20by%20County%20198602017.xlsx?CRP+Enrollment+and+Rental+Payments+History+by+County+2017.

———. "CRP Practices Library." Last accessed October 3, 2019. https://www.fsa.usda.gov/programs-and-services/conservation-programs/crp-practices-library/index.

———. "Fact Sheet: Emergency Forestry Conservation Reserve Program 2005 Hurricane Assistance." July 2006. https://www.fsa.usda.gov/Internet/FSA_File/efcrp06.pdf.

———. "Farmable Wetlands Program." Last accessed January 30, 2019. https://www.fsa.usda.gov/programs-and-services/conservation-programs/farmable-wetlands/index.

Farmers' Legal Action Group Inc. "Is Your Farm 'EQIPed' for Conservation?" 2007. http://www.flaginc.org/wp-content/uploads/2013/03/FGtoEQIP.pdf.

Farmland Information Center. "Farmland Directory." Last accessed August 13, 2019. https://www.farmlandinfo.org/directory.

———. "Farmland Information Center." Last accessed August 13, 2019. https://www.farmlandinfo.org.

Federation, The Natonal Wildlife. "Saving the Greater Sage-Grouse." Last accessed May 7, 2018, https://www.nwf.org/Our-Work/Wildlife-Conservation/Greater-Sage-Grouse.

Florida Fish and Wildlife Conservation Commission. "Florida Panther." Last accessed November 16, 2018, http://myfwc.com/wildlifehabitats/managed/panther/.

Georgia Forestry Commission. "Contact Us." Last accessed July 27, 2019. http://www.gfc.state.ga.us/about-us/contact-us/county-units/index.cfm.

———. "Cost Share & Incentive Programs." Last accessed July 26, 2019. http://www.gfc.state.ga.us/forest-management/private-forest-management/landowner-programs/other-landowner-programs/index.cfm.

———. "Forest Health in Georgia." Last accessed July 26, 2019. http://www.gfc.state.ga.us/forest-management/forest-health/index.cfm.

———. "Forest Legacy & Conservation Easements." Last accessed July 26, 2019. http://www.gfc.state.ga.us/forest-management/private-forest-management/landowner-programs/forest-legacy/index.cfm.

———. "Landowner Services." Last accessed July 26, 2019. http://www.gfc.state.ga.us/forest-management/private-forest-management/landowner-programs/.

"Grants.Gov." Last accessed August 11, 2019. https://www.grants.gov/.

Hall, Pete. "Wetland Reserve Easements (WRE) and Benefits to Landowners." Last accessed January 22, 2019. https://www.johnhallco.com/wetland-reserve-easements-and-the-practical-benefits-to-landowners/.

Intermountain West Joint Venture. "Intermountain West Joint Venture Factsheet." 2017. https://iwjv.org/sites/default/files/iwjv_fact_sheet.pdf.

———. "Mission & Scope." Last accessed March 27, 2019. https://iwjv.org/mission-scope.

———. "Partnering to Conserve Sagebrush Rangelands." Last accessed September 20, 2019. https://iwjv.org/sites/default/files/sagebrush_parntership_four_pages.pdf.

———. "Water 4 Initiative." Last accessed October 3, 2019. https://iwjv.org/water-4-initiative.

———. "Western Association of Fish & Wildlife Agencies and the IWJV." Last accessed March 28, 2019. https://iwjv.org/node/1028.

———. "Where We Work." Last accessed March 27, 2019. https://iwjv.org/where-we-work.

Jackass Honey Farms. "How to Get Qualified for an Ag Valuation in Texas." 2018. https://www.jackasshoneyfarms.com/bee-tax-exemption.

Kansas Department of Wildlife Parks and Tourism. "Fishing Impoundments and Stream Habitats (F.I.S.H.) Program." Last accessed July 25, 2019. https://ksoutdoors.com/Fishing/Special-Fishing-Programs-for-You/F.I.S.H.-Program.

———. "Habitat First Habitat Practices and Payments." January 2016. https://ksoutdoors.com/content/download/46850/478951/version/1/file/Habitat+Practices+and+Payments.pdf.

———. "Pond Management Program." Last accessed July 25, 2019. https://ksoutdoors.com/Fishing/Special-Fishing-Programs-for-You/Pond-Management-Program.

———. "Public Hunting Access & Your Land." 2018. https://ksoutdoors.com/content/download/39527/378058/version/2/file/WIHA+Flyer+-+2018+Update.pdf.

———. "Wildlife Biologists." Last accessed July 25, 2019. https://ksoutdoors.com/Services/Private-Landowner-Assistance/Wildlife-Biologists.

———. "Wildlife Biologists & Bio-Technicians." April 2019. https://ksoutdoors.com/content/download/26765/182561/version/11/file/Biologists+Counties_April_2019.pdf.

Land Conservation Assistance Network. "The Conservation Tax Center." Last accessed July 22, 2019. https://www.landcan.org/conservation-tax-center/.

———. "Federal & State Conservation Assistance Programs." Last accessed August 11, 2019. https://www.landcan.org/Grant-and-Assistance-Programs.

———. "Land Conservation Assistance." Last accessed August 11, 2019. https://www.landcan.org/.

Land Trust Alliance. "Estate Tax Incentives for Land Conservation: Keeping Land in the Family." Last accessed June 28, 2018. https://www.landtrustalliance.org/topics/taxes/estate-tax-incentives-land-conservation.

———. "Find a Land Trust." Last accessed July 28, 2019. https://www.findalandtrust.org/.

———. "Find a Land Trust: Kittery Land Trust." Last accessed March 21, 2019. https://www.findalandtrust.org/land_trusts/1138.

———. "Find a Land Trust: North Carolina Coastal Land Trust–an Accredited Land Trust." Last accessed March 22, 2019. https://www.findalandtrust.org/land_trusts/10371.

———. "Find a Land Trust: The Nature Conservancy–an Accredited Land Trust." Last accessed March 21, 2019. https://www.findalandtrust.org/land_trusts/263/national.

———. "Find a Land Trust: The Trustees of Reservations–an Accredited Land Trust." Last accessed March 21, 2019. https://www.findalandtrust.org/land_trusts/767.

———. "Income Tax Incentives for Land Conservation." Last accessed May 9, 2019. https://www.landtrustalliance.org/topics/taxes/income-tax-incentives-land-conservation.

———. "Tax Incentives." Last accessed June 29, 2018. http://www.landtrustalliance.org/issues-action/take-action/tax-incentives.

———. "What You Can Do." Last accessed August 11, 2019. https://www.landtrustalliance.org/what-you-can-do/conserve-your-land/getting-started.

LandCAN Land Conservation Assistance Network. "Get to Know the Land Conservation Assistance Network (Landcan)." Septemper 22, 2017. https://www.youtube.com/watch?time_continue=237&v=OapyBl6tW4U.

Lesser Prairie Chicken Initiative. "Partners." Last accessed January 14, 2019. http://lpcinitiative.org/about/partners/.

Maine Land Trust Network. "Foothills Land Conservancy." Last accessed March 21, 2019. http://www.mltn.org/trust_detail.php?t=1246.

McEowen, Roger A. "Deductibility of Soil and Water Conservation Expenses." March 9, 2017. https://lawprofessors.typepad.com/agriculturallaw/2017/03/deductibility-of-soil-and-water-conservation-expenses.html.

Miao, Ruiqing, David Hennessy, and Hongli Feng. "How Much Can the Sodsaver Provision in the 2014 Farm Bill Save Sod? It Depends on Crop Prices." *PolicyMatters*, January 19, 2017.

Migratory Bird Joint Ventures. "Joint Venture Directory." Last accessed March 25, 2019. https://mbjv.org/who-we-are/directory/.

———. "Joint Venture Timeline." Last accessed March 26, 2019. http://mbjv.org/who-we-are/jv-timeline/.

———. "Migratory Bird Joint Ventures Map." Last accessed March 26, 2019. https://mbjv.org/joint-venture-map/.

Miller, Laura. "Beekeeping Qualifies as an Agricultural Use Enterprise." April 13, 2012. Texas A&M AgriLife Extension. https://agrilife.org/tcch/2012/04/13/beekeeping-qualifies-as-an-agricultural-use-enterprise/

Mississippi Forestry Commision. "Forest Resource Development Program." Last accessed July 26, 2019. https://www.mfc.ms.gov/FRDP.

———. "Forest Stewardship Program." Last accessed July 26, 2019. https://www.mfc.ms.gov/Forest-Stewardship.

———. "Private Landowner Services." Last accessed July 26, 2019. https://www.mfc.ms.gov/private-landowner-services.

———. "Reforestation Tax Credit." Last accessed July 26, 2019. https://www.mfc.ms.gov/reforestation-tax-credit.

———. "Southern Pine Beetle Prevention Program." Last accessed July 26, 2019. https://www.mfc.ms.gov/SPB-Prevention.

Missoula County, Montana. "2006 Open Space Bond Program for Missoula County." Last accessed July 26, 2019. https://www.missoulacounty.us/government/community-development/community-planning-services/open-lands/open-space-bond-program#ad-image-0.

Mule Deer Foundation. "Conservation Projects." Last accessed July 30, 2019. https://muledeer.org/projects/.

———. "MDF in Your State." Last accessed October 5, 2019. https://muledeer.org/state/.

Munger, Sarah. "Coupling Up: Are Crop Insurance and Conservation Compliance Mated for Life?" April 5, 2018. http://www.farmbilllaw.org/2018/04/05/coupling-crop-insurance-conservation-compliance-mated-life/.

National Association of Conservation Districts. "2018 Farm Bill Breakdown: Conservation Stewardship Program." Janaury 15, 2019. https://www.nacdnet.org/2019/01/15/2018-farm-bill-breakdown-conservation-stewardship-program/.

———. "About NACD." Last accessed May 23, 2019. https://www.nacdnet.org/about-nacd/.

———. "Conservation District Directory." Last accessed August 11, 2019. https://www.nacdnet.org/general-resources/conservation-district-directory/.

National Audubon Society. "About Us." Last accessed October 2, 2019. https://www.audubon.org/about.

———. "Audubon Near You." Last accessed October 2, 2019. https://www.audubon.org/about/audubon-near-you.

———. "Conservation Ranching." Last accessed October 2, 2019. https://www.audubon.org/conservation/ranching.

———. "Guide to North American Birds: Red-Cockaded Woodpecker." Last accessed April 25, 2019. https://www.audubon.org/field-guide/bird/red-cockaded-woodpecker.

———. "Prairie Management Toolbox Program." Last accessed October 2, 2019. https://dakota.audubon.org/conservation/prairie-management-toolbox-program.

———. "Working Lands." Last accessed October 3, 2019. https://www.audubon.org/conservation/working-lands.

National Conservation Easement Database. "National Conservation Easement Database." Last accessed August 11, 2019. https://www.conservationeasement.us/.

———. "Resources." Last accessed August 11, 2019. https://www.conservationeasement.us/resources.

National Fish and Wildlife Foundation. "2017 Annual Report." Last Accessed October 5, 2019. https://www.nfwf.org/whoweare/mediacenter/Documents/2017_annual_report.pdf.

———. "2018 Annual Report." Last accessed October 5, 2019. https://www.nfwf.org/whoweare/mediacenter/Documents/2018-annual-report.pdf.

———. "About the National Fish and Wildlife Foundation." Last accessed February 15, 2019. https://www.nfwf.org/whoweare/Pages/home.aspx.

———. "Applicant Information." Last accessed February 15, 2019. https://www.nfwf.org/whatwedo/grants/applicants/Pages/home.aspx.

———. "Conservation Programs." Last accessed February 15, 2019. https://www.nfwf.org/whatwedo/programs/Pages/home.aspx.

———. "Gulf Environmental Benefit Fund." Last accessed July 2, 2019. https://www.nfwf.org/gulf/Pages/home.aspx.

———. "Gulf Environmental Benefit Fund Projects." Last accessed August 14, 2019. https://www.nfwf.org/gulf/Pages/gulf-projects.aspx.

———. "Hurricane Sandy Coastal Resiliency Competitive Grant Program." Last accessed February 1, 2019. https://www.nfwf.org/hurricanesandy/Pages/home.aspx.

———. "Pecos Watershed Conservation Initiative." Last accessed August 14, 2019. https://www.nfwf.org/swrivers/pecos/Pages/home.aspx.

———. "Pecos Watershed Conservation Initiative 2018 Grant Slate." Last accessed August 14, 2019. https://www.nfwf.org/swrivers/pecos/Documents/2018-grant-slate.pdf.

———. "Pecos Watershed Conservation Initiative 2019 Grant Slate." Last accessed August 14, 2019. https://www.nfwf.org/swrivers/pecos/Documents/2019grantslate.pdf.

———. "Pecos Watershed Conservation Initiative 2019 Request for Proposals." 2018. https://www.nfwf.org/swrivers/pecos/Documents/2019rfp.pdf

———. "What We Do." Last accessed February 15, 2019. https://www.nfwf.org/whatwedo/Pages/home.aspx.

National Marine Protected Areas Center. "Fact Sheet." July 2013. https://nmsmarineprotectedareas.blob.core.windows.net/marineprotectedareas-prod/media/archive/pdf/helpful-resources/mpa_center_fact_sheet071213.pdf.

National Oceanic and Atmospheric Administration Fisheries. "Biological Opinions." Last accessed August 1, 2019. https://www.fisheries.noaa.gov/national/endangered-species-conservation/biological-opinions.

———. "Critical Habitat." Last accessed August 1, 2019. https://www.fisheries.noaa.gov/national/endangered-species-conservation/critical-habitat#critical-habitat-designations,-maps,-and-gis-data.

National Oceanic and Atmospheric Administration Office for Coastal Management. "About Coastal and Estuarine Land Conservation Program." Last accessed July 18, 2019. https://coast.noaa.gov/czm/landconservation/about/.

———. "Coastal and Estuarine Land Conservation Program." Last accessed February 26, 2019. https://coast.noaa.gov/czm/landconservation/media/celpfactsheet.pdf.

———. "Coastal Resilience Grant Program Project Summaries." Last accessed February 26, 2019. https://www.coast.noaa.gov/resilience-grant/projects/.

———. "Coastal Resilience Grants for Coastal Communities." Last accessed February 26, 2019. https://www.coast.noaa.gov/data/resilience/factsheet-resilience-grants.pdf.

———. "NOAA Coastal Resilience Grants Program." Last accessed February 26, 2019. https://coast.noaa.gov/resilience-grant/.

———. "Preserving the Pristine." Last accessed July 18, 2019. http://noaa.maps.arcgis.com/apps/Cascade/index.html?appid=30535f92095d4a25b0ba26ec04f5af57.

National Sustainable Agriculture Coalition. "2018 Farm Bill by the Numbers." December 21, 2018. https://sustainableagriculture.net/blog/2018-farm-bill-by-the-numbers/.

———. "2018 Farm Bill Drilldown: Commodity Programs and Crop Insurance." December 14, 2018. https://sustainableagriculture.net/blog/2018-farm-bill-commodity-subsidies-crop-insurance/.

———. "A Closer Look at the 2018 Farm Bill: Lands Conservation Programs." January 14, 2019. http://sustainableagriculture.net/blog/a-closer-look-at-the-2018-farm-bill-working-lands-conservation-programs/.

———. "Conservation Reserve Program." October, 2016. http://sustainableagriculture.net/publications/grassrootsguide/conservation-environment/conservation-reserve-program/.

———. "Enforcement of Conservation Compliance for Highly Erodible Lands." June 5, 2018. http://sustainableagriculture.net/wp-content/uploads/2018/06/CFRA-NSAC-Conservation-compliance-special-report.pdf.

———. "Environmental Quality Incentives Program." Last accessed May 28, 2019. http://sustainableagriculture.net/publications/grassrootsguide/conservation-environment/environmental-quality-incentives-program/.

———. "Path to the 2018 Farm Bill: Conservation." March 14, 2017. http://sustainableagriculture.net/blog/path-to-2018-farm-bill-conservation/.

———. "Regional Conservation Partnership Program: 2018 Awards and Farm Bill Opportunities." February 8, 2018. http://sustainableagriculture.net/blog/rcpp-awards-farm-bill-fy2018/.

"National Timber Tax Website." Last accessed August 11, 2019. https://www.timbertax.org.

National Wild Turkey Federation. "In Your Area." Last accessed July 27, 2019. http://www.nwtf.org/about/nation.

———. "Manage Your Property." Last accessed July 27, 2019. http://www.nwtf.org/conservation.

———. "Save the Habitat. Save the Hunt." Last accessed July 27, 2019. http://www.nwtf.org/about/save-the-habitat-hunt.

National Wildlife Federation. "History & Heritage." Last accessed July 27, 2019. https://www.nwf.org/About-Us/History.

———. "Red-Cockaded Woodpecker." Last accessed April 25, 2019. https://www.nwf.org/Educational-Resources/Wildlife-Guide/Birds/Red-Cockaded-Woodpecker.

———. "Regional Centers & Affiliates." Last accessed July 27, 2019. https://www.nwf.org/About-Us/Regional-Centers-and-Affiliates.

———. "Southeast Forestry Program." Last accessed July 27, 2019. https://www.nwf.org/Our-Work/Our-Lands/Southeast-Forestry-Program.

Natural Capital Investment Fund. "Borrow from NCIFund." Last accessed July 26, 2019. https://www.ncifund.org/apply.

Natural Resources Conservation Servce. "Ecological Site Descriptions." Last accessed October 3, 2019. https://www.nrcs.usda.gov/wps/portal/nrcs/main/national/technical/ecoscience/desc/.

———. "Longleaf Pine Initiative." Last accessed October 3, 2019. https://www.nrcs.usda.gov/wps/portal/nrcs/detailfull/national/programs/initiatives/?cid=nrcsdev11_023913.

———. "Nine Major Resource Concern Categories and Their Subcategories." Last accessed October 4, 2019. https://www.nrcs.usda.gov/Internet/FSE_DOCUMENTS/nrcseprd1140206.pdf.

Natural Resources Conservation Servce Mississippi. "Healthy Forests Reserve Program." Last accessed August 13, 2019. https://www.nrcs.usda.gov/wps/portal/nrcs/main/ms/programs/easements/forests/.

Natural Resources Conservation Service. "2014–Farm Bill–VPA-HIP 2014 Funding Grantees." Last accessed February 8, 2019. https://www.nrcs.usda.gov/wps/portal/nrcs/detail/national/programs/farmbill/?cid=stelprdb1258890.

———. "2015 VPA-HIP Grantees." Last accessed February 8, 2019. https://www.nrcs.usda.gov/wps/portal/nrcs/detail/national/programs/farmbill/?cid=nrcseprd388655.

———. "Agricultural Water Enhancement Program." Last accessed February 8, 2019. https://www.nrcs.usda.gov/wps/portal/nrcs/main/national/programs/financial/awep/.

———. "Agricultural Water Enhancement Program (AWEP)." Last accessed February 8, 2019. https://www.nrcs.usda.gov/Internet/NRCS_RCA/reports/fb08_cp_awep.html.

———. "Apply for CSP." Last accessed January 16, 2019. https://www.nrcs.usda.gov/wps/portal/nrcs/detailfull/national/programs/financial/csp/?cid=nrcseprd1288620.

———. "Conservation Reserve Program." Last accessed May 22, 2019. https://www.nrcs.usda.gov/wps/portal/nrcs/detail/national/programs/?cid=stelprdb1041269.

———. "Conservation Reserve Program CP-25." Last accessed September 20, 2019. https://www.fsa.usda.gov/Assets/USDA-FSA-Public/usdafiles/FactSheets/2015/CRPProgramsandInitiatives/Practice_CP25_Rare_and_Declining_Habitat.pdf.

———. "Conservation Stewardship Program." Last accessed August 13, 2019. https://www.nrcs.usda.gov/wps/portal/nrcs/main/national/programs/financial/csp.

———. "Conservation Stewardship Program (CSP)." Last accessed June 21, 2019. https://www.nrcs.usda.gov/Internet/NRCS_RCA/reports/fb08_cp_cstp.html.

———. "Conservation Stewardship Program Producer Self-Screening Checklist." Last accessed January 16, 2019. https://www.nrcs.usda.gov/wps/PA_NRCSConsumption/download?cid=stelprdb1269861&ext=pdf.

———. "Cooperative Conservation Partnership Initiative." Last accessed February 8, 2019. https://www.nrcs.usda.gov/wps/portal/nrcs/main/national/programs/financial/ccpi/.

———. "CRP Practice CP25 (Idaho)." June 2010. https://www.nrcs.usda.gov/Internet/FSE_DOCUMENTS/nrcs144p2_042378.pdf.

———. "Environmental Quality Incentives Program." Last accessed August 13, 2019. https://www.nrcs.usda.gov/wps/portal/nrcs/main/national/programs/financial/eqip/.

———. "Environmental Quality Incentives Program (EQIP)." Last accessed June 21, 2019. https://www.nrcs.usda.gov/Internet/NRCS_RCA/reports/fb08_cp_eqip.html.

———. "EWP Floodplain Easement (EWP-FPE)." Last accessed February 20, 2019. https://www.nrcs.usda.gov/wps/portal/nrcs/detail/national/programs/financial/ewp/?cid=nrcs143_008225.

———. "Highly Erodible Land Conservation Compliance Provisions." Last accessed February 4, 2019. https://www.nrcs.usda.gov/wps/portal/nrcs/detail/national/programs/alphabetical/camr/?cid=nrcs143_008440.

———. "How to Apply for USDA-NRCS Conservation Programs." Last accessed August 13, 2019. https://www.nrcs.usda.gov/wps/portal/nrcs/detail/national/technical/nra/?cid=nrcs144p2_027058.

———. "How to Ask for Conservation Technical Assistance." Last accessed June 19, 2019. https://www.nrcs.usda.gov/wps/portal/nrcs/detail/national/programs/technical/cta/?cid=nrcs143_008232.

———. "Lesser Prairie-Chicken Initiative FY16-18 Conservation Strategy." 2015. https://www.nrcs.usda.gov/wps/PA_NRCSConsumption/download?cid=nrcseprd967006&ext=pdf.

———. "More Than 80 Years Helping People Help the Land: A Brief History of NRCS." Last accessed January 3, 2019. https://www.nrcs.usda.gov/wps/portal/nrcs/detail/national/about/history/?cid=nrcs143_021392.

———. "Natural Disaster Recovery Assistance through EQIP." Last accessed February 20, 2019. https://www.nrcs.usda.gov/Internet/FSE_MEDIA/nrcseprd1429025.pdf.

———. "Nine Step Conservation Planning Process." Last accessed June 18, 2019. https://www.nrcs.usda.gov/wps/portal/nrcs/detail/national/programs/financial/eqip/?cid=nrcs144p2_015695.

———. "NRCS Conservation Programs." Last accessed August 9, 2019. https://www.nrcs.usda.gov/Internet/NRCS_RCA/reports/cp_nat.html.

———. "NRCS Conservation Programs: Agricultural Conservation Easement Program (ACEP)." Last accessed June 23, 2019. https://www.nrcs.usda.gov/Internet/NRCS_RCA/reports/srpt_cp_acep.html.

———. "NRCS Conservation Programs: Farm and Ranch Lands Protection Program (FRPP)." Last accessed June 23, 2019. https://www.nrcs.usda.gov/Internet/NRCS_RCA/reports/fb08_cp_frpp.html.

———. "NRCS Conservation Programs: Grassland Reserve Program (GRP)." Last accessed June 23, 2019. https://www.nrcs.usda.gov/Internet/NRCS_RCA/reports/fb08_cp_grp.html.

———. "NRCS Conservation Programs: Healthy Forests Reserve Program (HFRP)." Last accessed June 23, 2019. https://www.nrcs.usda.gov/Internet/NRCS_RCA/reports/fb08_cp_hfrp.html.

———. "NRCS Conservation Programs: Wetlands Reserve Program (WRP)." Last accessed June 23, 2019. https://www.nrcs.usda.gov/Internet/NRCS_RCA/reports/fb08_cp_wrp.html.

———. "NRCS Field Office Technical Guide." Last accessed August 13, 2019. https://efotg.sc.egov.usda.gov.

———. "NRCS Funding Opportunities." Last accessed August 11, 2019. https://www.nrcs.usda.gov/wps/portal/nrcs/detail/national/programs/?&cid=stelprdb1048817.

———. "NRCS Regional Conservation Partnership Program (RCPP) 2014-2018 Project Summaries." Last accessed February 7, 2019. https://www.nrcs.usda.gov/wps/PA_NRCSConsumption/download?cid=nrcseprd1386891&ext=pdf.

———. "RCPP Partnership Stories." Last accessed February 8, 2019. https://www.nrcs.usda.gov/wps/portal/nrcs/detailfull/national/programs/farmbill/rcpp/?cid=stelprdb1253919.

———. "RCPP: Partner-Led Solutions." December 2016. https://www.nrcs.usda.gov/wps/PA_NRCSConsumption/download?cid=nrcseprd1309909&ext=pdf.

———. "Regional Boundaries, State Offices & Centers." Last accessed August 12, 2019. https://www.nrcs.usda.gov/wps/portal/nrcs/detail/national/about/org/?&cid=nrcs143_021421#state.

———. "Regional Conservation Partnership Program." Last accessed February 7, 2019. https://www.nrcs.usda.gov/wps/portal/nrcs/main/national/programs/farmbill/rcpp/.

———. "Sage Grouse Initiative 2.0 Investment Strategy, FY 2015-2018." August, 2015. http://www.sagegrouseinitiative.com/wp-content/uploads/2015/08/SGI2.0_Final_Report.pdf.

———. "State Technical Committees." Last accessed June 14, 2019. https://www.nrcs.usda.gov/wps/portal/nrcs/main/national/programs/technical/stc/.

———. "Technical Service Providers." Last accessed February 6, 2019. https://www.nrcs.usda.gov/wps/portal/nrcs/main/national/programs/technical/tsp/.

———. "Working Lands for Wildlife." Last accessed October 4, 2019. https://www.nrcs.usda.gov/wps/portal/nrcs/detailfull/national/programs/?cid=stelprdb1046975.

Natural Resources Conservation Service and J. Douglas Helms. "Brief History of the USDA Soil Bank Program." January 1985. https://www.nrcs.usda.gov/Internet/FSE_DOCUMENTS/stelprdb1045666.pdf.

Natural Resources Conservation Service and Jason Johnson. "USDA Helps Iraqi War Veteran Enhance Conservation on Farm." September, 2016. https://www.nrcs.usda.gov/wps/portal/nrcs/detail/national/programs/financial/csp/?cid=nrceprd1290732.

Natural Resources Conservation Service and Colette Kessler. "We Are the Brewers. We Are Conservationists." September, 2016. https://www.nrcs.usda.gov/wps/portal/nrcs/detail/national/programs/financial/csp/?cid=nrcseprd1292065.

Natural Resources Conservation Service and Dave Sanden. "Building Sustainability on a California Vineyard, Ranch." April 4, 2017. https://www.nrcs.usda.gov/wps/portal/nrcs/detail/national/newsroom/features/?cid=nrcseprd1323428.

Natural Resources Conservation Service Kansas State Office. "Lesser Prairie-Chicken Habitat in Jeopardy–a Conservation Plan Can Help." Last accessed June 17, 2019. https://www.nrcs.usda.gov/wps/portal/nrcs/detail/ks/technical/cp/?cid=nrcs142p2_033117.

Natural Resources Conservation Service Washington. "Partnerships Support Wildlife Programs for NRCS-WA." Last accessed February 7, 2019. https://www.nrcs.usda.gov/wps/portal/nrcs/detail/wa/newsroom/factsheets/?cid=nrcs144p2_036546.

NatureServe. "Conservation Status Assessment." Last accessed July 29, 2019. http://www.natureserve.org/conservation-tools/conservation-status-assessment.

North American Bird Conservation Initiative U.S. Committee. "State of the Birds." Last accessed August 12, 2019. http://nabci-us.org/how-we-work/state-of-the-birds/.

North Carolina Coastal Land Trust. "Coastal Land Trust Saves 'Site X.'" Last accessed March 22, 2019. https://coastallandtrust.org/sitex/.

———. "How We Save Land." Last accessed March 22, 2019. https://coastallandtrust.org/how/.

———. "Salmon Creek State Natural Area." Last accessed March 22, 2019. https://coastallandtrust.org/lands/salmon-creek-natural-area/.

———. "Who We Are." Last accessed March 22, 2019. https://coastallandtrust.org/who-we-are/.

North Carolina State Parks. "About the State Parks System: State Natural Areas." Last accessed March 22, 2019. https://www.ncparks.gov/more-about-us/about-state-parks-system/state-natural-areas.

Partners in Flight. "Partners in Flight Landbird Conservation Plan (2016)." Last accessed August 12, 2019 https://www.partnersinflight.org/resources/the-plan/.

"Partners in the Sage." Last accessed September 19, 2019. https://www.partnersinthesage.com/.

PBS. "The Loneliest Animals: Captive Breeding Success Stories." April 1, 2009. http://www.pbs.org/wnet/nature/the-loneliest-animals-captive-breeding-success-stories/4920/.

PennFuture. "Our Priorities." Last accessed July 27, 2019. https://www.pennfuture.org/our-priorities.

Pennsylvania Land Trust Association. "What Is a Land Trust?" Last accessed March 21, 2019. https://conservationtools.org/guides/150-what-is-a-land-trust.

Pheasants Forever. "Farm Bill Biologists." Last accessed June 14, 2019. https://pheasantsforever.org/getdoc/d68502f8-71ec-423c-a8a0-dbaf1e2851b2/Farm-Bill-Biologists.aspx.

———. "Find a Chapter." Last accessed July 27, 2019. https://pheasantsforever.org/Participate/Find-a-Chapter.aspx.

———. "Habitat Forever." Last accessed October 1, 2019. https://pheasantsforever.org/getdoc/c5e1e18a-5b7f-484e-8b5b-8d1eafb439f5/Habitat-Forever-,LLC.aspx.

———. "Unique Model." Last accessed July 27, 2019. https://pheasantsforever.org/getdoc/5dee6cc7-d6ca-475c-aa05-6195cdcd1942/Unique-Model.aspx.

———. "Why Habitat?" Last accessed July 27, 2019. https://pheasantsforever.org/Habitat/Why-Habitat.aspx.

———. "Why Participate?" Last accessed June 14, 2019. https://www.pheasantsforever.org/Participate/Why-Participate.aspx.

Phillips, Mary. "Pollinator Protection: Pledges, Policy and Proclamations." *National Wildlife Federation Blog*, June 10, 2019. https://blog.nwf.org/2019/06/pollinator-protection-pledges-policy-and-proclamations/

Pollinator Stewardship Council Inc. "Funding Programs for Pollinator Habitat." Last accessed July 26, 2019. http://pollinatorstewardship.org/index.php/funding-programs-for-pollinator-habitat/.

Prairie Pothole Joint Venture. "U.S. Fish and Wildlife Service's Small Wetland Acquisition Program." Last accessed July 10, 2019. http://ppjv.org/newsletters/summer-2016/us-fish-and-wildlife-services-small-wetland-acquisition-program.

Quality Deer Management Association. "Attend an Event." Last accessed July 27, 2019. https://www.qdma.com/get-involved/events/.

———. "Find a Branch." Last accessed July 27, 2019. https://www.qdma.com/get-involved/find-a-branch/.

———. "Mission & History." Last accessed July 27, 2019. https://www.qdma.com/about/mission-and-history/.

"Report to Missoula City Council: Projects for the 2018 Proposed County Wide General Obligation Bond and Proposed City Wide Conservation and Responsible Stewardship Mill Levy." 2018. https://www.ci.missoula.mt.us/DocumentCenter/View/46387/2018-06-25-Projects-for-the-2018-Proposed-County-Wide-GO-Bond.

Rocky Mountain Elk Foundation. "Chapter Web Sites." Last accessed July 27, 2019. http://www.rmef.org/NewsandMedia/PressRoom/AboutRMEF/Chapters/RMEFChapterMap.aspx.

———. "Eastern Elk Initiative." Last accessed July 27, 2019. http://www.rmef.org/Conservation/Initiatives/EasternelkInitiative.aspx.

———. "RMEF Mission Statement & Fast Facts." Last accessed July 27, 2019. http://www.rmef.org/NewsandMedia/PressRoom/AboutRMEF/MissionFastFacts.aspx.

Roth, Caroline. "Red-Cockaded Woodpecker Populations Begin Rebound." March 28 2016. https://www.usda.gov/media/blog/2016/03/28/red-cockaded-woodpecker-populations-begin-rebound

Roulson, Leanne. "The Culture of Conservation and the Arctic Grayling: A State Wildlife Management Success Story." July 14, 2018. http://greatergallatin.org/the-culture-of-conservation-and-the-arctic-grayling-a-state-wildlife-management-success-story/.

Ruffed Grouse Society & American Woodcock Society. "Chapter Info." Last accessed July 28, 2019. https://ruffedgrousesociety.org/chapter-info/.

———. "History of RGS." Last accessed July 28, 2019. https://ruffedgrousesociety.org/history-of-rgs/.

Sage Grouse Conservation Initiative. "Partner List." Last accessed January 14, 2019. https://www.sagegrouseinitiative.com/about/partners/.

Sage Grouse Initiative. "1,800 Ranchers and Counting!" December 18, 2018. https://www.sagegrouseinitiative.com/1800-ranchers-and-counting/.

———. "Collaborative Conservation Is Paying Off for the Sagebrush Ecosystem." August, 2015. https://www.sagegrouseinitiative.com/collaborative-conservation-paying-off-sagebrush-ecosystem/.

———. "#Fridaysonthefarm for Future Generations of Ranchers and Sage Grouse." November 2, 2018. https://www.sagegrouseinitiative.com/fridaysonthefarm-for-future-generations-of-ranchers-and-sage-grouse/.

———. "#Habitathero Mike Fenn: Leaving a Legacy for Wildlife in Wyoming." September 25, 2017. https://www.sagegrouseinitiative.com/habitat-hero-mike-fenn-wyoming/.

———. "New Paradigm." Last accessed January 14, 2019. https://www.sagegrouseinitiative.com/about/new-paradigm/.

———. "Round-up of Ranching Success Stories for Sage Grouse." May 20, 2015. https://www.sagegrouseinitiative.com/round-up-of-ranching-success-stories-for-sage-grouse/.

San Bruno Mountain Watch. "San Bruno Mountain Watch Oral History Project: Bette Higgins: The Queen of the Environmental Movement on San Bruno Mountain." February 25, 2011. http://www.mountainwatch.org/sbmw-oral-history-project/2011/2/25/bette-higgins.html.

———. "San Bruno Mountain Watch Oral History Project: The Story of the Breakup of the CSSBM and the Evolution of the HCP." February 25, 2011. http://www.mountainwatch.org/sbmw-oral-history-project/2011/2/25/the-story-of-the-breakup-of-the-cssbm-and-the-evolution-of-t.html.

———. "San Bruno Mountain Watch Oral History Project: Who Is Bette Higgins?" February 25, 2011. http://www.mountainwatch.org/sbmw-oral-history-project/2011/2/25/who-is-bette-higgins.html

Schewel, Steve, and David Proper. "Cities: Building an Urban Greenspace and Trail in Durham, North Carolina." March 11, 2019. https://www.conservationfund.org/blog/cities/2005-building-an-urban-greenspace-and-trail-in-durham-north-carolina.

Sentinel Landscapes. "About." Last accessed July 16, 2019. https://sentinellandscapes.org/about/.

South Dakota Department of Game Fish and Parks. "2019 Walk-in Area Program Information." 2019. http://habitat.sd.gov/WIA%20INFO%20PACK%202019.pdf.

South Dakota Habitat Pays. "Dense Nesting Cover." Last accessed July 25, 2019. http://habitat.sd.gov/resources/nestingcover.aspx.

———. "Habitat Advisors." May 28, 2019. http://habitat.sd.gov/advisors/docs/habitat-advisor-map.pdf.

———. "Habitat Resources for Landowners." Last accessed July 25, 2019. http://habitat.sd.gov/resources/default.aspx.

———. "Hunting Access Programs." Last accessed July 25, 2019. http://habitat.sd.gov/resources/access.aspx.

———. "Pollinator Habitat Assistance." Last accessed July 25, 2019. http://habitat.sd.gov/resources/pollinator-habitat.aspx.

———. "SD Habitat Pays." Last accessed July 27, 2019. http://habitat.sd.gov/.

———. "Working Grassland Habitat Programs." Last accessed July 25, 2019. http://habitat.sd.gov/resources/workinggrassland.aspx.

St. Mary's Soil Conservation District. "History of Soil Conservation Districts." Last accessed May 23, 2019. https://stmarysscd.com/about-us/history-of-soil-conservation-districts/.

Tenneessee Wildlife Federation. "Habitat Conservation." Last accessed July 27, 2019. https://tnwf.org/conservation/#tab-id-1.

Texas Parks & Wildlife. "2014 Farm Bill Programs: Healthy Forest Reserve Program (HFRP)." Last accessed June 7, 2019. https://tpwd.texas.gov/landwater/land/private/farmbill/hfrp/.

———. "Find a Wildlife Biologist." Last accessed July 30, 2019. https://tpwd.texas.gov/landwater/land/technical_guidance/biologists/.

———. "Landowner Incentive Program (LIP)." Last accessed July 30, 2019. https://tpwd.texas.gov/landwater/land/private/lip/.

———. "Landowner Request for Technical Guidance." 2013. https://tpwd.texas.gov/publications/pwdforms/media/pwd_0153_w7000_landowner_request_technical_guidance.pdf.

———. "Private Lands and Habitat Program." Last accessed July 30, 2019. https://tpwd.texas.gov/landwater/land/private/.

———. "Texas Farm and Ranch Lands Conservation Program." Last accessed July 30, 2019. https://tpwd.texas.gov/landwater/land/private/farm-and-ranch/.

The Conservation Fund. "Cities: Green Infrastructure." Last accessed March 20, 2019. https://www.conservationfund.org/our-work/cities-program.

———. "Coastal Headwaters Forest–Longleaf Conservation and Restoration." Last accessed February 8, 2019. https://www.conservationfund.org/projects/coastal-headwaters-forest-longleaf-conservation-and-restoration.

———. "Conservation Loans: Flexible Financing and Tailored Support to Advance Conservation." Last accessed September 20, 2019. https://www.conservationfund.org/images/resources/Conservation_Loans_Program.pdf.

———. "Contact Us." Last accessed March 20, 2019. https://www.conservationfund.org/about-us/contact-us.

———. "Financials." Last accessed March 20, 2019. https://www.conservationfund.org/about-us/financials.

———. "The Florida Panther Conservation Bank, II." Last accessed September 20, 2019. https://www.conservationfund.org/images/programs/files/Florida-Panther.pdf.

———. "Hand-Built in Asheville." Last accessed March 20, 2019. https://www.conservationfund.org/projects/hand-built-in-asheville.

———. "Lavender Helps a Family Farm Bloom." Last accessed March 20, 2019. https://www.conservationfund.org/projects/lavender-helps-a-family-farm-bloom.

———. "Milwaukee River Watershed Conservation Partnership." Last accessed March 20, 2019. https://www.conservationfund.org/projects/milwaukee-river-watershed-conservation-partnership.

———. "Natural Capital Investment Fund: Business Finance." Last accessed March 20, 2019. https://www.conservationfund.org/our-work/natural-capital-investment-fund.

———. "Supporting the Sage Grouse and Family Ranches in the West." Last accessed March 21, 2019. https://www.conservationfund.org/projects/support-for-wyoming-family-ranches-and-the-sage-grouse.

———. "Working Forest Fund: Forestry." Last accessed March 20, 2019. https://www.conservationfund.org/our-work/working-forest-fund.

The Longleaf Alliance. "The Longleaf Alliance." Last accessed October 3, 2019. https://www.longleafalliance.org/.

The Nature Conservancy. "BirdReturns Bird Response." Last accessed July 26, 2019. http://birdreturns.org/bird-response/.

———. "BirdReturns Program Overview." Last accessed July 26, 2019. http://birdreturns.org/home-page/about/about/.

———. "Conservation Easements: All About Conservation Easements." Last accessed June 28, 2018. https://www.nature.org/about-us/private-lands-conservation/conservation-easements/all-about-conservation-easements.xml.

———. "Conservation Easements: The Evolution of Conservation Easements." Last accessed June 28, 2018. https://www.nature.org/about-us/private-lands-conservation/conservation-easements/evolution-of-conservation-easements.xml.

———. "Contact Us." Last accessed February 21, 2019. https://www.nature.org/en-us/connect/contact-us/.

———. "Florida Panthers: Crossing the Caloosahatchee." Last accessed November 16, 2018, https://www.nature.org/en-us/about-us/where-we-work/united-states/florida/stories-in-florida/florida-panther-kittens-north-of-caloosahatchee/.

———. "The Matador Ranch." Last accessed February 25, 2019. https://www.nature.org/en-us/get-involved/how-to-help/places-we-protect/matador-ranch/.

The Pioneers Alliance. "The Pioneers Alliance." Last accessed March 21, 2019. http://www.pioneersalliance.org/.

The Roost Stop. "Red-Cockaded Woodpecker Artificial Cavity Insert." Last accessed April 25, 2019. https://www.therooststop.com/Red-cockaded-Woodpecker-insert-p/red-cockaded%20woodpecker%20insert.htm.

Theodore Roosevelt Conservation Partnership. "Sportsmen's Priorities for Conservation and Access in the 2018 Farm Bill Voluntary Public Access & Habitat Incentive Program (VPA-HIP)." 2018. http://www.trcp.org/wp-content/uploads/2018/02/Public-Access-fact-sheet.pdf.

Travis County Texas. "The Balcones Canyon Conservation Plan: Participating." Last accessed August 3, 2019. https://www.traviscountytx.gov/tnr/bccp/participating.

———. "The Balcones Canyonlands Conservation Plan (BCCP)." Last accessed August 3, 2019. https://www.traviscountytx.gov/tnr/bccp.

———. "Species Covered under the Balcones Canyonlands Conservation Plan." Last accessed September 20, 2019. https://www.traviscountytx.gov/images/tnr/Docs/species_covered.pdf.

Trout Unlimited. "About Trout Unlimited." Last accessed September 16, 2019. https://www.tu.org/about/.

———. "Find Your Chapter." Last accessed September 16, 2019. https://www.tu.org/find-your-chapter/.

———. "Fishing, Conservation and Science, Community: Trout Unlimited 2017 Annual Report." Last accessed October 4, 2019. https://www.tu.org/wp-content/uploads/2019/02/Trout_Unlimited_Annual_Report_2017-1.pdf.

U.S Army Corps of Engineers. "Regulatory in-Lieu Fee and Bank Information Tracking System." Last accessed June 19, 2018. https://ribits.usace.army.mil.

U.S. Air Force. "Readiness and Environmental Protection Integration [REPI] Program Project Profile: Dare County Bombing Range: North Carolina." 2017. https://www.repi.mil/Portals/44/Documents/Current%20Year%20Fact%20Sheets/DareCounty.pdf.

U.S. Department of Agriculture. "Service Center Locator." Last accessed August 12, 2019. https://offices.sc.egov.usda.gov/locator/app.

———. "USDA Local Offices." Last accessed June 12, 2019. https://www.outreach.usda.gov/USDALocalOffices.htm.

U.S. Department of Agriculture National Agroforestry Center. "National Agroforestry Center." Last accessed August 11, 2019. https://www.fs.usda.gov/nac/index.shtml.

U.S. Fish and Wildlife Service. "1939 through 2019 Wildlife Restoration Apportionments." 2019. https://wsfrprograms.fws.gov/Subpages/GrantPrograms/WR/WRApportionmentsHE-1939-2019.xlsx.

———. "1952 through 2019 Sport Fish Restoration Apportionments." 2019. https://wsfrprograms.fws.gov/Subpages/GrantPrograms/SFR/SFRApportionments-1952-2019.xlsx.

———. "2019 National Coastal Wetland Conservation Grants Project Summaries." 2019. https://www.fws.gov/coastal/CoastalGrants/pdfs/2019-Coastal-Wetlands-Grants-Project-Summaries-Final.pdf.

———. "Candidate Conservation Agreements." Last accessed May 30, 2018, https://www.fws.gov/endangered/esa-library/pdf/CCAs.pdf.

———. "Candidate Species: Section 4 of the Endangered Species Act." October 2017. https://www.fws.gov/endangered/esa-library/pdf/candidate_species.pdf.

———. "Critical Habitat for Threatened & Endangered Species [USFWS]." Last accessed August 1, 2019. https://fws.maps.arcgis.com/home/webmap/viewer.html?webmap=9d8de5e265ad4fe09893cf75b8dbfb77.

———. "Endangered Species." Last accessed July 29, 2019. https://www.fws.gov/endangered/.

———. "Endangered Species Act Listings: Frequently Asked Questions." July 2013. https://www.fws.gov/greatersagegrouse/factsheets/ESA%20Landowner%20Fact%20Sheet_080713.pdf.

———. "Environmental Conservation Online System: Habitat Conservation Plans." Last accessed August 3, 2019. https://ecos.fws.gov/ecp0/conservationPlan/region/summary?region=9&type=HCP.

———. "Environmental Conservation Online System: Red-Cockaded Woodpecker (*Picoides borealis*)." Last accessed October 3, 2018, https://ecos.fws.gov/ecp0/profile/speciesProfile?spcode=B04F.

———. "Environmental Conservation Online System: Section 7 Consultation Issued Biological Opinions." Last accessed August 1, 2019. https://ecos.fws.gov/ecp/report/biological-opinion.html.

———. "Environmental Conservation Online System: Species with 4d Rules." Last accessed July 31, 2019. https://ecos.fws.gov/ecp0/reports/species-fourd-report.

———. "Environmental Conservation Online System: USFWS Threatened & Endangered Species Active Critical Habitat Report." Last accessed August 1, 2019. https://ecos.fws.gov/ecp/report/table/critical-habitat.html.

———. "ESA Basics: 40 Years of Conserving Endangered Species." 2013. https://www.fws.gov/endangered/esa-library/pdf/ESA_basics.pdf.

———. "Farm Bill Endangered Species Tax Deduction." March 2010. https://www.fws.gov/endangered/esa-library/pdf/ES_TaxCredit2a.pdf.

———. "Financial Assistance Home." July 8, 2015. https://www.fws.gov/grants/.

———. "For Landowners." Last accessed August 11, 2019. https://www.fws.gov/endangered/landowners/index.html.

———. "The Greater Sage Grouse: Facts, Figures and Discussion." Last accessed September 20, 2019. https://www.fws.gov/greatersagegrouse/factsheets/GreaterSageGrouseCanon_FINAL.pdf.

———. "Items Taxed to Support Wildlife and Sport Fish Restoration in America." January, 2018. https://wsfrprograms.fws.gov/subpages/aboutus/ItemsTaxedJan2018.pdf.

———. "Migratory Bird Joint Ventures." Last accessed August 13, 2018, https://www.fws.gov/birds/management/bird-conservation-partnership-and-initiatives/migratory-bird-joint-ventures.php.

———. "National Coastal Wetlands Conservation Grants." August 20, 2019. https://www.fws.gov/coastal/CoastalGrants/index.html.

———. "North American Wetlands Conservation Act: Protecting, Restoring and Enhanging Wetland Habitats for Birds." Last accessed October 4, 2019. https://www.fws.gov/birds/grants/north-american-wetland-conservation-act.php.

———. "ODFW Columbian-Sharp Tailed Grouse Candidate Conservation Agreement with Assurances." Last accessed October 17, 2018. https://ecos.fws.gov/ecp0/conservationPlan/plan?plan_id=854

———. "Red-Cockaded Woodpecker Recovery." December 19, 2016. https://www.fws.gov/rcwrecovery/rcw.html.

———. "Safe Harbor Agreements for Private Landowners." October, 2017. https://www.fws.gov/endangered/esa-library/pdf/harborqa.pdf.

———. "State Wildlife Grant Program Apportionment History." 2019. https://wsfrprograms.fws.gov/Subpages/GrantPrograms/SWG/SWGAppt2002-2019.pdf.

———. "Statistical Data Tables for Fish & Wildlife Service Lands (as of 9/30/2018)." https://www.fws.gov/refuges/land/PDF/2018_Annual_Report_of_Lands_Data_Tables.pdf.

———. "Wetland Management Districts of the National Wildlife Refuge System." 2007. https://www.fws.gov/refuges/smallwetlands/WPAs/WMDs2008.pdf.

———. "Who's on the Lek: A Guide to Players." Last accessed August 11, 2017. https://www.fws.gov/mountain-prairie/species/birds/sagegrouse/Primer2SGWhoOnTheLek.pdf.

———. "Working Together: Tools for Helping Imperiled Wildlife on Private Lands." 2005. https://digitalmedia.fws.gov/digital/collection/document/id/887/.

U.S. Fish and Wildlife Service Coastal Program. "Coastal Conservation: Key to a Thriving Economy and a Healthy Environment." Last accessed September 20, 2019. https://www.fws.gov/coastal/pdfs/Coastal_Program_Brochure-Online_Version.pdf.

———. "Regional Coastal Contacts." July 24, 2019. https://www.fws.gov/coastal/contacts.html.

———. "Strategic Plan." February 28, 2019. https://www.fws.gov/coastal/strategic_plan.html.

———. "What Is the Coastal Program?" August 5, 2019. https://www.fws.gov/coastal/about.html.

U.S. Fish and Wildlife Service Division of Realty. "Grassland Easements." March 31, 2010. https://www.fws.gov/mountain-prairie/realty/grassesmt.htm.

U.S. Fish and Wildlife Service Environmental Conservation Online System. "Gopher Tortoise (Gopherus Polyphemus)." Last accessed July 15, 2019. https://ecos.fws.gov/ecp0/profile/speciesProfile?sId=6994.

U.S. Fish and Wildlife Service Estuary Restoration Act Council. "Estuary Restoration Act (ERA) of 2000." February 28, 2019. https://www.fws.gov/coastal/era.html.

U.S. Fish and Wildlife Service National Coastal Wetlands Conservation Grant Program. "National Coastal Wetlands Conservation Grants." May 8, 2019. https://www.fws.gov/coastal/CoastalGrants/index.html.

U.S. Fish and Wildlife Service National Wildlife Refuge System. "The Small Wetlands Program: A Half Century of Conserving Prairie Habitat." November 7, 2012. https://www.fws.gov/refuges/whm/fwn.html.

———. "Waterfowl Production Areas." October 15, 2015. https://www.fws.gov/refuges/about/wpas.html.

U.S. Fish and Wildlife Service Oregon Fish & Wildlife Office. "Oregon's First Conservation Bank." July 22, 2008. https://www.fws.gov/oregonfwo/LandAndWater/ConservationPlanning/ConservationBank.asp.

U.S. Fish and Wildlife Service Partners for Fish and Wildlife Program. "Background and History." Last accessed July 8, 2019. https://www.fws.gov/mountain-prairie/refuges/partners.php#.

———. "Partners for Fish and Wildlife Program." Last accessed August 10, 2018. https://www.fws.gov/partners/aboutus.html.

———. "Partners for Fish and Wildlife Program: Program History." Last accessed December 18, 2018. https://www.fws.gov/partners/programHistory.html.

———. "Regional and State Contact Information." Last accessed August 11, 2019. https://www.fws.gov/partners/contactUs.html.

U.S. Fish and Wildlife Service Wildlife & Sport Fish Restoration Program. "Multistate Conservation Grant Program–Overview." February 2, 2018. https://wsfrprograms.fws.gov/Subpages/GrantPrograms/MultiState/MS.htm.

U.S. Forest Service. "By the Numbers." November 2013. https://www.fs.fed.us/about-agency/newsroom/by-the-numbers.

———. "Forest Taxation." Last accessed August 11, 2019. https://www.fs.fed.us/spf/coop/programs/loa/tax.shtml.

———. "Grants." Last accessed August 11, 2019. https://www.fs.fed.us/working-with-us/grants.

———. "How the Community Forest Program Works." Last accessed August 14, 2019. https://www.fs.fed.us/managing-land/private-land/community-forest/program.

———. "Our History." Last accessed July 3, 2019. https://www.fs.fed.us/learn/our-history.

———. "Treesearch." Last accessed August 15, 2019. https://www.fs.usda.gov/treesearch/.

U.S. Sustainability Alliance. "Conservation Compliance." Last accessed February 4, 2019. https://thesustainabilityalliance.us/conservation-compliance/.

United States Department of Agriculture Economic Research Service. "Wetland Reserve Program Expenditures." January 15, 2009. https://web.archive.org/web/20090115231303/http://www.ers.usda.gov/Briefing/ConservationPolicy/images/wrpexpenditures.htm.

United States Department of Agriculture National Institute of Food and Agriculture. "Cooperative Extension History." Last accessed July 28, 2019. https://nifa.usda.gov/cooperative-extension-history.

———. "Land-Grant University Website Directory." Last accessed July 28, 2019. https://nifa.usda.gov/land-grant-colleges-and-universities-partner-website-directory.

United States Environmental Protection Agency. "319 Grant Program for States and Territories." Last accessed July 18, 2019. https://www.epa.gov/nps/319-grant-program-states-and-territories.

———. "Wetlands Compensatory Mitigation." Last accessed October 4, 2019. https://www.epa.gov/sites/production/files/2015-08/documents/compensatory_mitigation_factsheet.pdf

United States Forest Service Research and Development. "About R&D." August 13, 2018. https://www.fs.fed.us/research/about/.

United States Geological Survey Cooperative Research Units. "About the Cooperative Research Units Program." Last accessed July 1, 2019. https://www.coopunits.org/Headquarters/About_Us/.

———. "History of the Unit Program." Last accessed July 1, 2019. https://www.coopunits.org/Headquarters/History/.

University of New Hampshire Cooperative Extension Land & Water Conservation Program. "Preparing a Conservation Plan." March 2013. https://extension.unh.edu/resources/files/Resource001227_Rep1568.pdf.

USDA National Agroforestry Center. "About the Center." Last accessed July 3, 2019. https://www.fs.usda.gov/nac/about/index.php.

———. "Tools." Last accessed July 3, 2019. https://www.fs.usda.gov/nac/resources/tools/index.shtml.

———. "What Is Agroforestry?" August 2012. https://www.fs.usda.gov/nac/assets/documents/workingtrees/infosheets/WhatIsAgroforestry07252014.pdf.

Venture, Intermountain West Joint. "Priority Species List." Last accessed October 5, 2019. https://iwjv.org/priority-species-list.

Ventures, Migratory Bird Joint. "Migratory Bird Joint Ventures Fact Sheet: Conserving Birds and Their Habitats Throughout North America." 2017. https://www.fws.gov/migratorybirds/pdf/management/national_jv_factsheet.pdf

Walmart Acres for America. "2014 Grant Fact Sheet Coastal Headwaters Forest, Longleaf Conservation and Restoration (AL, FL)." November 12, 2014. https://www.nfwf.org/acresforamerica/Documents/acres2014longleaf.pdf.

Western Association of Fish and Widlife Agencies. "WAFWA Report Documents Greater Sage-Grouse Population Rebound." August 17, 2015. https://www.wafwa.org/news/e_1606/News/2015/8/_WAFWA_Report_Documents_Greater_Sage-Grouse_Population_Rebound.htm.

Western Association of Fish and Wildlife Agencies. "Greater Sage-Grouse Population Trends: An Analysis of Lek Count Databases 1965-2015." August, 2015. http://www.wafwa.org/Documents%20and%20Settings/37/Site%20Documents/News/Lek%20Trend%20Analysis%20final%208-14-15.pdf.

Wild Farm Alliance. "Farm Bill 2018 Issue Brief: Conservation Stewardship Program." 2018. https://d3n8a8pro7vhmx.cloudfront.net/wildfarmalliance/pages/302/attachments/original/1531441423/csp-issue-brief_FinalWeb.pdf?1531441423.

Wilson, Mel. "Corporate Sustainability: What Is It and Where Does It Come From?" *Ivey Business Journal*, 2003, https://iveybusinessjournal.com/publication/corporate-sustainability-what-is-it-and-where-does-it-come-from/

Wisconsin Department of Natural Resources. "DNR Service Foresters." Last accessed July 26, 2019. https://dnr.wi.gov/topic/ForestLandowners/dnrForesters.html.

———. "Forest Crop Law." Last accessed July 26, 2019. https://dnr.wi.gov/topic/ForestLandowners/fcl/.

———. "Managed Forest Law." Last accessed July 26, 2019. https://dnr.wi.gov/topic/forestlandowners/mfl/.

———. "Professional Forestry Assistance." Last accessed July 26, 2019. https://dnr.wi.gov/topic/ForestLandowners/assist.html.

———. "Wisconsin Forest Landowner Grant Program (WFLGP)." Last accessed July 26, 2019. https://dnr.wi.gov/Aid/ForestLandowner.html.

Other Sources

Bean, Michael J. "Endangered Species Act Safe Harbor Agreements: An Assessment." Working Paper. Madison, WI: Sand County Foundation/Environmental Policy Innovation Center, 2017.

———. "Habitat Exchanges: A New Tool to Engage Landowners in Conservation." Working Paper. Madison, WI: Sand County Foundation/Environmental Policy Innovation Center, 2017.

———. "Landowner Assurances under the Endangered Species Act." Working Paper. Madison, WI: Sand County Foundation/Environmental Policy Innovation Center, 2017.

Black's Law Dictionary. Tenth edition. 2014.

Faulkner, Jerome, and Jessica Groves. Presentation at the Association of Fish and Wildlife Agencies 109th Annual Meeting. September 24, 2019.

Fearer, Todd. Migratory Bird Joint Ventures. Presentation at the Bird Conservation Alliance 116th Congress Legislative Briefing, Washington, DC, April 2, 2019.

Gillenwater, Michael. "What Is Additionality? Part 1: A Long Standing Problem: Discussion Paper No. 001." Silver Spring, MD: Greenhouse Gas Management Institute, 2012.

Harris, Michael. "Federal Tools for Conservation of at-Risk Species on Private Working Forests in the Southeast." Paper presented at the Southeastern Association of Fish and Wildlife Agencies Annual Conference, Mobile, AL, October 22 2018.

"Journals of the Lewis & Clark Expedition, March 2, 1806." Last accessed August 11, 2017, https://lewisandclarkjournals.unl.edu/item/lc.jrn.1806-03-02#n28030201.

Natural Resources Conservation Service. "Working Lands for Wildlife Obligations, All Years All Species." 2019.

Pheasants Forever. "Farm Bill and Conservation Technical Assistance Summary." May 2019.

Pitman, Jim, Jon Ungerer, and Rod Winkler. "Lesser Prairie-Chicken Conservation Plan & Mitigation Framework. (Slideshow)." https://ksoutdoors.com/content/download/34160/243327/file

"Report to the Sage-Grouse Task Force, Greater Sage-Grouse Compensatory Mitigation." December 2016.

Riley, Steve. Presentation at the Association of Fish and Wildlife Agencies 109th Annual Meeting. September 24, 2019.

Tran, Huong N., and Liu Chuang. "State Conservation District Laws Development and Variations." Natural Resources Conservation Service, Working Paper No. 3, 1996.

U.S. Fish and Wildlife Service. "20 Years of the Safe Harbor Program: An Interview with Michael Bean." 2015. https://www.youtube.com/watch?time_continue=622&v=htd4xCljx9I.

U.S. Internal Revenue Service. 990 Form Filed by Ducks Unlimited. 2017.

———. 990 Form Filed by Mule Deer Foundation. 2018.

———. 990 Form Filed by National Audubon Society. 2017.

———. 990 Form Filed by National Wild Turkey Federation. 2017.

———. 990 Form Filed by National Wildlife Federation. 2017.

———. 990 Form Filed by Pheasants Forever. 2017.

———. 990 Form Filed by Quality Deer Management Association. 2017.

———. 990 Form Filed by Rocket Mountain Elk Foundation. 2017.

———. 990 Form Filed by Ruffed Grouse Society. 2018.

———. 990 Form Filed by the Conservation Fund. 2017.

———. 990 Form Filed by the Nature Conservancy. 2018.

———. 990 Form Filed by Trout Unlimited. 2017.

Working Lands for Wildlife. "2018 Fact Sheet."

Index

About the Author

Lowell E. Baier is an attorney and a legal and environmental historian and author. Baier holds a B.A. from Valparaiso University, a J.D. from Indiana University, and has received three honorary doctorates. He received an honorary Doctor of Legal Letters (L.L.D.) awarded in 2010 (Rocky Mountain College), a Doctor of Humane Letters (L.H.D.) awarded in 2015 (Indiana University), and a Doctor of Public Service awarded in 2019 (Valparaiso University). He's worked in Washington, DC, throughout his fifty-five-year career as a tireless advocate for natural resources and wildlife conservation. Throughout his career, he has observed and documented wildlife and its habitats on extensive treks and expeditions in the mountains and wilderness regions across the North American Continent, the Pamirs and Caucasus of Russia, and Mongolia's Gobi Desert and Altai Mountains, providing him with firsthand observations of wildlife and man's interactions across the globe. He was recognized as the Conservationist of the Year by the National Fish and Wildlife Foundation in 2008, and again in 2010 by both the Association of Fish and Wildlife Agencies and *Outdoor Life* magazine, and in 2016 by the National Wildlife Federation. In 2018 he was one of four judges chosen to select the 2019 Duck Stamp image by the Department of Interior's U.S. Fish and Wildlife Service.

www.ingramcontent.com/pod-product-compliance
Ingram Content Group UK Ltd.
Pitfield, Milton Keynes, MK11 3LW, UK
UKHW021905150926
475560UK00003B/198

9 781538 139370